Lecture Notes in Electrical Engineering

Volume 969

The book series *Lecture Notes in Electrical Engineering* (LNEE) publishes the latest developments in Electrical Engineering—quickly, informally and in high quality. While original research reported in proceedings and monographs has traditionally formed the core of LNEE, we also encourage authors to submit books devoted to supporting student education and professional training in the various fields and applications areas of electrical engineering. The series cover classical and emerging topics concerning:

- Communication Engineering, Information Theory and Networks
- Electronics Engineering and Microelectronics
- Signal, Image and Speech Processing
- Wireless and Mobile Communication
- Circuits and Systems
- Energy Systems, Power Electronics and Electrical Machines
- Electro-optical Engineering
- Instrumentation Engineering
- Avionics Engineering
- Control Systems
- Internet-of-Things and Cybersecurity
- Biomedical Devices, MEMS and NEMS

For general information about this book series, comments or suggestions, please contact leontina.dicecco@springer.com.

To submit a proposal or request further information, please contact the Publishing Editor in your country:

China

Jasmine Dou, Editor (jasmine.dou@springer.com)

India, Japan, Rest of Asia

Swati Meherishi, Editorial Director (Swati.Meherishi@springer.com)

Southeast Asia, Australia, New Zealand

Ramesh Nath Premnath, Editor (ramesh.premnath@springernature.com)

USA, Canada

Michael Luby, Senior Editor (michael.luby@springer.com)

All other Countries

Leontina Di Cecco, Senior Editor (leontina.dicecco@springer.com)

**** This series is indexed by EI Compendex and Scopus databases. ****

Liheng Wang · Yirong Wu · Jianya Gong
Editors

Proceedings of the 8th China High Resolution Earth Observation Conference (CHREOC 2022)

High Resolution Earth Observation: Wide Horizon, High Accuracy

Springer

Editors
Liheng Wang
China Aerospace Science and Technology
Beijing, China

Yirong Wu
Aerospace Information Research Institute
Chinese Academy of Sciences
Beijing, China

Jianya Gong
Wuhan University
Wuhan, China

ISSN 1876-1100 ISSN 1876-1119 (electronic)
Lecture Notes in Electrical Engineering
ISBN 978-981-19-8201-9 ISBN 978-981-19-8202-6 (eBook)
https://doi.org/10.1007/978-981-19-8202-6

This Springer imprint is published by the registered company Springer Nature Singapore Pte Ltd.
The registered company address is: 152 Beach Road, #21-01/04 Gateway East, Singapore 189721, Singapore

Contents

先进航天制造技术

Research on Digital Satellite Design Featuring On-Board Software in the Loop for on Orbit Management 3
Yang Tongzhi, Dang Jiancheng, Zhou Jun, Dong Fang, and Zhang Hai

High-Precision Directional Management Technology of Fengyun-4 Satellite with Multi-load and Multi-line of Sight 15
Dong Yaohai, Bian Zhiqiang, Wang Tianye, Zeng Qing, and Wang Hao

Anti-interference of Direct Sequence Spread Spectrum System Based on Wavelet and Eigenvalue Decomposition 37
Zhao Tengfei, Shen Zhao, Shang Jie, Li Junyao, and Liu Peijie

Orbit Optimization Technology for Multi-satellites Network Earth Observation 46
Hui Yu, Jun Zhao, Jun Liu, Tiening Nie, and Bingguang Zhan

临近空间系统与应用技术

Thermal Characteristics Analysis of Airship Based on Multi-node Model 59
Yao Xuan, Geng Baogang, Lu Du, Ye Hu, and Li Yunfei

Analysis of Space Contamination Effect on the Detecting Data from “XY-1” Satellite 72
Yuan Xiao-xue, Yang Dong-sheng, Zang Wei-guo, and Y. U. Qian

Thermosphere Neutral Densities at Dusk/Dawn Derived from Space-Borne Atmospheric Density Detectors 79
Yegui Wang, Xiaofang Zhang, Qifa Cai, and Guangming Chen

对地精确感知技术

Semi-automatic Annotation Method for Semantic Segmentation of Synthetic Aperture Radar Images 95
Nai-Rong Zheng, Zi-An Yang, Hong Yang, Yue Sun, and Feng Wang

Dominate Tree Species Classification on Large-Scale Mountainous Areas Using Voting Strategy-Based Ensemble Learning Method 102
Pengfei Zheng, Panfei Fang, Peiwei Liu, Weiheng Xu, Guojun Wang, and Leiguang Wang

A Radar Clutter Model Selection Method Based on Multi-layer Perceptron 116
Shu-Qi Lei, Dong-Xiao Yue, and Feng Wang

Urban Surface Solid Waste Detection Based on UAV Images 124
Yang Liu, Peng Gou, Wei Nie, Nuo Xu, Tianyu Zhou, and Yalan Zheng

Image-Based Sliding Mode Attitude Control Research on Small Video Satellites 135
Mengmeng Wang, Chao Song, Caizhi Fan, and Yonghe Zhang

Optimization and Application of Lucy-Richardson Filter on Optical Remote Sensing Image Restoration 150
Yunsen Wang, Chengzhi Ma, Panfeng Wu, Weiqiang Li, and Qi Song

Hyperspectral Image Data Construction and Expansion Method of Ground Object 161
Zhao Jiale, Zhou Bing, Wang Guanglong, Ying Jiaju, Deng Lei, and Wang Qianghui

Quad-Polarimetric Reconstruction with Adaptive Volume Scattering Model from Compact PolSAR Data 170
Xiaopeng Zhao, Rui Guo, Hui Wang, and Liang Guo

A High-Precision Monolithic 3D Modeling Method Based on UAV Tilt Photography 178
Changliang Ha, Guangrui Yu, Lili Yu, and Shuqi Wang

SAR Image Classification by Combining Transformer and Convolutional Neural Networks 193
Zi-An Yang, Nai-Rong Zheng, and Feng Wang

Simulation of a Typical Meiyu Case Over the Yangtze-Huai River in a Record-Breaking Meiyu Period of 2020 201
J. I. N. Yang, Wang Ying, and J. I. N. Meibing

Hyperspectral Image Classification Based on 3D Calibrated Attention 216
Menglin Zhang and Feng Gao

协同融合应用技术

Fine Registration of SAR and Optical Image Based on Improved Hausdorff Distance ... 229
Yifan Xu, Aifang Liu, Hui Xu, Long Huang, and Fan Wang

High Resolution Imagery-Based Statistical Analysis for Urban Rivers Extraction and Quality Classification—Case Study in Beijing District of China ... 242
Dongsheng Liu and Ling Han

Design and Implementation of a Target Characteristics Recommender System for Image Interpretation ... 264
L. V. Shouye, C. E. N. Pengrui, and Zheng Mei

Application Research on Water Body Extraction of Gaofen-3 Polarimetric SAR Based on Deep Learning ... 274
Zhang Wei, Wang Guanghui, Qi Jianwei, and Zhang Tao

A Study on Buildings Liveability Index Based on High Resolution Land Cover Data ... 284
Jie Wang, Guanghui Wang, Jianwei Qi, Yu Liu, and Tao Zhang

Radiometric Cross-Calibration of GF1-B/C/D Satellite with Landsat-8 OLI ... 293
Hengyang Wang, Hongzhao Tang, Honggeng Zhang, and Chen Hui

Validation of ZY1-02D Satellite Multispectral Sensor on-Orbit Radiometric Calibration Accuracy ... 303
Hongzhao Tang, Honggeng Zhang, and Chen Hui

Vibrational Frequency Analysis for the Ziyuan 3-02 Satellite Platform ... 313
Mo Fan, Xie Junfeng, Li Xiang, Tang Hongzhao, and Zhao Jianyi

Land Cover Change Information Extraction of Danjiangkou Wetland Based on GF_1 Data and Softmax Algorithm ... 322
Xue Longqin

Deep Learning-Based Matching of Sar Images with Optical Images ... 331
Dehua Li, Canhai Li, Hao Han, and Hui Qu

A Fast Verification Method of Small Building Samples Using Deep Learning ... 341
Zhongwu Wang, Shucheng You, Lei Du, Zhipan Wang, Yuhang Gan, Zhengyu Luo, and Lirong Liu

Template-Based Universal Adversarial Perturbation for SAR Target Classification ... 351
Junfan Zhou, Hao Sun, and Gangyao Kuang

Semantics-Based Analysis for Multi-source Remote Sensing Image Retrieval 361
Xiaolu Su, Lilu Zhu, Yang Wang, and Yanfeng Hu

A High-Precision Orographic Dataset Suitable for Global Numerical Weather Prediction Spectral Models Was Developed Based on Earth Observation Data 380
Yingjie Wang, Jianping Wu, and Dazheng Liu

PolSAR Image Classification Based on Deep Features 392
Ping Han, Lunan Men, and Zetao Chen

Research on Lake Changes in China Over the Past 50 Years Based on Satellite Remote Sensing Technology 404
Lei Du, Xu Han, Shucheng You, Ke Liu, Zhongwu Wang, Yun He, and Xinglin Mu

Author Index 411

先进航天制造技术

Research on Digital Satellite Design Featuring On-Board Software in the Loop for on Orbit Management

Yang Tongzhi, Dang Jiancheng(✉), Zhou Jun, Dong Fang, and Zhang Hai

Shanghai Institute of Satellite Engineering, Shanghai, China
363884916@qq.com

Abstract. With the development of satellite autonomy, intelligence and systematization, the traditional verification means can no longer meet the needs of operation, maintenance and application. It is urgent to adopt digital means and build digital satellites to provide users with comprehensive and real tools for task planning verification, daily operation and maintenance and effectiveness evaluation. The mission level system effectiveness simulation system for equipment planning and index demonstration has the problems of non fidelity of on-board software, large simulation granularity and insufficient process detail simulation, which is difficult to be applied to daily business scenarios such as mission deduction and replay, on orbit operation and maintenance management and so on. To solve this problem, the on-board software in the loop simulation is realized by introducing virtual computer technology, and digital satellite is realized by co-simulation of virtual computer and multi-system models based on various flows such as satellite information flow, control flow, energy flow and data flow, which runs on co-simulation platform based on DDS(data distributed service). Digital satellite can effectively simulate the real operation and response, replay and deduction task process, assist in on orbit operation and maintenance, and improve the service efficiency of equipment.

Keywords: Software in the loop · Virtual computer · Co-simulation · System model

1 Introduction

With the increasing complexity, autonomy and intelligence of satellite system, the difficulty and requirements of test verification are becoming higher and higher. There are some problems in the verification of real satellite test, such as high cost, low efficiency, poor flexibility and inability to support boundary simulation [1]. It is urgent to build digital satellite, which comprehensively simulates satellite functions and serves large-scale system interface verification, autonomous task planning verification, on orbit health management [2]. Based on digital satellite, ensure the coordination and consistency between satellite and ground, and the correctness and accuracy of mission application.

L. Wang et al. (Eds.): CHREOC 2022, LNEE 969, pp. 3–14, 2023.
https://doi.org/10.1007/978-981-19-8202-6_1

At the same time, digital satellite is the basis for realizing digital twin satellite, mirror system and parallel system, and it is also the trend of future development. Gartner, a world-famous consulting company, has listed digital twin technology as one of the top ten strategic science and technology development trends for three consecutive years [3]. As early as the 1970s, digital simulation technology began to be applied to aerospace technology. The earliest application should be the Apollo 13 lunar landing simulator. On April 13, 1970, the Apollo 13 spacecraft had an accident in space. The rescue plan was simulated in real time through the ground simulator, and the astronauts returned to the ground safely according to the operation instructions formed by the ground drill [4]. NASA and the U.S. Air Force laboratory put forward the example of digital twin for future aircraft, and are committed to using digital twin technology to improve production efficiency and improve the ability of health management and task application [5]. The traditional satellite system simulation is mainly used for system demonstration and index allocation. The simulation level is high, but the on-board software is not fidelity, the simulation granularity is large, and the support capacity of on orbit operation management is insufficient. Therefore, it is necessary to build a digital satellite of on-board software in the loop for deduction and replay of on orbit missions. Through multi-disciplinary and multi system collaborative simulation, digital satellite can verify the operation and task scheduling process.

2 System Design of Digital Satellite

The digital satellite adopts the "virtual computer+co-simulation" architecture. The virtual computer is used to simulate the real on-board integrated electronic computer. It can load and run the real satellite business software (including service software, task planning software, attitude and orbit control software, etc), so as to support the on-board software in the loop simulation. Based on satellite design, multi-disciplinary and multi system satellite models are constructed through simulation modeling tools. Through the co-simulation of virtual computer and various system models, digital satellite is completed which truly restores and deduces the task process.

The system design of digital satellite is shown in Fig. 1. Real-time DDS(data distributed service) is used as the basic co-simulation facility for object management, time management and other functions [6], so as to realize the flexible integration, deployment and operation of heterogeneous models. The 1553B bus, CAN bus, RS422, A/D, D/A, command pulse and other hardware interface are simulated and replaced by DDS bus. The interaction between various software modules is also realized by DDS network. All model software modules can be deployed on one high-performance host or distributed on multiple hosts. Digital satellite has the functions of virtual-real data interaction and state synchronization. Digital satellite receives the data input of on orbit satellite, extracts the initial state (such as attitude angle, orbit information, satellite time, single machine on-off state, sail illumination angle, antenna pointing angle, battery soc and other state parameters), automatically configures each model parameter, and completes the state synchronization of digital satellite to on orbit satellite. The simulation model is gradually corrected by virtual-real comparison.

The main components of digital satellite simulation system are as follows:

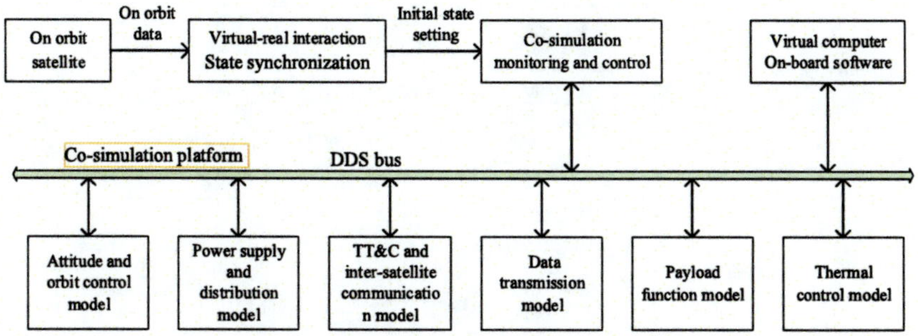

Fig. 1. System design of digital satellite

(1) Virtual computer: realize the virtualization of integrated electronic computer, simulate the operation environment of real on-board computer, run real satellite software, and realize the adaptive interface with other system simulation models.
(2) Simulation models: cover the main satellite systems, including attitude and orbit control model, power supply and distribution model, TT&C and inter-satellite communication model, thermal control model, data transmission model, payload function model. The models realize the adaptive interface with virtual computer and other simulation models.
(3) Co-simulation platform: integrate various professional simulation models, run multi-system model co-simulation. It adopts DDS standard which is used for data communication between different heterogeneous applications. It can control the timing of the overall simulation, and support multiple nodes to publish and subscribe in the LAN at the same time, so as to provide end-to-end transmission services between model software.

As shown in Fig. 2, digital satellite is built in layers and is divided into five layers. The basic model and data layer provides modeling data and basic model library. The simulation modeling layer realizes the multi-disciplinary and multi-system modeling. The simulation operation layer is a co-simulation platform, which is responsible for model integration, management and operation of various models. The virtual-real interaction layer interacts with the real satellite, extracts the model state from on orbit data, and synchronizes the virtual-real state. At the same, through the comparison of virtual and real states, the model correction is assisted. The system application layer is used to support users' business functions, such as autonomous task planning verification, on orbit fault inversion and disposal, test evaluation, etc.

Digital satellite supports real software in the loop. The satellite information scheduling is basically consistent with the real satellite information flow. Based on the basic component model & data, the satellite system model is constructed in the simulation modeling layer to form the configuration file and interface control document, which are used to service the assembly of digital satellite. In the simulation operation layer, digital satellite is loaded according to the configuration file. According to the task scenario or instructions, the simulation engine is controlled to conduct digital satellite co-simulation,

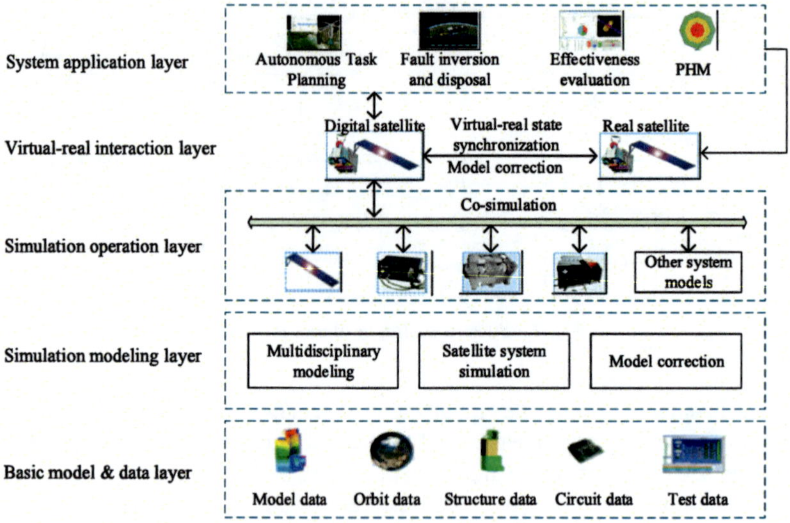

Fig. 2. Digital satellite layered architecture

and the simulation variable results are processed and presented in the user interface. The virtual-real interaction layer provides model parameters synchronized with the real satellite state, and assists model correction through virtual-real comparison. In the system application layer, simulation tasks and failure modes can be configured to generate task scenarios, and task scenarios are sent to the simulation operation layer for collaborative simulation.

The information flow of digital satellite external interaction is shown in Fig. 3, which is mainly divided into two categories: real satellite external interface (such as telemetry, command, navigation, inter-satellite communication, data transmission, etc.) and simulation monitoring interface (such as simulation initialization, simulation state parameter monitoring, simulation process control, etc.).

3 Digital Satellite Function Design

3.1 Virtual Computer for Software in the Loop

The virtual computer simulates the operation environment of on-board embedded software, such as satellite service software, autonomous task planning software and attitude and orbit control software. Let the on-board software run as if it were on a real on-board computer. The virtual of integrated electronic computer is realized in three steps:

(1) Complete the simulation of CPU core and CPU on-chip peripherals.
(2) Complete the simulation of RS422, 1553B bus, CAN bus, AD, DA, pulse command and other peripherals.
(3) Load and run the on-board operating system, satellite business software on virtual CPU.

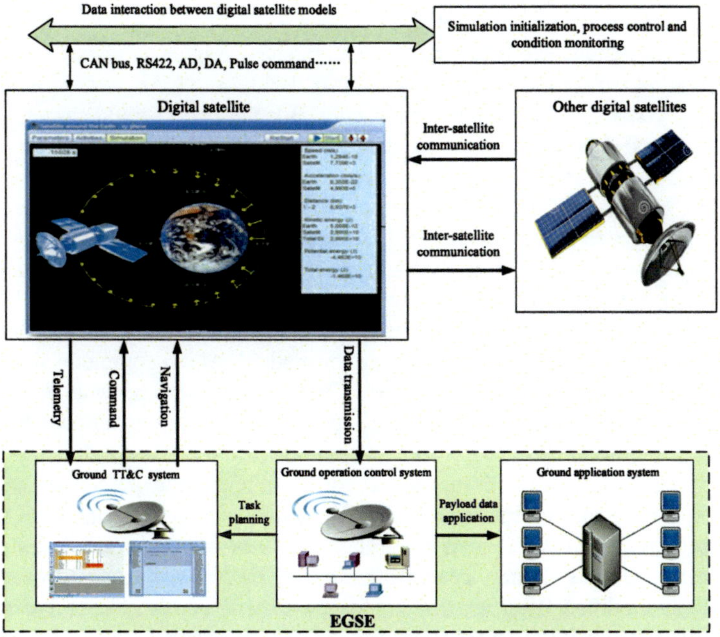

Fig. 3. Digital satellite external interactive information

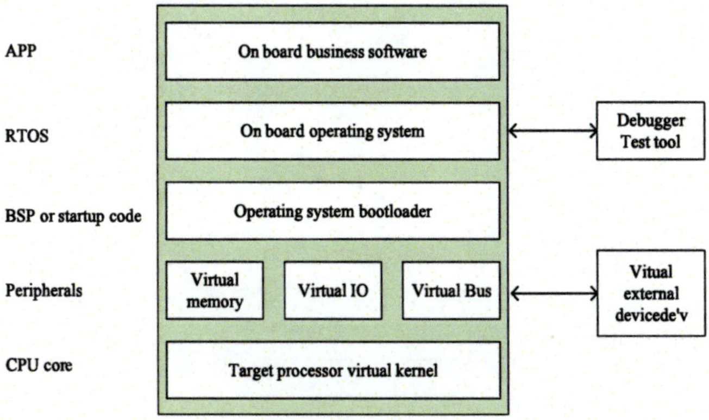

Fig. 4. Integrated electronic computer virtualization

The composition of the virtual integrated electronic computer is shown in Fig. 4. The virtual computer supports the on-board CPU instruction set. Instruction simulation adopts the method of simulating processor execution, which takes the instructions from the simulation memory, analyzes the instructions (including arithmetic instructions, logic instructions, jump instructions, etc.) according to the CPU instruction set specification. The virtual processor is used to simulate and execute the corresponding behavior, and the results are written back to the corresponding virtual register.

After the virtualization of onboard CPU, then 1553B bus, CAN bus, RS422, AD acquisition, instructions and other integrated computer peripherals are simulated. Bus peripherals simulation completes the functions of 1553B and CAN bus chips, bus registers, memory space and interrupts on the chip. Bus peripheral simulation completes bus messages and data flow simulation. RS422, AD, instructions and other peripherals also simulate the corresponding registers, memory space, interrupts, data streams, etc. The satellite business software runs on on-board operating system or multi task micro kernel. Software code execution call the CPU instruction set through operating system conversion, and call CAN bus, RS422 and other peripherals for simulation operation through driver interface and BSP conversion.

The virtual computer synchronizes the timing of each component in the system. The timing synchronization adopts the event trigger method, which completes the coordinated and synchronous operation of the virtual CPU core, computer peripherals, on-board operating system and on-board software through synchronous event callback function,. The user instructions are executed by the processor in turn according to the time benchmark. The virtual computer loads the object code file or binary file of on-board software, and customizes the interface module according to the hardware interface and address space of each functional unit, so as to digitally simulate the functions, data flow and communication process of interface chips. The execution codes of on-board software do not need to be modified. They are directly loaded and run in this system. The running process is consistent with the results in the physical environment.

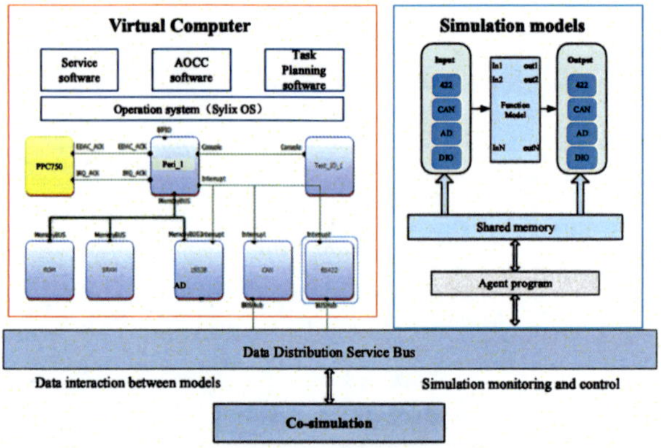

Fig. 5. Co-simulation of virtual computer and system model

As shown in Fig. 5, under the scheduling of the co-simulation platform, the virtual computer and virtual peripheral environment support the integration and execution of various professional simulation models. Co-simulation of virtual computer and system model realize the closed-loop simulation operation in the all digital mode, and form satellite high-precision digital satellite system.

3.2 Simulation Model Based on Various Flows

The real on-board software is simulated in the loop, and the on-board software is scheduled according to satellite information flow. Therefore, the external interaction interface of the system model needs to be consistent with the real satellite. Based on satellite information flow, energy flow, control flow, data flow and heat flow, as shown in Fig. 6, the system model interface can be divided into TM&TC interface, power interface, dynamic interface, payload data interface and thermal interface. According to the system function design, the corresponding model is developed. Satellite system engineers and product designers at all levels can obtain digital models of historical products in information world and establish specific digital models of current spacecraft [7]. Support professional model modeling and adaptation such as MATLAB / Simulink, ADAMS, AMESim, STK, C/C++ [8]. The co-simulation platform provides open model interface adapter, build a distributed co-simulation environment through configuration and message publishing and subscription mode.

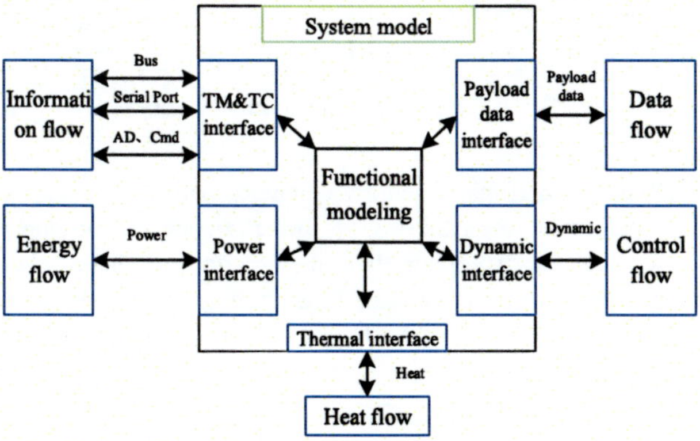

Fig. 6. System model interface based on flow

The composition and information interaction of digital satellite model are shown in Fig. 7, including six system models: attitude and orbit control model, power supply and distribution model, thermal control model, TT&C and inter-satellite communication model, data transmission model and payload function model. The interaction between models is based on satellite information flow, energy flow, data flow, control flow and thermal flow.

Information flow simulation is completed through the interaction between virtual computing and various system models in Fig. 7. Control flow simulation is completed through attitude and orbit control model. As shown in Fig. 8, the attitude and orbit control model simulates attitude and orbit control loop. The whole control loop involves on-board computer, measurement sensor, actuator, dynamics and environmental perturbation [9]. The on-board computer uses the virtual integrated electronic computer to run the real on-board control software. The actuator receives the actuator control command from the virtual computer, generates control torque and thrust, and adjusts the position and attitude

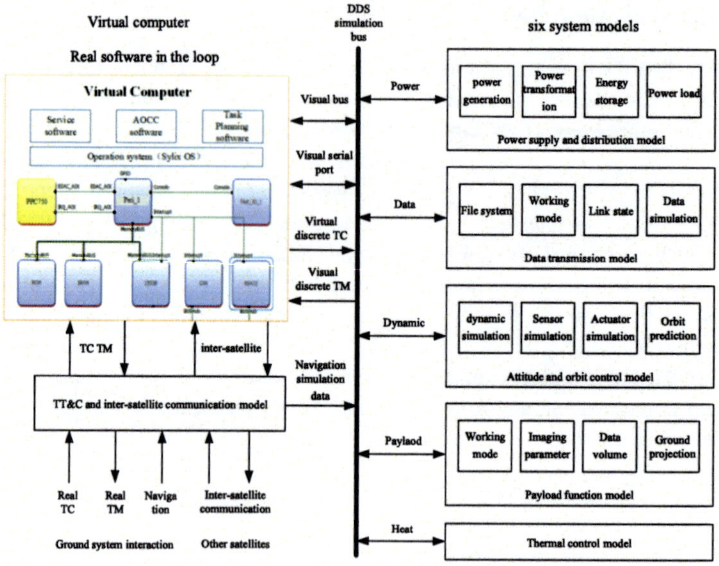

Fig. 7. Digital satellite system simulation model

of the satellite. Dynamics and space environment perturbation complete the attitude and orbit dynamics calculation of the satellite body under various environmental interference moments. The measurement sensor simulates the functions of all attitude determination and orbit determination sensors on the satellite, including the construction of star sensor, gyroscope, digital sun sensor, analog sun sensor, 0-1 sun sensor, magnetometer, etc.

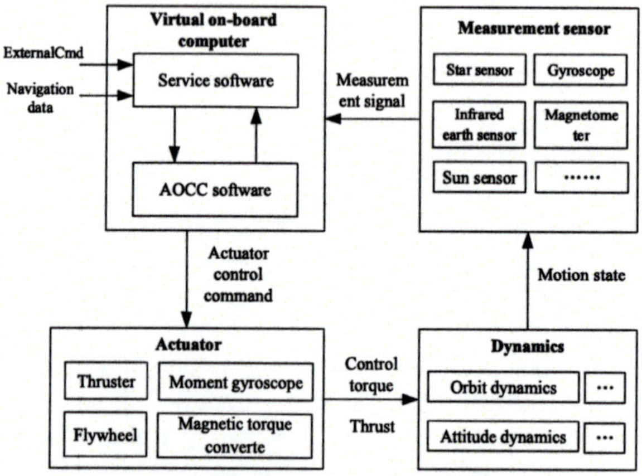

Fig. 8. Attitude and orbit control simulation

Energy flow simulation is completed through the power supply and distribution model is shown in Fig. 9, which simulates power supply, change, distribution and load.

It realizes the analysis and evaluation of satellite power supply and energy balance, and completes satellite energy closed-loop dynamic simulation [10]. At the same time, the evolution model realizes the accurate SOC monitoring of space power supply, on orbit SOH prediction and joint estimation of multiple parameters, analyzes the service capacity of power supply under different working conditions, and improves on-orbit intelligent perception ability [11].

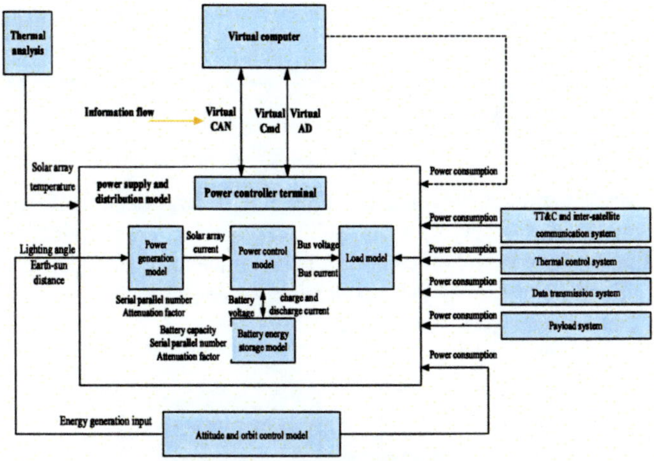

Fig. 9. Power simulation

Data flow simulation is completed by data transmission model and payload function model in Fig. 10. Data transmission model simulates various working modes and data links of data transmission and builds a solid state storage file system model which receives and records load data, simulate and manage payload data files. Payload function model simulates various imaging modes, and calculates the imaging width, resolution, ground projection and imaging data volume under the corresponding mode. The imaging data is generated to the data transmission model to complete the data flow simulation.

TT&C and inter satellite model mainly completes the function and channel simulation, navigation data simulation, and supports satellite information flow simulation. Thermal flow simulation is completed through thermal control model. According to satellite work mode and orbit, calculate internal and external heat flow. Then based on satellite heat dissipation design, heating design, heating unit and heat capacity design, satellite heat flow closed-loop simulation and heat prediction are realized. According to the mission needs and satellite design, build the corresponding system model. The functional models are assembled and integrated into a digital satellite on the co-simulation platform.

3.3 Co-simulation Platform Based on DDS

As shown in Fig. 11, the co-simulation platform adopts DDS (Data Distribution Service) mechanism for information exchange between models, which can realize the flexible

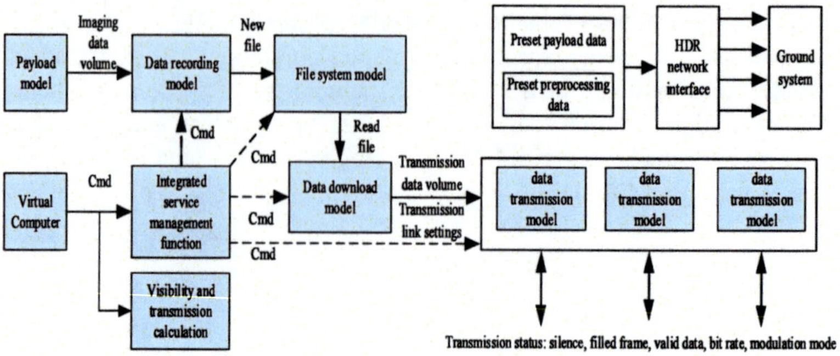

Fig. 10. Data transmission simulation

integration, configuration, deployment and operation of large-scale heterogeneous simulation systems. The bottom layer adopts "publish/subscribe" communication to provide real-time, reliable, flexible and efficient data communication. The whole platform construction is guided by the simulation application for end users and based on the open architecture that provides good system scalability.

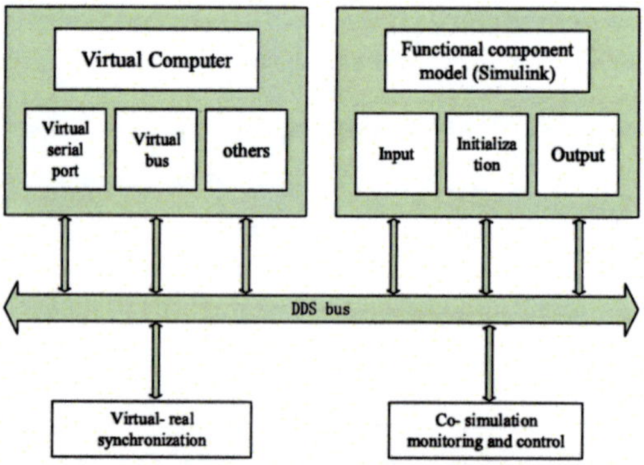

Fig. 11. Collaborative simulation platform

According to the input and output ports parsed by the model, the configuration of data cross-linked publish/subscribe relationship is generated. As shown in Fig. 12, this interface is used to configure the data interaction relationship. On the left is the publishing end, which lists the data published by all output ports to the simulation data bus. In the middle is the data center, which is used to present the name of interactive data. At the same time, you can also set the alias of data. On the right is the subscriber. Each data will have a subscriber window, which can configure the data interaction relationship.

The co-simulation platform completes the loading and operation management of digital satellite simulation engineering. The description file and startup script of each model are configured through model configuration management. All simulation models can be loaded and started through interface. Through the co-simulation engine, the

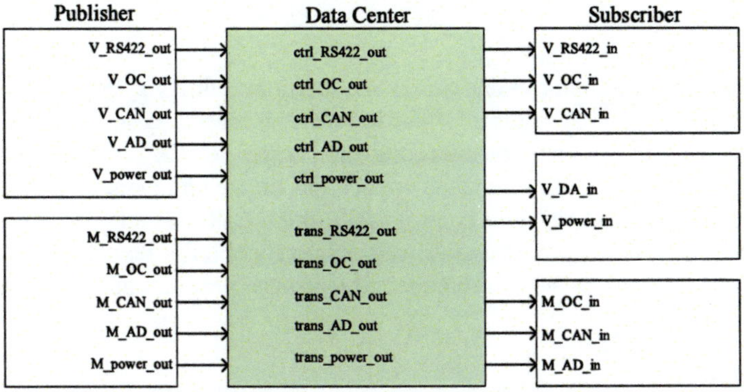

Fig. 12. Data interactive subscription between models

running time sequence of each node is coordinated to adjust the running speed of the system.

The timing of each simulation entity is controlled synchronously, and the synchronous command and response mechanism is introduced for all simulation entities connected to the DDS bus. Send synchronization commands to each system model after the execution of each control cycle. After receiving the synchronization response from all nodes, broadcast start commands to promote the continuous operation of each node. When all the simulation entities are executed in a control cycle and do not receive the start command of the co-simulation and timing control software, the blocking waiting state is maintained to ensure synchronous timing control of digital satellite simulation model. As shown in Fig. 13, the simulation operation is monitored through the co-simulation platform, which can set the simulation start and stop, simulation acceleration ratio and the fault or task scene. During operation, visually monitor the simulation parameters and present the satellite operation scene in three dimensions [12].

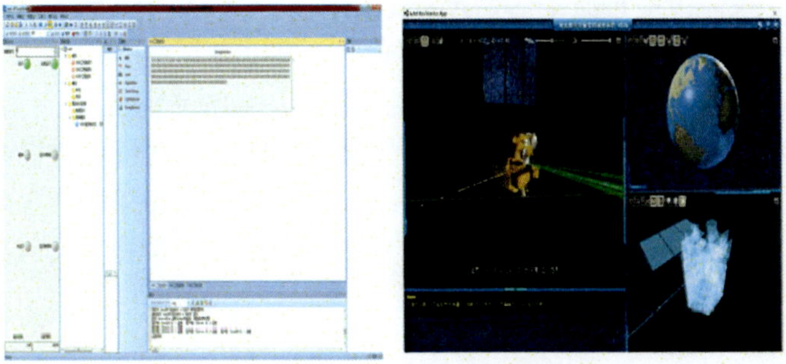

Fig. 13. Co-simulation monitoring

4 Summary

Through on-board computer virtualization and multidisciplinary system model collaborative simulation, digital satellite of software in the loop is designed. The simulation system runs real on-board business software, complete task deduction and replay, fault inversion and deduction, command verification, on orbit programming verification, attitude control verification, orbit control verification, task planning and other operational verification, which provides technical support for on orbit fault disposal and task scheduling, and improves the on orbit operation management capability of satellites.

References

1. Cao, Y.H.: Zhang Bei Shi Yan She Ji Yu Ping Gu. National Defense Industry Press, Beijing (2019)
2. Ouyang, G.X., Li, X., Dong, X. et al.: All-digital simulation technique with application to on-orbit spacecraft maintenance and servicing. J. Syst. Simul. **28**(4), 981–990 (2016)
3. Liug, D.T., Guo, K., Wang, B.K. et al.: Summary and perspective survey on digital twin technology. Chin. J. Sci. Instr. **28**(4), 1–10 (2018)
4. Meng, S.H., Ye, Y.M., Yang, Q., et al.: Digital twin and its aerospace applications. Acta Aeronautica et Astronautica Sinica **41**(9), 1–12 (2020)
5. Dong, Y.F., Li, Z., Lei, M.: Research on concept of digital satellite. Aerosp. Shanghai **38**(1), 1–12 (2021)
6. Gao, X.L., Ma, J.X., Li, J.: HLA-DDS distributed simulation system of telecommunication satellite. Foreign Electron. Measur. Technol. **36**(2), 89–95 (2017)
7. Wang, J.J., Xiang, Y.Q., He, Z.W.: Models and implementation of digital twin based spacecraft system engineering. Comput. Integr. Manuf. Syst. **25**(6), 1348–1360 (2019)
8. Wang, P., Yang, M., Zhu, J.C. et al.: Dynamic data driven modeling and simulation method for digital twin. Syst. Eng. Electron. **42**(12), 2779–2786 (2020)
9. Zhao, Q., Guo, X.Y., Zhao, Y.: Exploration of digital twin application in satellite constellation system. Foreign Electron. Measur. Technol. **40**(2), 121–127 (2020)
10. Liu, Z.G., Xia, N., Du, Q., et al.: Application research of spacecraft electrical power system digital accompanying flight technology. Spacecr. Eng. **29**(5), 135–141 (2020)
11. Zhu, K., Chen, J., Lv, T.L. et al.: Digital twin system for space power-sources. Aerosp. Shanghai **38**(3), 197–206 (2021)
12. Su, H.X., Dong, Z.H., Yang, F. et al.: Visual simulation analysis platform for satellite load based on cesium. Comput. Eng. **46**(10), 193–198 (2020)

High-Precision Directional Management Technology of Fengyun-4 Satellite with Multi-load and Multi-line of Sight

Dong Yaohai[1], Bian Zhiqiang[2,3](✉), Wang Tianye[2], Zeng Qing[2], and Wang Hao[2]

[1] Shanghai Academy of Spaceflight Technology, Shanghai 201109, China
[2] Shanghai Institute of Satellite Engineering, Shanghai 201109, China
bianzhiqiang2003@163.com
[3] Nanjing University of Aeronautics and Astronautics, Nanjing 211106, China

Abstract. The application background of high-precision directional management technology in Fengyun-4 satellite is introduced in this paper, in order to meet the requirement of "accurate pointing, stable attitude and quiet platform" for multiple high-precision remote sensing loads, the dynamic characteristic of the moving parts, the characteristic of space disturbance torque and the dynamic model of the whole satellite are studied, The attitude control system of Fengyun-4 satellite is introduced. In order to realize the precise directional control of the LOS of several remote sensing instruments, a series of directional management measures such as high-precision measurement and control of platform benchmark, geometric calibration for LOS of remote sensing instruments are put forward. It has broken through many key technologies such as precise planning and testing of satellite dynamics spectrum, high-precision measurement and determination, high-precision attitude control, image navigation and registration, micro-vibration measurement and suppression, etc. After in-orbit flight verification, all indexes meet the design requirements, and high-precision directional management of multi-view axes of onboard loads has been realized.

Keywords: Fengyun-4 · Load line of sight · Directional management · Attitude determination · Torque compensation · Image navigation and registration

1 Summary

Fengyun-4 satellite (FY-4 for short) is the second generation geostationary orbit meteorological satellite in China, which adopts the three-axis stability control scheme to replace Fengyun-2 satellite [1, 2]. FY-4A was successfully launched on December 11th, 2016. It has completed the five-year assessment life in orbit, and is currently in overdue service. FY-4B was successfully launched on June 3rd, 2021. Through the joint observation of FY-4A and FY-4B, the continuous and stable operation has greatly improved the detection level of geostationary meteorological satellites in China. FY-4 satellite carries three main payloads, among which the advanced geostationary radiation imager (AGRI) acquires a 15-band image of the Earth's disk [1, 3] every 15 min, with the highest spatial

L. Wang et al. (Eds.): CHREOC 2022, LNEE 969, pp. 15–36, 2023.
https://doi.org/10.1007/978-981-19-8202-6_3

resolution of 500 m in the visible band and 2 km in the infrared band. The Geostationary interferometric infrared sounder (GIIRS) can realize the infrared radiation detection in 1650 bands, and obtain the vertical distribution of high-frequency sub-atmospheric temperature and humidity [4] from the geostationary orbit. The geostationary high-speed imager (GHI) is capable of making observations from GEO orbit for 2000 km × 2000 km regions with spatial resolution up to 250 m in 1 min intervals. The coordinated observation of the three payloads has the real-time comprehensive observation ability of "multispectral two-dimensional imaging+hyperspectral three-dimensional detection+small area high-frequency observation", which provides strong support for disaster and environmental monitoring, weather modification and space weather research.

In order to give full play to the quantitative remote sensing capability of meteorological payloads, the three loads are required to point to the target observation position accurately and stably. While ensuring the accurate positioning of the image, the platform is required to provide a quiet observation environment [5]. In order to achieve this goal, this paper focuses on the high-precision directional management technology of Fengyun-4 satellite with multiple loads and multiple sight axes (Fig. 1).

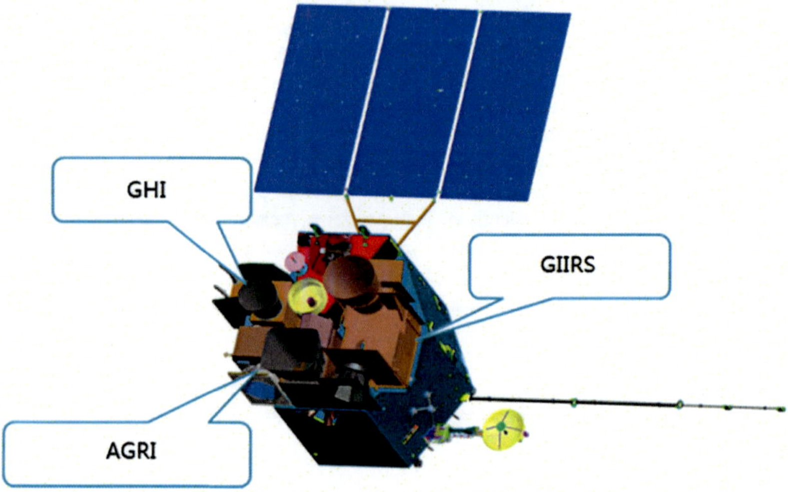

Fig. 1. Fengyun-4 satellite

2 Research Status at Home and Abroad

The development period of American geostationary meteorological satellites is long, the platform stabilization mode is from spin stabilization to three-axis stabilization, the attitude measurement technology is from horizon to high-precision star sensor and gyro, and the satellite high-precision directional management technology is becoming more and more mature. In November 2016, the third-generation geostationary orbit meteorological satellite GOES-R was successfully launched. The payload has been upgraded from the original Imager to the Advance Baseline Imager (ABI). Compared with the previous generation of meteorological satellites, the spatial resolution, spectral coverage and resolution, time refresh rate and radiation sensitivity of the GOES-R satellite's earth observation have been significantly improved. The satellite adopts advanced attitude determination scheme, the satellite is equipped with three high-precision star sensors (output frequency of 20 Hz) and two inertial measurement units (each of which contains four hemispherical resonator gyros and two accelerometers, output frequency of 200 Hz). The satellite's precise attitude information is output at 100 Hz through 6th-order Kalman filtering. The satellite has improved the management scheme of pointing control, through comprehensive management and control of satellite orbit drift, flexible vibration, satellite thermal deformation in orbit, instrument servo error, attitude control system noise and moving parts on board, the positioning accuracy of ABI is better than 21 μrad [7, 8].

Japan's new generation of geostationary meteorological satellites Himawari-8 and Himawari-9 were successfully launched in 2014 and 2016 respectively. The satellite adopts three-axis stable attitude control, and the loads are advance high-precision imager (AHI) and space environment data acquisition monitor (SEDA). AHI has 16 observation channels with the highest spatial resolution of 0.5 km. The image positioning of Himawari-8 satellite mainly depends on the ground application system. Therefore, the satellite attitude determination system requires high precision and high bandwidth. In order to shorten the transmission path of measurement error, the star sensor, inertial reference unit, angular vibration sensor and accelerometer are installed on the optical bench in common reference with the load. In addition, in order to reduce the influence of micro-vibration on imaging, the satellite uses a low disturbance reaction flywheel [9, 22].

Japan launched a sun-synchronous orbit-Advanced Land Observing Satellite [10] (ALOS) in 2016. A comprehensive solution was designed to meet the directional management requirements, which surpassed the traditional attitude and orbit control system and covered various satellite systems and ground systems. At the satellite system level, these methods include low thermal deformation structure, registration management, dynamics management, interference management, time management, high-bandwidth directional measurement and directional data processing. A precise star tracker, gyro, accelerometer and high-frequency angular vibration sensor are equipped to form a 0–500 Hz broadband attitude determination system, and the attitude determination accuracy on the satellite can reach 0.0003°. Dual-frequency GPS receiver can measure pseudo-range and carrier phase, so that the accuracy of ground position determination can reach sub-meter level. Through attitude and position estimation, the accuracy of geographical position determination can reach 6 m, and supplemented by high-bandwidth measurement, the accuracy can reach 3 m (Figs. 2, 3 and 4).

Gaofen-4 satellite is a high-resolution optical remote sensing satellite in China's geostationary orbit. The satellite adopts the three-axis stability control scheme, and is equipped with four large torque and momentum wheels of 0.5 Nm. It has the ability of rapid maneuver, can complete the maneuver and stability of 0.64° within 30 s, and can respond to the observation mission of China and surrounding areas in time. The attitude determination method of star sensor and gyro is adopted, The attitude stability is better than 0.0005°/s, and the image positioning accuracy is 4 km [11, 12].

Fig. 2. GOES-R satellite

Fig. 3. Himawari-8 satellite

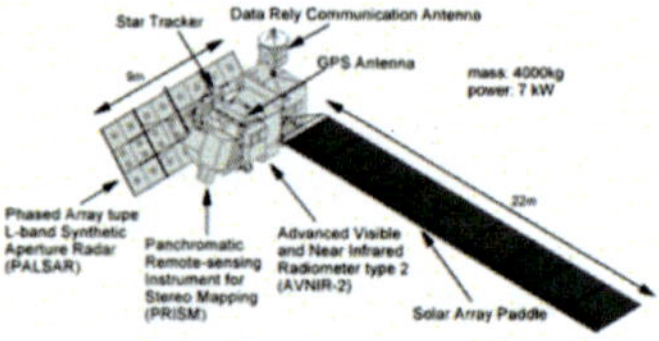

Fig. 4. ALOS satellite

Summing up the demand and realization of high-precision positioning of satellite loads at home and abroad is based on the determination of high-precision attitude datum, and through the design and management of load layout and installation, thermal control management, structural deformation suppression and calibration, combined with the data processing of satellite and ground, the high-precision directional management of load sight axis is realized.

3 Requirement of Directional Management Technology for Payloads

In view of the requirements of "high stability, high pointing accuracy and high positioning accuracy" of multiple remote sensing loads, high-precision directional management focuses on the pointing of loads, establishes the relative transformation relationship between multiple loads and satellite attitude measurement datum, and solves the measurement and control of low-frequency, medium-low-frequency and high-frequency pointing changes of satellite remote sensing loads, rather than just that attitude control of the satellite platform. Therefore, high-precision directional management not only depends on attitude control system, but also needs to be combined with several subsystems [6] such as satellite thermal control, structure, payloads, attitude and orbit control and ground processing. Through the optimization design of the satellite system, Comprehensive control of multi-source errors such as structural deformation, long-period thermal deformation, micro-vibration, disturbance of moving parts and on-board time reference management of three loads; Through the satellite-ground integration design, the systematic deviation and long-period deviation of remote sensing load pointing are corrected, and the load imaging effect is further improved (Fig. 5).

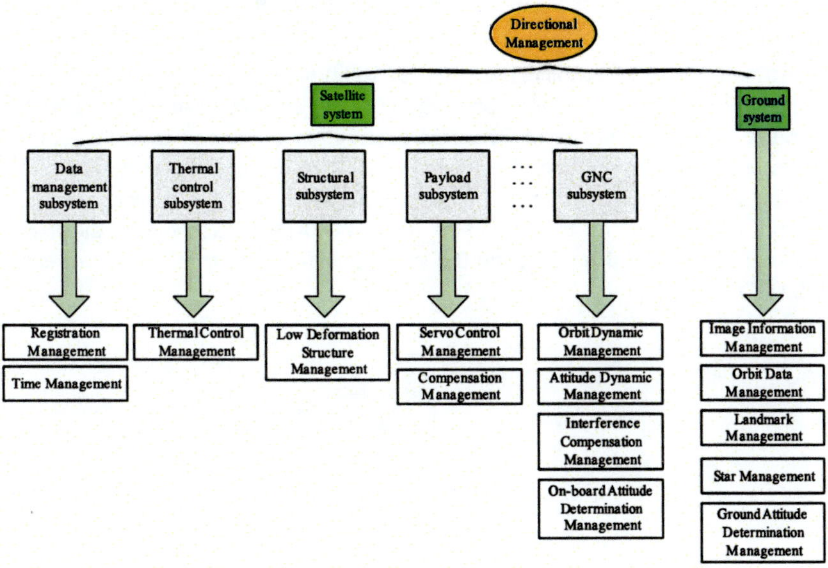

Fig. 5. Satellite directional management architecture

3.1 Accurate Directional Requirement

The demand of FY-4 satellite payloads for directional control is mainly reflected in three aspects: accurate pointing, stable attitude and quiet platform.

(1) "Pointing accuracy" represents the absolute pointing control accuracy of the load sight axis. The AGRI generates a global disk image every 15 min. In order to ensure the accuracy of the geographic information corresponding to the payload data, it is required that each pixel be accurately mapped with the geographic latitude and longitude, that is, the positioning accuracy of the original image downloaded from the satellite is better than that of one pixel [12].This puts a high demand on absolute pointing control, which requires the attitude control system of satellite platform to overcome the disturbance of space environment disturbance torque, moving parts on board, etc., and break through the angular-second-level high-precision attitude

determination technology to meet the requirements of pixel-level image navigation (Fig. 6).

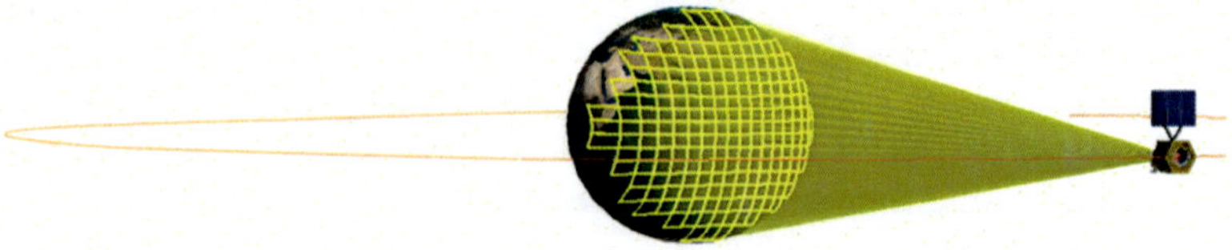

Fig. 6. Schematic diagram of image navigation

(2) "Attitude stability" means that the attitude stability of the satellite platform should meet the requirements of load observation [13]. Fengyun-4 satellite requires high image navigation accuracy. On-board observation of stars is used for real-time geometric calibration. Take the loaded GHI as an example, its spatial resolution is high. In order to ensure that the star recognition accuracy reaches the sub-pixel level, it is required that the jitter of the star image in the field of view of the detector should not exceed 1/6 pixel, and the attitude change within 1s should not exceed 0.0005° (Fig. 7).

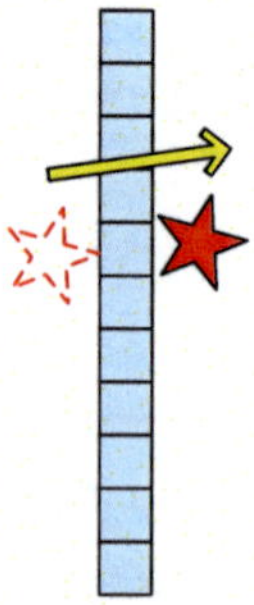

Fig. 7. Star sensitivity diagram of GHI

(3) "Quiet platform" means that all kinds of high-frequency micro-vibration interference are isolated during load operation. Take GIIRS as an example, the instrument is a typical spectral payload. When working in orbit, there is not only a strict correspondence between objects and images, but also the change of spectral position and interferogram contrast [14]. The micro-vibration on the satellite will induce the resonance of the moving mirror system of the GIIRS, The optical path difference is out of tolerance, and the spectrogram cannot be inverted. In order to ensure the normal operation of the load, the micro-vibration of the vertical detector near the modal frequency of the moving mirror is required to be less than 1 mg (Fig. 8).

In order to realize the high-precision directional management, the above three aspects are targeted for design, and the long-period error compensation, middle-low frequency

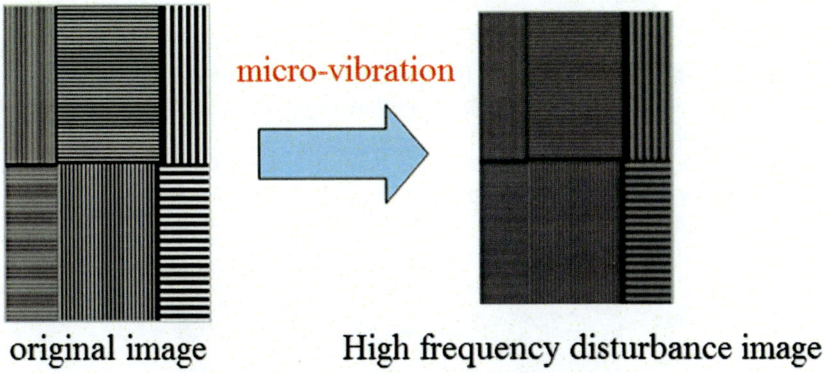

Fig. 8. Influence of micro-vibration

jitter control and high-frequency micro-vibration suppression of loads pointing should be comprehensively solved, which provides an ideal observation platform for payloads (Table 1).

Table 1. Requirements of payloads on satellite platform

Load name	Maximum angular resolution (μrad)	Index	Request	Remarks
AGRI	14	Pointing stability	Less than 0.001° in one second	During the observation of stars
		Pointing accuracy	0.01°	
GIIRS	28	Pointing stability	Less than 0.0005 in one second	The requirement for pointing stability is the highest during ground gaze detection
		Pointing accuracy	0.01°	
		Micro-vibration requirement	1 mg @ (within 200 Hz)	
GHI	7	Pointing stability	Less than 0.0005° in one second	During the observation of stars
		Pointing accuracy	0.01°	

3.2 Satellite Platform Design

(1) The load and attitude sensor are installed in the same benchmark.
Through comprehensive optimization design of onboard load and attitude sensors, ground verification and evaluation are carried out, and on-orbit calibration scheme is formulated, so as to realize effective control and measurement of benchmark changes under the coupling effect of force and heat (Fig. 9).

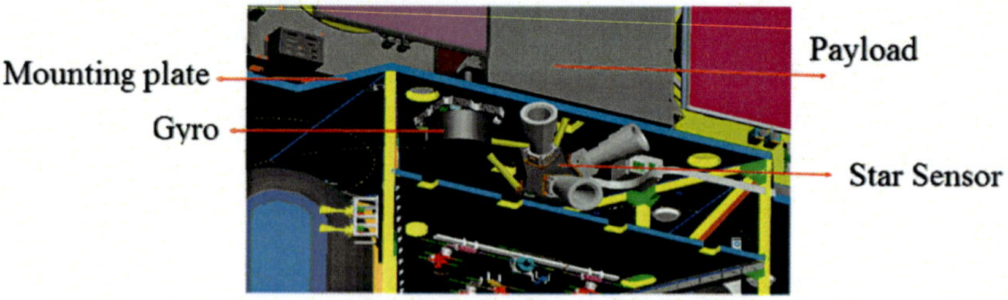

Fig. 9. Schematic diagram of payload and attitude sensor installation

The star sensor, gyro and high-frequency angular vibration sensor are installed integrally with three loads to reduce the relative deformation between the attitude sensor and the load sight axis. Multi-head star sensors are installed on the same star sensor bracket to reduce the influence of deformation between star sensors.

(2) Precise temperature control of star sensor
FY-4 satellite is equipped with three high-precision star sensors, which are mounted on the same bracket, and the mounting surface and bracket are precisely temperature-controlled, and the temperature of the mounting surface of the star sensors is controlled at 20 °C $\pm$ 0.2 °C, so as to minimize the influence of space thermal environment change on multiple star sensors. After the satellite is in orbit, According to the measurement data of the multi-head star sensor, the change relation of the relative reference is synchronously corrected, and the measurement reference of the star sensor is unified.

(3) Configuration of attitude and orbit control subsystem
The attitude and orbit control subsystem of FY-4 satellite includes three parts: controller, sensor and actuator, as shown in Fig. 10. The controller is realized by the attitude and orbit control computer. The attitude sensor includes two sets of three floating gyro components, one set of fiber optic gyro, three high-precision star sensors, one set of digital sun sensor and one infrared earth sensor. The actuator includes 6 momentum wheels, 18 thrusters and a 490 N engine, which can realize the high-precision control of attitude and orbit.

The detailed indexes of core measuring equipment is showed in the following table (Tables 2 and 3).

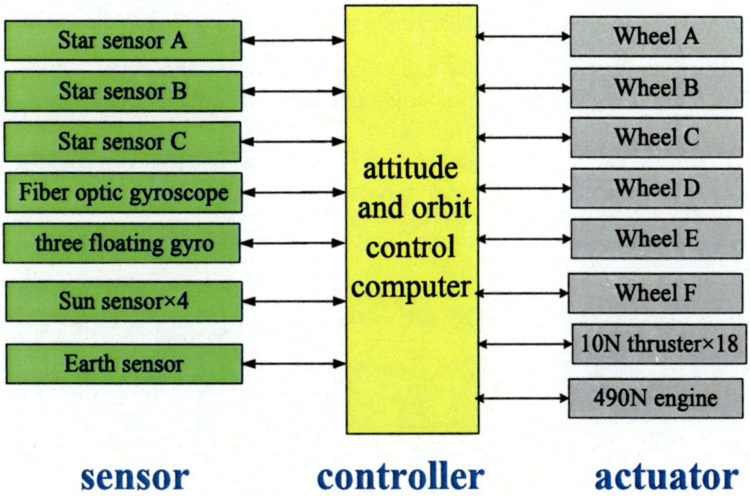

Fig. 10. Control system composition diagram

Table 2. FY-4 Satellite Gyro Index

Serial number	Index parameter	Index requirements
1	Pulse equivalent	0.2″
2	Random drift	0.01°/h(3σ)
3	Angular random walk coefficient	$0.0005°/h^{1/2}$

Table 3. FY-4 Satellite Star Sensor Index

Serial number	Index parameter	Index requirements
1	Update frequency	4Hz
2	Random noise	3″(3σ)
3	Low frequency error	3″(20 °C ± 1 °C)

4 Key Technology and Application of High Precision Directional Management

In view of the application requirement of "accurate pointing, stable attitude and quiet platform", high-precision directional management includes five key technologies such as precise planning and testing of satellite dynamics spectrum, high-precision attitude measurement and determination, high-precision attitude control, image navigation and registration, micro-vibration measurement and suppression.

4.1 Precision Planning and Test Technology of Satellite Dynamic Spectrum

FY-4 satellite is affected by various environmental disturbance torques, which brings great difficulties to the high-precision pointing control of the satellite. The cut-off frequency of satellite attitude control is designed to be below 0.01 Hz. Disturbances below this frequency are called low-frequency disturbances, and disturbances above this frequency are called medium-high-frequency disturbances. Tables 4 and 5 make a statistical summary of all kinds of disturbances (Fig. 11).

A) momentum wheel b) refrigerator c) load scanning mechanism

Fig. 11. Schematic diagram of FY-4 satellite moving parts

Through Analyzing the dynamic environment such as internal and external interference torque of the satellite, the dynamic planning of the whole satellite is completed [16]. It can be seen from the dynamic planning diagram that there are five interference transmission paths, and corresponding measures should be taken to reduce the influence on the satellite platform and load LOS:

(1) The space disturbance torque is a low-frequency disturbance, which will affect the pointing accuracy and stability index, and it is controlled within the index range by the attitude control system;
(2) Attitude motion, orbital motion and thermal deformation motion of satellite rigid body affect the LOS orientation of load, which is controlled within the index range by on-board real-time compensation;
(3) The low-frequency disturbance of onboard load can be controlled within the index range by attitude control;
(4) The risk of coupling resonance between the scanning frequency of the mirrors of the scanning radiometer and the frequency of the flexible accessories may have a serious impact on the satellite attitude stability and the load imaging stability. In order to solve this problem, firstly, frequency avoidance is carried out, secondly, the function of feedforward torque compensation is added, and the function of on-orbit injection to modify the load working mode is designed to ensure that resonance does not occur.
(5) The high-frequency disturbance of momentum wheel, refrigerator and other components can be transmitted to the payloads through the satellite structure, which affects the imaging stability. The flywheel flexible bracket is designed to restrain the influence of micro-vibration on load vibration and ensure the stability index of the GIIRS (Fig. 12).

Table 4. Statistics of moving parts (high frequency disturbance) on the satellite

Serial number	Disturbance source	Form of action	Frequency (Hz)	Interference amplitude
1	Reciprocating motion of east and west mirrors of AGRI	Moment around the yaw direction	0.125–2.5 discrete frequency, the main working frequency is 0.43478	Within 0.16Nm
2	Step detection by scanning mirror of GIIRS	Moment around the yaw direction of the star	The main working frequency is 0.02564	Within 0.1Nm
3	Moving mirror reciprocating motion of GIIRS	Three axis	1.33 or 0.77	Very small, negligible
4	Scan motion of GHI	Moment around the yaw direction	The main working frequency is 0.0387 or 0.0774	0.026Nm max
5	Windsurfing drive	Moment around the pitching direction	0.164 and its frequency doubling	0.03Nm
6	Mechanical refrigerator of GIIRS	Disturbance torque along star rolling direction	50 and its frequency doubling	Very small, negligible
7	Mechanical refrigerator of GHI	Disturbance torque along the rolling direction	50 and its frequency doubling	-
8	Momentum wheel vibration	Three axis	13–80, including several discrete frequencies and noise bands	It can be ignored after using vibration damping bracket

4.2 High Precision Attitude Measurement and Determination Technology

Accurate attitude measurement is the basis of high-precision directional management. FY-4 satellite adopts the attitude determination scheme of gyros and star sensors. The low-frequency measurement information of gyro is used to filter and suppress the high-frequency measurement noise of star sensors, and the absolute measurement benchmark of star sensors is used to accurately estimate gyro drift. Through multi-sensor information fusion, achieve accurate measurement of satellite attitude angle and angular velocity [17].

(1) High-precision star sensor combination
Compared with other satellites, FY-4 satellite is equipped with a multi-probe star sensor, which uses a single processor to process the image information from multiple probes, and outputs the three-axis attitude [18] with the same precision through

Table 5. Statistics of spatial disturbance torque (low frequency disturbance)

Serial number	Disturbance source	Form of action	Frequency (Hz)	Interference amplitude
1	Solar pressure moment	Moment in rolling or yawing direction	1.16×10^{-5}(orbital period)	Within 4.5×10^{-4} Nm
2	Single-wing deflation [15]	Early stage of orbit	–	Within 0.037 Nm
3	Geomagnetic moment	Three axis	1.16×10^{-5}(orbital period)	1.0×10^{-6} Nm
4	Earth oblateness, solar light pressure and so on cause orbital drift	Orbit drift motion	1.16×10^{-5}(orbital period)	Causing a drift of 0.25°
5	Thermal deformation	Structural deformation leads to pointing accuracy error of remote sensing instrument	Main frequency: 1.16×10^{-5}, thermal environment change frequency band: 0.0001 ~ 0.001	Within 1000 μrad

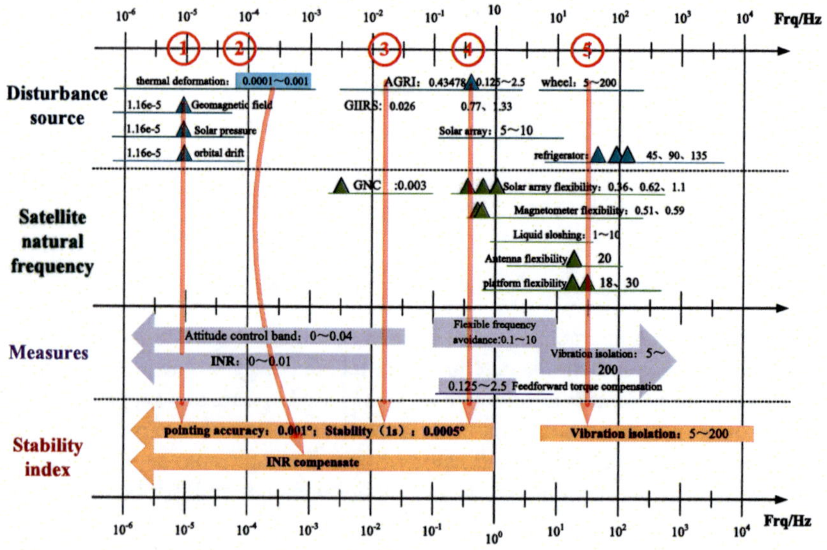

Fig. 12. Precision planning results of satellite dynamic spectrum

appropriate star map fusion algorithm, and the optical axis accuracy can be improved by 30% compared with the single-probe star sensor.

When the star sensor is observing the star, because the flying speed of the satellite in the inertial space is not parallel to the star vector, the actual observation optical axis will be far away from the speed vector, that is, it will rotate the angle $\Delta\theta$ around the plane normal composed of the star vector and the speed V, and the error is the aberration of the star sensor observing the star. The optical axis pointing error angle $\Delta\theta$ can be approximately written as:

$$\Delta\theta = \frac{v}{c} \sin\theta \tag{1}$$

Where c is the speed of light, θ is the angle between the satellite flight speed and the star vector; V is the speed of the satellite in the inertial system, which is the combination of the earth's speed around the sun and the satellite's speed around the earth. For high-orbit satellites, the calculation error of the load line-of-sight direction caused by aberration is about 19″–24″ [19], which seriously affects the absolute attitude measurement accuracy and needs to be corrected.

(2) High precision gyro
The high-precision gyro configured by FY-4 satellite can accurately estimate the attitude determination error caused by gyro drift by determining the noise covariance matrix Q of the gyro, and its analytical formula is as follows:

$$Q = \begin{bmatrix} E[f_{i+1}^2] & E[f_{i+1}g_{i+1}] \\ E[f_{i+1}g_{i+1}] & E[g_{i+1}^2] \end{bmatrix} = \begin{bmatrix} \sigma_v^2 T + \sigma_u^2 T^3/3 & -\sigma_u^2 T^2/2 \\ -\sigma_u^2 T^2/2 & \sigma_u^2 T \end{bmatrix} \tag{2}$$

Among them, σ_v Is the gyro angle random walk coefficient, σ_u Is the random walk coefficient of gyro angular rate.

The accuracy result of relative attitude determination obtained by using gyro alone is shown in Fig. 13. Compared with the commonly used star sensors, the gyro's relative attitude measurement accuracy is higher, and the total gyro error is less than 1.2″ within 50 s. However, the gyro output has a drift characteristic, and the error accumulates with time. For example, after 300 s, the accumulated error exceeds 3″, so it needs to be combined with the star sensor.

(3) Joint attitude determination of gyro and star sensor.
Gyro has high short-term accuracy and is not restricted by environment, but there is accumulated error with time. The star sensor has no accumulated error and can calibrate the gyro drift, while the gyro low-frequency attitude information can restrain the high-frequency noise of the star sensor and improve the attitude determination accuracy of the system. Given that statistical characteristic of gyro random error and star sensor measurement noise, kalman filtering algorithm is used to predict and update the attitude estimation, and finally the analytical formula of attitude determination error is obtained. After reaching the steady state of filtering, the attitude determination accuracy can be better than 2″. The formula for calculating the standard deviation of attitude error is:

$$\sigma_\infty = T^{\frac{1}{4}} \sigma_n^{\frac{1}{2}} (\sigma_v^2 + 2\sigma_u \sigma_v T^{\frac{1}{2}})^{\frac{1}{4}} \tag{3}$$

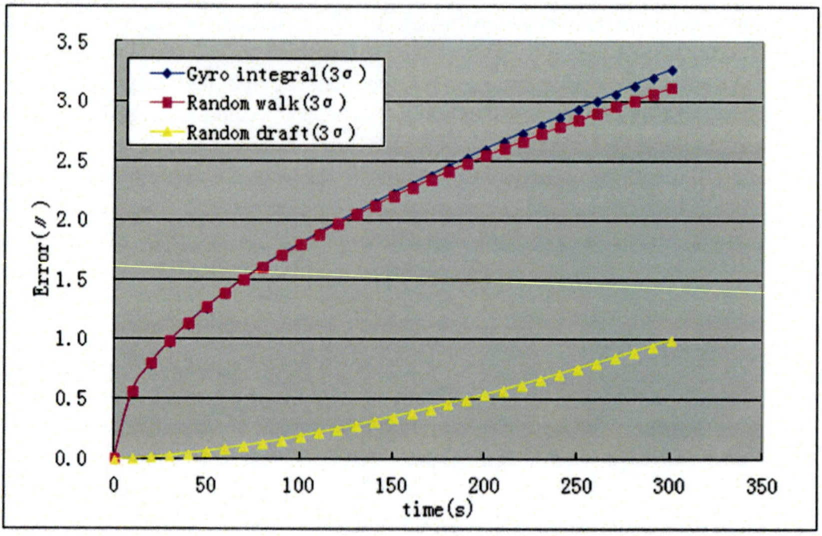

Fig. 13. Integral error angle curve of fiber optic gyroscope in different time

Among them, σ_v Is the gyro angle random walk coefficient, σ_u Is the random walk coefficient of gyro angular rate, σ_n Is star sensor noise, T Is the gyro integration time, σ_∞ Determine the error for the integrated attitude.

(4) Identification of long-period thermal deformation
Long-period thermal deformation identification is the key problem of satellite attitude determination with high accuracy. During the 24-h observation of the satellite, the satellite body, attitude sensor and load are all affected by different degrees of thermal deformation alternately, which makes the payload benchmark and attitude sensor benchmark have long-period deviation. To solve this problem of long-period thermal deformation, a geometric deformation calibration method of observing stars is proposed as continuous and high-precision absolute reference. The long-period thermal deformation law is identified and predicted by observing stars simultaneously with payload and star sensor (Fig. 14).

4.3 High Precision Attitude Control Technology

Attitude control technology is the key link of high-precision directional management, which is used to overcome the disturbance of various moving parts and the disturbance torque of space environment, and realize the high pointing accuracy and stability of the satellite.

(1) Feedforward torque compensation technology
The three loads of FY-4 satellite can detect the global disk by driving the two-dimensional scanning mirror. However, the interference torque generated during scanning will cause the satellite platform to shake, which will seriously affect the imaging effect and the line-of-sight pointing. The feedforward torque compensation

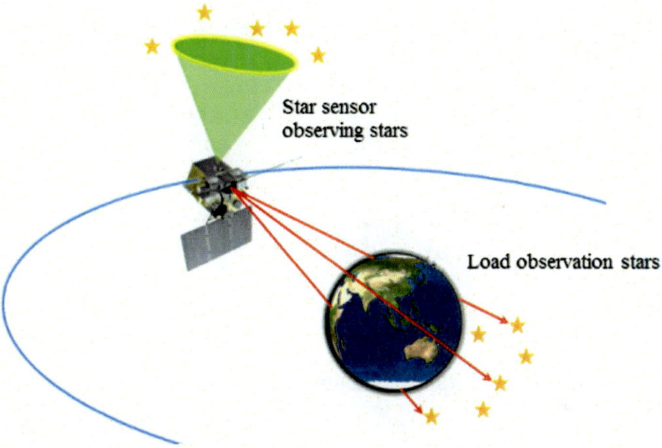

Fig. 14. Identification method of long-period thermal deformation

scheme is designed, and the angular momentum change trend of the satellite platform is calculated according to the law of scanning movement. Then, the momentum wheels are used to compensate the angular momentum change every 32 ms to ensure the stable imaging of the load and the accurate orientation of the LOS (Fig. 15).

The on-orbit test results show that the attitude stability performance of the satellite is improved from 0.0011°/s to 0.00035°/s after the feedforward torque compensation is turned on (Fig. 16).

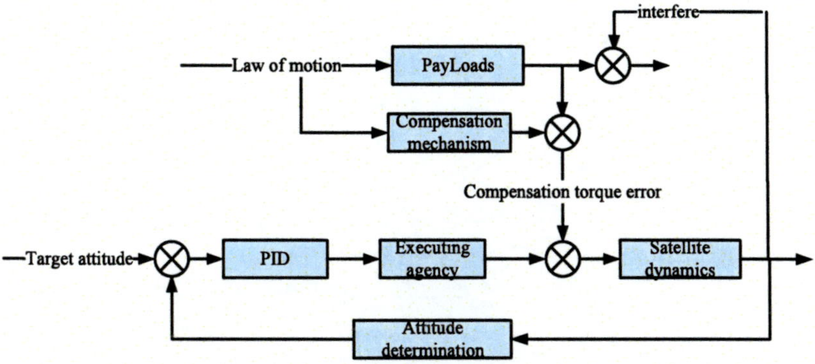

Fig. 15. Schematic diagram of feedforward torque compensation

(2) Zero-space angular momentum management technology
FY-4 satellite platform is equipped with six 50 Nms momentum wheels, which are in a pentagonal pyramid configuration. The control system adopts the zero-space angular momentum management mode, so that the actual rotational speed and angular momentum of the momentum wheel are consistent with the expected rotational speed and angular momentum [20, 21]. Compared with the traditional tri-orthogonal

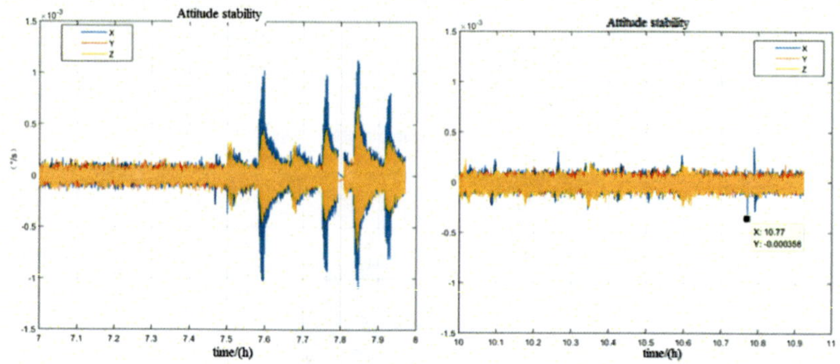

Fig. 16. Influence of feedforward torque compensation on attitude stability

configuration method, the zero-space angular momentum management technology has the following advantages: first, the zero-space angular momentum management can prevent the momentum wheel from rotating over zero, improve the accuracy and stability of attitude control; Secondly, through the optimized configuration design of the pentagonal pyramid, it is possible to ensure that the momentum wheel combination has the maximum angular momentum and torque output capacity; Third, the zero-space angular momentum management can redistribute the three-axis control torque to avoid the influence on satellite stability caused by the control failure or capability degradation of a wheel. In-orbit flight shows that, through zero-space angular momentum management, the satellite attitude control accuracy is better than 0.003° (Figs. 17 and 18).

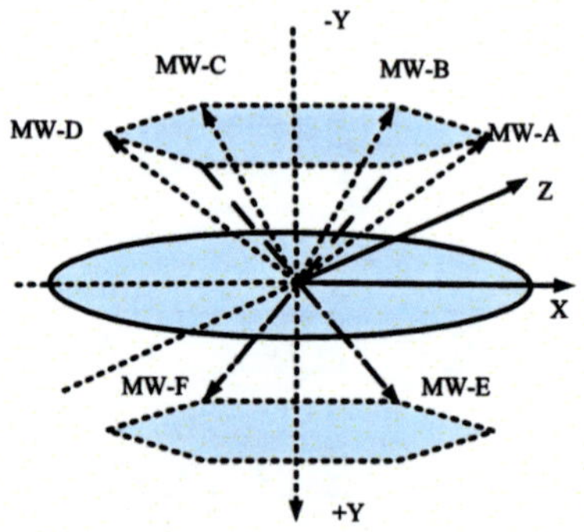

Fig. 17. Schematic diagram of wheel angular momentum vector

(3) Autonomous unloading technology of angular momentum
FY-4 satellite adopts the single-wing sail configuration, the angular momentum caused by solar light pressure increases by 35 Nms every day, and the control system adopts six momentum wheels (4 working and 2 cold standby) to absorb the angular momentum change caused by solar pressure. At the same time, in order to

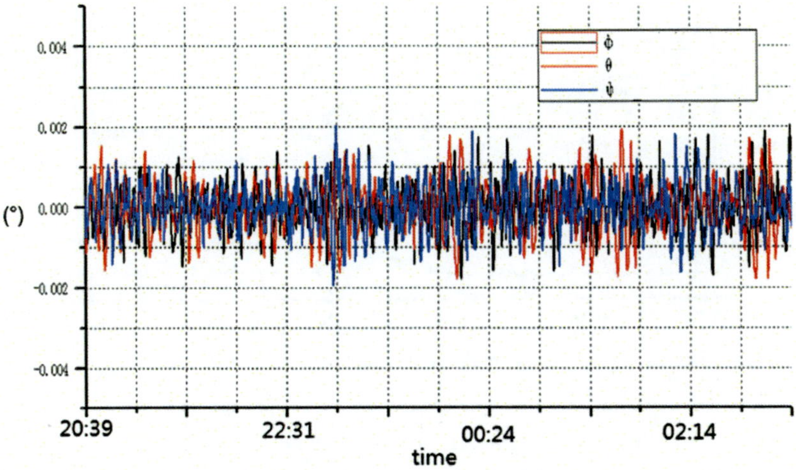

Fig. 18. Measured results of attitude control in orbit

avoid the angular momentum saturation of momentum wheel, the satellite control system adopts the autonomous jet unloading scheme, and the angular momentum unloading lasts for 10 min at midnight every day (Figs. 19 and 20).

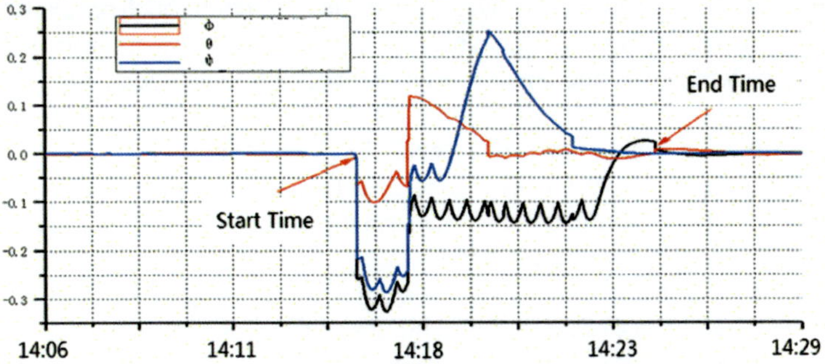

Fig. 19. Attitude change curve during unloading

(4) Autonomous fault diagnosis technology
In order to ensure the normal on-orbit operation of the satellite, the control system has backed up the measuring sensors and actuators. When a single momentum wheel fails, a star sensor fails, a gyro fails or other abnormal situations, the satellite can independently complete the fault diagnosis and reconstruction operation in orbit, and in the case of single fault, the reconstruction process will not affect the normal business operation of the load.

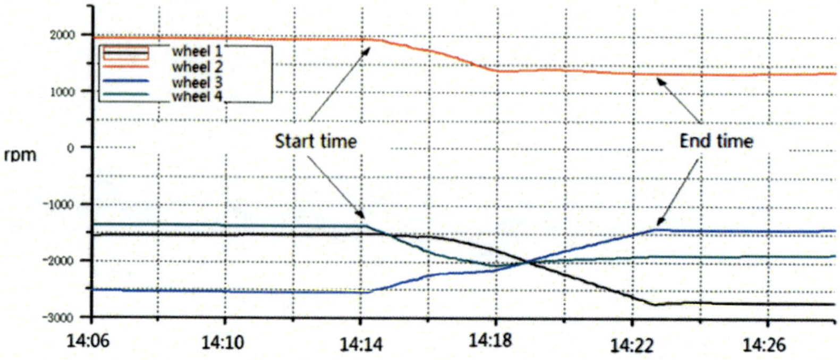

Fig. 20. Speed change curve of momentum wheel during unloading

4.4 Micro-vibration Measurement and Suppression Technology

Micro-vibration measurement and suppression technology is used to solve the high-precision directional management under the influence of high-frequency vibration. With the increasing complexity of satellite functions and more and more rotating parts, the micro-vibration of satellite platform is inevitable. The micro-vibration of FY-4 satellite (generally 1.0×10^{-5}g–1.0×10^{-2}g, frequency 0–200 Hz) is mostly caused by the motion of momentum wheel, refrigerator and other moving parts and structural coupling vibration. It has great influence on interference optical sensitive load. Micro-vibration measurement and suppression technology is dedicated to solving such problems, mainly including micro-vibration simulation analysis and impact assessment, micro-vibration suppression system design and micro-vibration on-orbit measurement [14].

(1) Micro-vibration simulation analysis and impact assessment
The key of micro-vibration simulation analysis and impact assessment lies in the establishment of system coupling model and the selection of disturbance assessment model. The former is to estimate the micro-vibration environment of the load and determine its frequency range and its corresponding magnitude. The latter is to evaluate the influence of micro-vibration disturbance on sensitive load.
(2) Design of micro-vibration suppression system
It is necessary to evaluate the vibration reduction efficiency of commonly used vibration suppression materials and investigate their applicability. On this basis, select the appropriate vibration isolation technology to design the vibration isolation system (Fig. 21).
(3) On-orbit measurement of micro-vibration
Micro-vibration measurement system focuses on solving the difficulties of system background noise suppression, micro-signal amplification and anti-interference design. Through on-orbit measurement, the micro-line vibration near FY-4 satellite load is less than 1mg, which meets the working requirements of the load (Fig. 22).

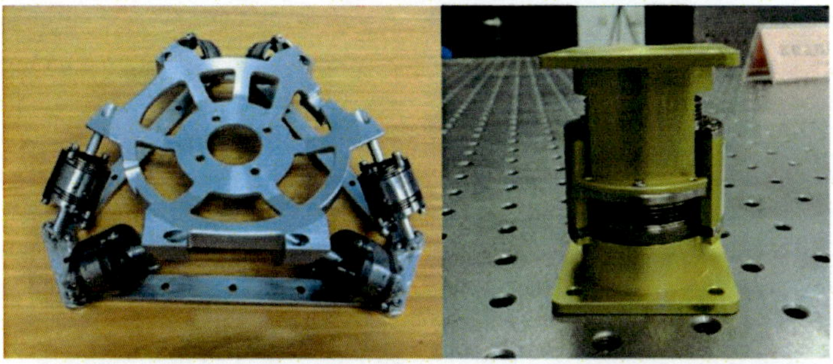

Fig. 21. Momentum wheel vibration isolation bracket and liquid damping vibration isolation unit

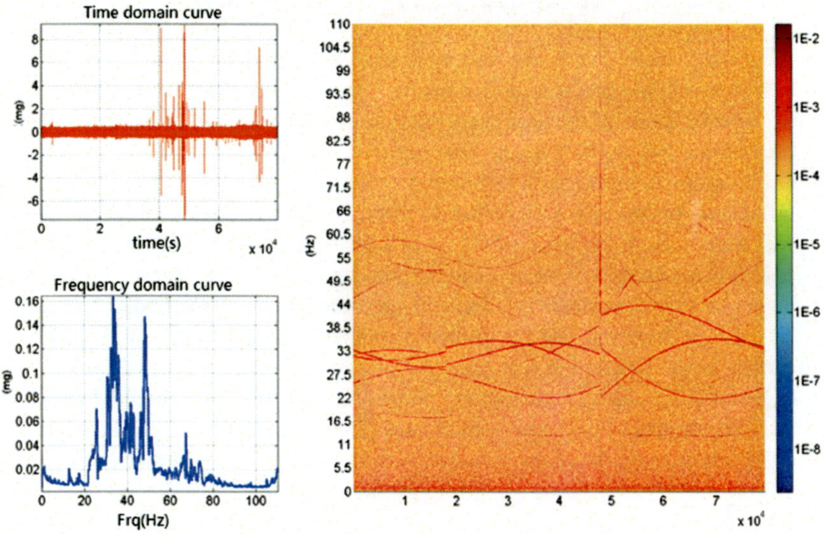

Fig. 22. On-orbit micro-vibration data

4.5 Image Navigation and Registration Technology

In order to ensure the accurate pointing of the three loads of FY-4 satellite, it is necessary to control several benchmarks. Image navigation and registration (INR) technology can solve the problem of inconsistent imaging datum of multiple loads. Each load is taken as an independent imaging datum, and the conversion relationship between each load datum and satellite attitude measurement datum is established. Independent compensation is made for the actuator by the scanning mirror inside the loads, and each load is guided to scan and image the earth according to the predetermined trajectory in real time, so that the imaging benchmark of each load on the satellite is unified with the attitude measurement benchmark. According to the analysis of disturbance sources on board, image navigation and registration include attitude motion compensation, orbit motion compensation and long-period thermal deformation compensation.

(1) Attitude motion compensation technology
The satellite attitude control system is constrained by the system bandwidth, so it is difficult to realize the precise pointing control at the arc-second level. Therefore, a real-time high-bandwidth attitude motion compensation scheme on the satellite is designed, which utilizes the satellite attitude angle determined by the high-precision star sensor and gyro, combined with the rotation of the load scanning mirrors, On-board real-time high-frequency (5 ms) calculation is needed to offset the attitude deviation. The attitude control residual of satellite platform is eliminated by adding compensation amount to the load scanning mirrors. The formula for calculating the two-dimensional compensation of on-board attitude is as follows:

$$\begin{aligned} \Delta\varepsilon &= (\psi \sin 2\eta + \theta \cos 2\eta)/2 \\ \Delta\eta &= [-\varphi + (\theta \sin 2\eta - \psi \cos 2\eta) \tan 2\varepsilon]/2 \end{aligned} \tag{4}$$

Among them, $\Delta\varepsilon$ Is the east-west attitude compensation amount, $\Delta\eta$ Is the east-west attitude compensation amount, ε To load the east-west mirror angle, η Is the rotation angle of the load north-south mirror, θ Is the satellite pitch angle, φ Is the satellite roll angle, ψ Is the yaw angle.

(2) Orbit motion compensation technology
Influenced by the space perturbation force, the actual flight orbit of the satellite will deviate from the ideal orbit to a certain extent, which will cause the LOS of payloads to change constantly, and the projection of ground objects in the satellite image will move accordingly. The real-time compensation method of orbit change on board is designed, and the on-board computer calculates the currently planned surface target position in real time. According to the current actual orbit position of the satellite, the angle increment of the scanning mirror required by the current satellite pointing to the target is calculated. The load line-of-sight motion caused by orbital motion is eliminated by adding compensation amount to the load scanning mirror.

(3) Long-period thermal deformation compensation technology
When the satellite is working in orbit, the dynamic trend of thermal deformation changes in the day can be obtained to the greatest extent by star observation, and it can be transformed into the change curve of equivalent parameters. Fourier series is used to fit the changing trend of each parameter, and the characteristics of the instrument thermal deformation with time are described, so as to judge the thermal deformation trend in advance. When the instrument scans the earth, an appropriate increment is added on the scanning mirror angle to offset the influence of thermal deformation on the load line of sight.

Through the real-time compensation on the satellite, the imaging datum of the three loads on the satellite are all unified with the attitude measurement datum, which is close to the ideal state. The down-transmitted images have been registered to the fix grid of the earth, and the image navigation accuracy is better than one pixel (Figs. 23 and 24).

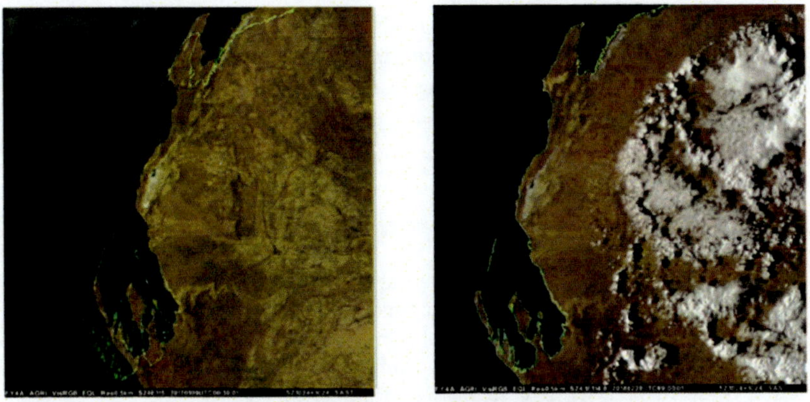

A) before compensation b) after compensation

Fig. 23. INR performance of AGRI

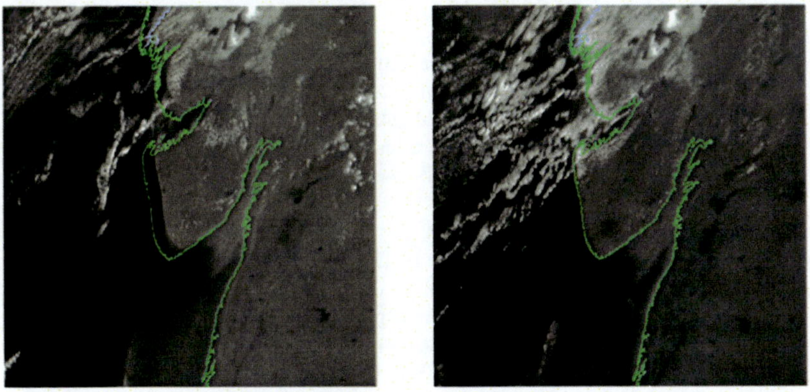

A) before compensation b) after compensation

Fig. 24. INR performance of GIIRS

5 Summary

In order to meet the demand of quantitative remote sensing of FY-4 satellite's high-precision payloads, the high-precision directional management technology is put forward, which breaks through the key technologies such as satellite dynamic spectrum precise planning, high-precision attitude measurement and determination, high-precision attitude control, micro-vibration measurement and suppression, image navigation and registration, etc. The corresponding solutions for high frequency, medium-low frequency and long-period disturbances on board are given, and the benchmark unification and independent compensation control of all loads are realized. After in-orbit flight verification, all control indexes meet the use requirements, which provides strong support for quantitative observation of remote sensing loads.

References

1. Yaohai, D.: FY-4 meteorological satellite and its application prospect. Aerosp. Shanghai **33**(2), 1–8 (2016)
2. Zhiqing, Z.: FY-4 geostationary meteorological satellite and product application introduction: seminar on application technology of new rainstorm detection data [C]. Wuhan (2011)
3. Ganquan, W., Xia, S., Weicheng, W., et al.: On board adaptation of temperature of the FY-4 meteorological satellite radiation imager. J. Infrared Millim. Waves **37**(5), 540–544 (2018)
4. Chen, G.K., Xiaowei, W., Siyu, Z., Jianwen, H., Dinglei,: Application of FY-4 atmospheric vertical sounder in weather forecast. J. Infrared Millim. Waves **38**(3), 285–289 (2019)
5. Shen Yili, L., Yongjiang, W.Y., et al.: Research of imager scan mirror's thermal distortion model for image navigation and registration. Aerosp. Shanghai **31**(2), 26–30 (2014)
6. Iwata, T.: Precision attitude and position determination for the advanced land observing satellite. Proc. SPIE 34–50 (2005)
7. NASA: GOES-R Series Data Book (2018)
8. GOES-R Program: GOES-R Series Concept of Operations (2017)
9. European Space Agency: Himawari-8 and 9 (2017)
10. Kang, L., Guodong, Q., Xinyan, L.: Highlights of attitude ans orbit control system(AOCS) for Gaofen-4 satellite with on-orbit Verification. Aerosp. Control Appl. **42**(6), 1–8 (2016)
11. Fengjing, L., Guo, L., Dengyun, Y. et al.: Gaofen-4 satellite and its application overview. Satell. Appl. (2018)
12. Lyu, W., Wang, T., Dong, Y., Shen, Y.: Imaging navigation and registration for geostationary imager. IEEE Geosci. Remote Sens. Lett. **14**(12), 2175–2179 (2017)
13. Zhiqiang, B., Cai Chensheng, L., Wang, et al.: High accuracy and high stability control technology of remote sensing satellites. Aerosp. Shanghai **34**(4), 24–34 (2017)
14. Yaohai, D., Bin, Z.X., Junfeng, S. et al.: Study on micro-vibration suppression technology of FY-4 satellite. Aerosp. Shanghai **31**(3), 20–27 (2014)
15. Zhiqiang, B., Wangxin, L.S. et al.: Analysis of disturbance torque caused by air-bleed of single solar wing. Spacecr. Eng. **27**(1), 67–74 (2018)
16. Wang, L., Mingjiang, X., Yili, S., et al.: Dynamic frequency spectrum planning of remote sensing satellite. Aerosp. Control Technol. Appl. **4**(2), 12–16 (2015)
17. Zhiqiang, B., Weiqiang, C., Xiaobu, X., et al.: Satellite attitude determination algorithm based on gyro and star-sensor. Spacecr. Eng. **20**(2), 29–34 (2011)
18. Xiong Yazhou, W., Yanpeng, C.H.: Attitude determination accuracy of multi-head star tracker based on star-image fusion. J. Chin. Inert. Technol. **10**(5), 612–617 (2016)
19. Miao, L., Lianwen, Z., Yikang, H. et al.: A method of star sensor aberration correction on-orbit. Navig. Position. Timing **1**(1), 60–63 (2018)
20. Yongjun, L., Li, Y.: Momentum envelope analysis and momentum management for spacecraft with a flywheel array. Chin. Space Sci. Technol. **37**(6), 1–9 (2017)
21. Yanning, G., Guangfu, M., Chuanjiang, L.: Design and analysis of torque distribution strategy for redundant flywheel configuration. Acta Aeronautica et Astronautica Sinca **31**(11), 2259–2265 (2010)
22. Bessho, K., Hayashi, M., Ikeda, A. et al.: An introduction to Himawari-8/9—Japan's new-generation geostationary meteorological satellites. J. Meteorol. Soc. Japan Ser. II **94**(2), 151–183 (2016)

Anti-interference of Direct Sequence Spread Spectrum System Based on Wavelet and Eigenvalue Decomposition

Zhao Tengfei[1,2](✉), Shen Zhao[1,2], Shang Jie[1,2], Li Junyao[1,2], and Liu Peijie[2]

[1] State Key Laboratory of Astronautic Dynamics, Xi'an 710043, China
zhaotengfei9527@126.com

[2] Xi'an Satellite Control Center, Xi'an 710043, China

Abstract. Limited by the processing gain of the spread spectrum system, the high power interference in the satellite space-to-earth communication cannot be suppressed effectively. It will bring high Bit Error Rate(BER) to the telemetry、track and command(TT&C) link. Based on the analysis of the Direct Sequence Spread Spectrum (DSSS) system in anti-interference, a method of anti-interference in DSSS is proposed in this paper. Combining the advantages of Wavelet Transform(WT) in singularity detection and Eigenvalue Decomposition(ED) in blind source separation, the method based on WD and ED is proposed. The simulation results indicate that the method can suppress the interference effectively and will improve the anti-interference ability of DSSS.

Keywords: Wavelet transform · Eigenvalue decomposition · Direct sequence spread spectrum · Bit error rate

1 Introduction

The interference is core problem in communication. In order to improve the communication quality, the infuluence of interference must be reduced or eliminated. Because of the strong anti-interference performance, multiple access capability, security, anti-multipath and other functions the spread spectrum communication system is widely used in military and civil communication, especially in the field of aerospace TT&C. DSSS TT&C can provide better communication link for satellite and ground equipment [1]. With the change of modern war, electromagnetic spectrum warfare has been paid more and more attention. The change results in complex electromagnetic environment for the satellite and ground equipment. The more active interferences will seriously affect the link between satellite and ground equipment. Although the DSSS TT&C has certain anti-interference ability, the limited spread spectrum gain can not ensure the safety of the link between satellite and ground equipment when the directional and high power interference occur during the war. The method of anti-interference in DSSS is divided into time domain and frequency domain. The method of anti-interference in time domain makes use of obvious difference in correlation between the interference

L. Wang et al. (Eds.): CHREOC 2022, LNEE 969, pp. 37–45, 2023.
https://doi.org/10.1007/978-981-19-8202-6_4

and spread spectrum signal. The method of anti-interference in frequency domain makes use of the features in frequency domain between the interference and spread spectrum signal. The literature [2] performed interference detection and FIR filter design in the frequency domian to complete interference cancellation in the time domain. The literature [3, 4] proposed an anti-interference method based on overlapping windowing, which can reduce frequency leakage and damage to useful signal. In literature [5], when the interference signal exists in one side and the other side is not disturbed, the amplitude of spectrum of the symmertric position corresponding to the interference is replaced to achieve interference suppression.

As an effective tool in linear algebra, ED has been successfully applied in signal processing field. In the aspect of interference suppression, some researchs have been carried out and good effect has been achieved[6]. WT is a powerful tool for signal singularity detection, because it can analyze the position and strength of singularity in signal.

Combining the advantages of WT in signal singularity detection and ED in blind source separation, an anti-interference method used in DSSS TT&C based on WT and ED is proposed in this paper. In view of difference between interference and DSSS TT&C signal after ED, using WT to find out the singular position corresponding to the eigenvalue sequence, interference subspace can be constructed to achieve the purpose of interference suppression.

2 DSSS System

The basic principle of DSSS: During the transmitting time, the useful information is multiplied by the pseudo noise(PN) sequence, after which the signal spectrum is broadened and the power spectrum density is reduced, and then carrier modulation is carried out. During the receiving time, the received signal is multiplied by the PN sequence synchronized with the transmitter, and then demodulated to obtain the useful information. The diagram is shown in Fig. 1.

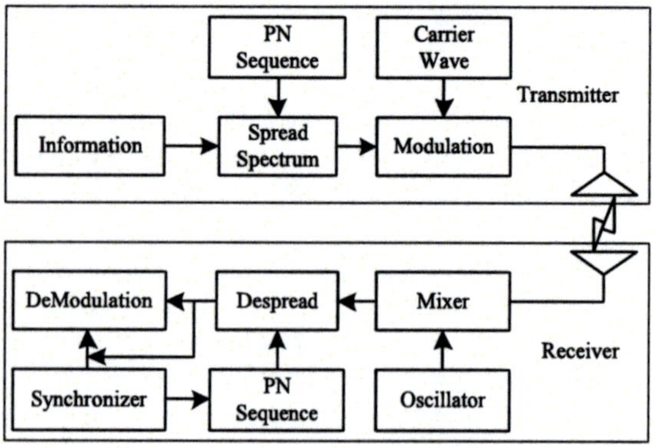

Fig. 1. DSSS diagram

In spread spectrum system, the spread spectrum processing gain and the interference tolerance are important anti-interference indicators. The calculation method is given in literature [1]. The DSSS processing gain G_p is the ratio of the bandwidth of the spread spectrum signal to the transmitted information bandwidth, or the ratio of the PN sequence rate R_c to the information rate R_a. The G_p is also known as the spreading factor, generally expressed in decibels:

$$G_p = R_c/R_a \tag{1}$$

Interference tolerance M_j refers to the number of decibels by which the receiver can withstand the interference exceeding the useful signal under the premise of ensuring the normal operation of the system.

$$M_j = G_p - \left[L_s + (S/N)_o\right] \tag{2}$$

where, L_s denotes the system demodulation loss, $(S/N)_o$ denotes the minimum output SNR required for the normal operation of the system.

Due to the limitation of communication bandwidth and spreading factor, spread spectrum gain cannot be amplified indefinitely. When the interference received by the receiver exceeds the interference tolerance of the system, a large number of error codes will be generated in the communication link, which will affect the normal operation of the whole DSSS TT&C system. In order to improve the survival and stability of the system in wartime, it is necessary to carry out the research on anti-interference in DSSS TT&C.

3 Signal Model

Affected by the interference, the spread spectrum signal input by the receiver consists of three parts:

$$r(t) = Ad(t)c(t)\cos(2\pi f_0 t + \varphi_0) + N(t) + J(t) \tag{3}$$

where, $d(t)$ denotes the signal that the information after encoded, $c(t)$ denotes the PN sequence, A denotes the amplitude, f_0 denotes the carrier frequency, φ_0 denotes the random phase, $N(t)$ denotes the thermal noise and $J(t)$ denotes the interference.

The single-frequency interference or narrow-band inteference has a very serious influence on the DSSS TT&C system, especially the artificial interference form the enemy. When the receiver receives the signal with the interference, the base band will be affected by the interference and produce a high BER. Then, the normal operation of the DSSS TT&C system is damaged [1]. For the narrow-band interference, showing a certain peak in the frequency domain, which can be approximated by the superposition of sinusoidal signals:

$$J(t) = \sum_{j=1}^{P} A_i \cos\left(2\pi f_j t + \varphi_j\right) \tag{4}$$

where, A_j, f_j and φ_j denote the amplitude, frequency and phase of the j-th single-frequency interference, and P denotes the number of the interferences.

4 Anti-interference Algorithm

4.1 Eigenvalue Decomposition

The basic idea on the ED: Using the principle that the interference and data signal are orthogonal, the correlation matrix of the original data is eigendecomposed. According to the size distribution of eigenvalues, the corresponding eigenvectors are divided into two categories. The main eigenvalues (significantly larger eigenvalues) correspond to the frequency component of strong interference, and its eigenvectors is constructed as the interference subspace I. The other eigenvecors are constructed as signal subspace G, corresponding to the remain signal components. Since I and G are orthogonal in space, the original data is projected into the interference subspace I, and the projected result is subtracted form the original data to obtain the filtered data.

The process of the anti-interference based on ED is as follows:

Suppose that $\boldsymbol{x} = [x_1, \cdots, x_M]^T$ is the vector in a certain period of time after being disturded, and x_m denotes the m-th sampling point($m = 1, \cdots, M$).

(1) Using the data vector $\boldsymbol{x}$ to construct a Hankel matrix with $L \times K, K = M - L + 1$:

$$X = \begin{bmatrix} x(1) & x(2) & \cdots & x(K) \\ x(2) & x(3) & \cdots & x(K+1) \\ \vdots & \vdots & \ddots & \vdots \\ x(L) & x(L+1) & \cdots & x(M) \end{bmatrix} \tag{5}$$

$$x_k = [x(k), \cdots, x(k+L-1)]^T, k = 1, 2, \cdots, K$$

(2) Estimate the covariance correlation matrix of X:

$$\hat{\boldsymbol{R}} = \frac{1}{M}\sum_{k=1}^{K} \boldsymbol{x}_k \boldsymbol{x}_k^T \tag{6}$$

(3) The eigenvalues of $\hat{\boldsymbol{R}}$ are obtained after ED, and the eigenvalues are arranged in descending order $\lambda_1 > \lambda_2 > \cdots > \lambda_L$. The eigenvectors corresponding to the eigenvalues are $\boldsymbol{u}_1, \boldsymbol{u}_2, \cdots, \boldsymbol{u}_L$.

(4) If

$$\frac{\lambda_r}{\lambda_{r+1}} >> \frac{\lambda_{r+1}}{\lambda_{r+2}} \tag{7}$$

Then, r is the number of main eigenvalues, and the interference subspace is established by the corresponding eigenvectors:

$$\boldsymbol{F} = span\{\boldsymbol{u}_1, \boldsymbol{u}_2, \cdots, \boldsymbol{u}_r\}, \qquad 1 \leq r < L \tag{8}$$

(5) By projecting $\boldsymbol{x}_k$ into the interference subspace, the interference data can be obtained:

$$\boldsymbol{I}_k = \sum_{i=1}^{r} \boldsymbol{u}_i^T \boldsymbol{x}_k \boldsymbol{u}_i \tag{9}$$

(6) Subtract the interference from the original data:

$$\hat{\boldsymbol{x}}_k = \boldsymbol{x}_k - \boldsymbol{I}_k \tag{10}$$

(7) Use $\hat{\boldsymbol{x}}_k$ to reconstructed the new data matrix $\hat{\boldsymbol{X}}$. Rearrange the data matrix $\hat{\boldsymbol{X}}$ and get the signal vector $\boldsymbol{x}'$ after interference suppression.

4.2 Wavelet Transform

Mutation signals, also known as siggular signals, often carry important information and are one of the important characteristics of signals. WT can analyze the position of signal singularity and strength of singularity, and it has the advantages that the Fourier transform does not have for local mutation detection of signal.

The WT of the signal in discrete time can be expressed as:

$$X_{WT} = \sum_{n=0}^{N-1} x(n)\phi_k^*(n) \tag{11}$$

where, $x(n)$ denotes the original signal, ϕ_k denotes the discrete orthogonal wavelet basis function, which is obtained by stretching and translating the mother wavelet, N denotes the length of signal, and $*$ denotes the complex conjugate operation.

$$\phi_k(n) = \frac{1}{a^k}\phi\left(\frac{n-m}{a^k}\right) \tag{12}$$

The WT maps the one-dimensional time domain signal to the two-dimensional time and scale. By adjusting the stretch factor a and the shift factor m, the wavelet of different frequency width can be obtained to match any position of the original signal, thus achieving the purpose of time-frequency localization analysis of the signal. The inverse WT can reconstruct the two-dimensional signal back to the original one-dimensional signal.

4.3 Algorithm Process

ED is a non-parametric method, which does not need to estimate the parameters of the interference. According to the orthogonality of signal and interference, the signal and interference are projected into different subspace to filter the interference. The most important part of the ED is to determine the number of principal eigenvalues, based on which to construct an accurate interference subspace to ensure the accuracy of data projection. The detection of the number principal eigenvalues can be attributed to the detection of signal singularity, which can be accurately detected by WT. Daubechies wavelet is commomly used in signal singularity detection. In this paper, Daubechies4 wavelet is used for singularity detection of principal eigenvalues sequence [7].

The detection method of principal eigenvalues number based on WT is to perform multi-scale decomposition of the eigenvalues sequence. Then, the corresponding high-frequency and low-frequency coefficients are obtained. The process is shown in Fig. 2.

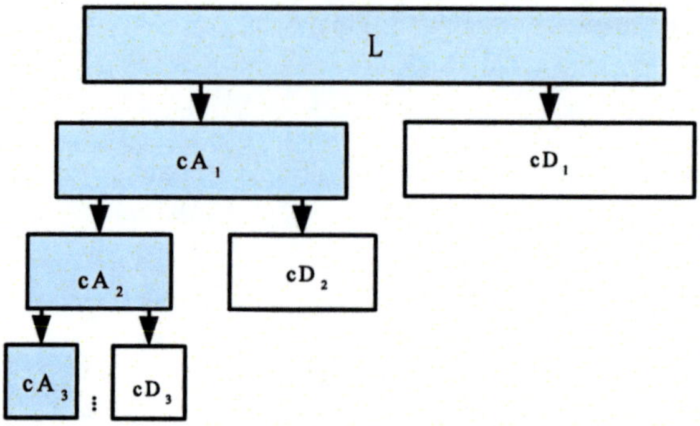

Fig. 2. WT

Where, *L* denotes the eigenvalues sequence after ED(descending order), cA and cD denote low-frequency and high-frequency coefficients, and the subscript of the coefficients denotes how many levels of decomposition. Decompose the low-frequency coefficients after the first level decomposition, and so on. The decomposition structure of signal at level i-th is $[cA_i cD_i cD_{i-1} \cdots cD_1]$. The detail of eigenvalues are contained in the hgih-frequency cofficients, which are used to reconstruct the signal to obtain the detail signals. Then, the number of the main eigenvalues can be determined by calculating the serial number corresopnding to the maximum value in the detail signals. The interference subspace will be constructed accurately by this method.

The interference suppression method based on WT and ED is consistent with the classical ED, but the difference is that the WT is used to calculate the number of principal eigenvalues, which can avoid the error loss caused by the incorrect construction of the interference subspace.

5 Simulation

This section verifies the effect of the proposed interference suppression method on DSSS TT&C. The simulation parameters are shown in Table 1.

Table 1. Simulation parameters

Parameters	Value
Carrier	50 MHz
Information rate	1000 bps
PN sequence rate	1.023 Mcps
PN sequence length	1023
Interference frequency	49.75 MHz, 50 MHz, 50.5 MHz

The results are shown in Figs. 3, 4 and 5. Figure 3 shows the spectrum of spread spectrum signal without interference, and the spectrum is relatively smooth. Figure 4 shows the spectrum disturbed by the narrow-band peaks in its frequency band. After interference suppression by using the method proposed in this paper, as shown in Fig. 5 , narrow-band interference is effectively suppressed.

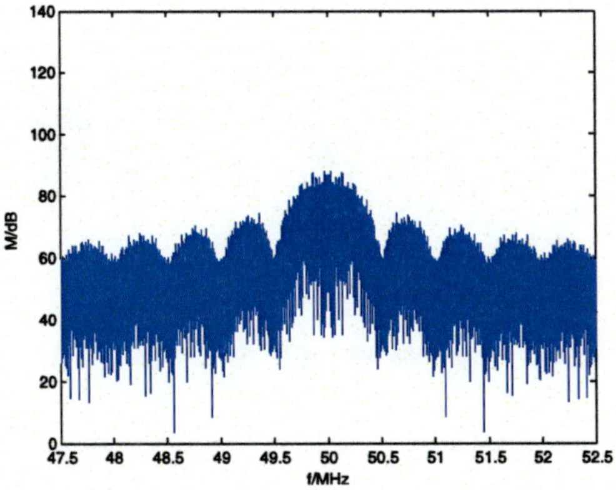

Fig. 3. Spectrum without interference

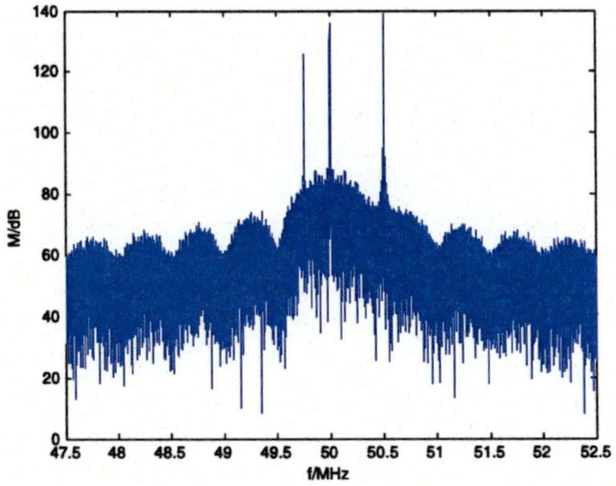

Fig. 4. Spectrum with interference

In the aspect of BER, the performance of the proposed method used in DSSS TT&C is analyzed. Figure 6 shows the BER curve of the system before and after interference suppression when it is interfered by three narrow-band interference, whose interference frequency is 49.75 MHz, 50 MHz, 50.5 MHz and interference to signal rate is 15 dB, 20 dB, 25 dB. The solid blue line in the figure represents the theoretical BER curve of the DSSS TT&C passing through a Gaussian channel. The green circle line indicates the BER of simulation, which roughly coincides with the theoretical curve. The blue rhombus line shows the BER curve with narrow-band interference, the SNR loss is large and BER increases obviously. The red deltoid line represents the BER curve after interference suppression, which basically returns to the same level as the simulation.

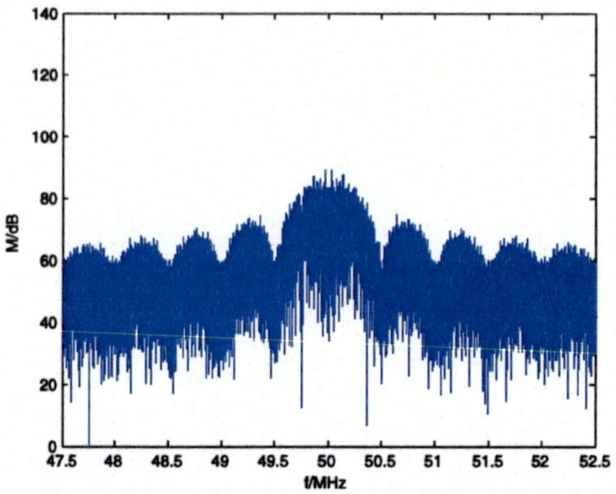

Fig. 5. Spectrum after interference suppression

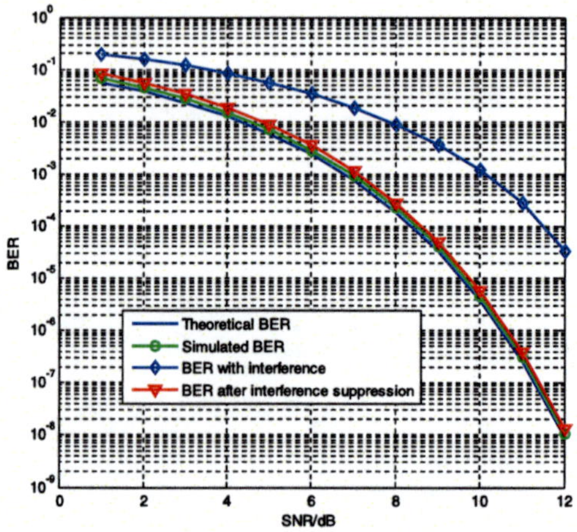

Fig. 6. BER curve

6 Conclusion

The interference tolerance of DSSS TT&C cannot be expanded infinitely, which is limited by the communication bandwidth and spread factor. Combined with the advantages of WT in singlualr point detection and ED in blind source separation, a method based on WT and ED is proposed to improve the anti-interference ability in this paper. Through theoretical analysis and simulation verification, the method can effectively improve the anti-interference ability of DSSS TT&C.

References

1. Xingwen, Z., Naian, L., Xianpu, S.: Spread Spectrum Communication and Multiple Access Technology. Xidian University Press, Xian (2004)

2. Weifeng, T., Wang Xuebin, D., Jiao.: Method for improving anti-jamming capacity of satellite direct spread spectrum TT&C systems. Commun. Technol. **51**(12), 2836–2840 (2018)
3. Zou Ning, X., Songtao, L.M., et al.: An overlap window-based frequency-domain narrowband interference suppression algorithm and its analysis. Mod. Defence Technol. **3**, 152–158 (2010)
4. Xiaowei, G.: Research on narrowband interference suppression in DSSS system. Radio Commun. Technol. **3**, 7–11 (2020)
5. Xiaolin, Z., Lili, G.: Narrow-band interference suppression in DSSS by side correlation replacement algorithm. J. Electron. Inf. Technol. **30**(1), 86–88 (2008)
6. Zhou, F., Wu, R., Xing, M., et al.: Eigensubspace-based filtering with application in narrow-band interference suppression for SAR. Geosci. Remote Sens. Lett. IEEE **4**(1), 75–79 (2007)
7. Defeng, Z.: MATLAB Wavelet Analysis. China Machine Press, Beijing (2009)

Orbit Optimization Technology for Multi-satellites Network Earth Observation

Hui Yu(✉), Jun Zhao, Jun Liu, Tiening Nie, and Bingguang Zhan

State Key Laboratory of Astronautic Dynamics, Xi'an, Shaanxi 710043, China
13630219579@139.com

Abstract. It is very difficult to achieve effective coverage of hot-spot targets in a short period of time by earth perturbation for orbit adjustment, because the coverage gap of the temporary multi-satellites network is too large in certain periods. In this paper, it is assumed that the satellite payload sensor beam type is a simple conic, the beam center of the satellite payload under the maximum lateral pendulum is calculated based on the precise ephemeris, and according to the projection ellipse of the payload on the ground, the coverage ability of the satellite payload on the ground target direction is obtained. Then the satellite is designed to encounter the ground target in the nodal period between the satellite passing the intersection of the same latitude circle of the ground target. The orbit adjustment control variables are calculated from the change of nodal period, and the optimal orbit adjustment control variables are selected from a specified time period. At last, the effective coverage of the ground target within a specified time period is achieved, and the maximum revisit interval is greatly shortened.

Keywords: Earth observation · Orbit adjustment · Projection ellipse · Nodal period · Control variables

1 Introduction

Earth observation satellites are generally retrogressive orbits [1, 2]. The designed regression orbit enables the satellite-borne microwave, visible light, multispectral, infrared, synthetic aperture and other payloads to repeatedly observe the fixed area on the ground.

As in the hotspot of events, such as local environmental pollution and dynamic monitoring of disasters, it is often necessary for multiple satellites to form an emergency temporary network to jointly conduct short-term observation of ground attention targets with multiple loads [5].

Before the network is formed, due to the different observation areas of each satellite, the observation coverage gap of the new target will be too large in certain periods after the temporary network is formed [3, 6]. In order to shorten the maximum revisit interval for the new target, a short period of time is required. It is necessary to complete the orbit adjustment of some networked satellites, so that these satellites can cover the observation target within the specified time period.

L. Wang et al. (Eds.): CHREOC 2022, LNEE 969, pp. 46–55, 2023.
https://doi.org/10.1007/978-981-19-8202-6_5

Since coverage of hot spot targets observation needs to be completed within a short period of time, and the fuel carried by the satellite is limited, saving fuel is also the preferred way of orbit adjustment under constraints.

It is difficult to achieve coverage and observation of hot targets in a short time by means of long-term orbit adjustment with the earth's perturbation force. In this paper, by calculating the longitude difference between the satellite load and the ground target at the same latitude, the beam projection of the satellite load and the ground target of interest are designed to meet each other. The increment of the intersection period is used to describe the orbit control variables of the satellite, and the most fuel-saving orbit adjustment control variables under the constraint condition is selected.

2 Load Beam Center

In order for satellites to jointly observe ground targets in a specific period of time through temporary networking, the prerequisite is that each satellite-borne payload must effectively cover ground target in a specified period of time. To achieve effective coverage, it is necessary to quickly grasp the coverage ability of the satellite to the ground area, and the key is to accurately calculate the intersection of the satellite payload beam center and the earth. In this paper, the normal vector of the orbit plane is calculated by using the satellite data under earth-centered, earth-fixed frame (ECEF). The center direction of the load beam is further determined by the normal vector and the geometric relationship between the satellite and the earth center line. The elevation error is close to zero by iteration, which greatly reduces the elevation error by using the average radius of the earth.

If the position vector $\boldsymbol{s} = [x_s, y_s, z_s]$ and velocity vector $\boldsymbol{v} = [d_x, d_y, d_z]$ of satellite S under ECEF, the earth center is **o**, $\boldsymbol{h}$ is the normal vector of the orbit plane, SN is the direction vector of the satellite load beam center, E is the intersection point of the satellite load beam center and the earth surface (as shown in Fig. 1), and let the rolling direction lateral pendulum angle of the satellite survey be $\boldsymbol{\beta}$, then $\angle OSN = \beta$, SN and h intersect at N, and the unit normal vector perpendicular to the orbit plane $\boldsymbol{h_o}$ can be expressed as follows [7, 8]:

$$h_o = \frac{h}{|h|} = \frac{s \times v'}{|s \times v'|} \tag{1}$$

In the formula, $v' = v + v_{\omega e}$, $v_{\omega e}$ is the direction vector of the earth's rotation speed under ECEF frame.

When the satellite geocentric distance is $\boldsymbol{r}$, the vector $\boldsymbol{ON}$ can be expressed as $r \tan \beta h_o$, and the coordinates of point N under ECEF frame (x_n, y_n, z_n) can be obtained. Then the beam center point $E(x_e, y_e, z_e)$ to be solved is on $\boldsymbol{SN}$, that is, the three points of satellite S, E and N are collinear, satisfying $SE = kSN$ (k is the proportional coefficient), and the distance between E and geocentric o is the earth radius R_e [7], namely:

$$\begin{cases} x_e - x_s = k(x_n - x_s) \\ y_e - y_s = k(y_n - y_s) \\ z_e - z_s = k(z_n - z_s) \end{cases} \tag{2}$$

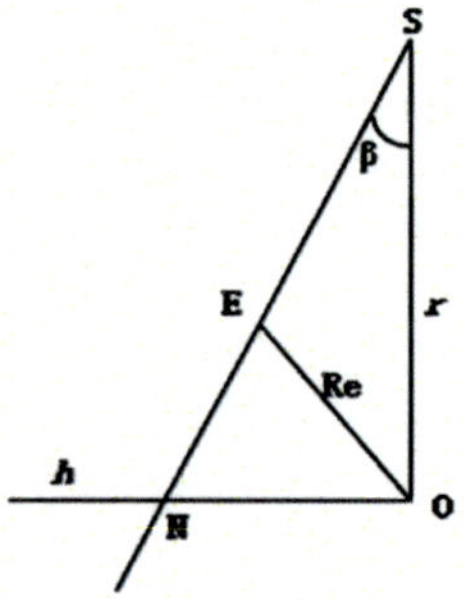

Fig. 1. Geometric schematic of beam center

$$\sqrt{x_e^2 + y_e^2 + z_e^2} = R_e \tag{3}$$

Combined solution:

$$k = \frac{-l_1 \pm \sqrt{l_1^2 - l_2 l_0}}{l_2} \tag{4}$$

In formula:

$$\begin{cases} l_0 = x_s^2 + y_s^2 + z_s^2 - R_e^2 \\ l_1 = x_s(x_n - x_s) + y_s(y_n - y_s) + z_s(z_n - z_s) \\ l_2 = (x_n - x_s)^2 + (y_n - y_s)^2 + (z_n - z_s)^2 \end{cases}$$

To ensure that the rolling lateral swing angle β and the beam center point E are on the same side of the satellite, the value of k in formula (4) should be taken smaller and β should be smaller than the earth radius angle ρ [7–9]:

$$\rho = arcsin(\frac{R_e}{\|r\|}) \tag{5}$$

Set the termination condition $\left|\sqrt{x_e^2 + y_e^2 + z_e^2} - R_e\right| \leq \varepsilon$, the beam center point $E(x_e, y_e, z_e)$can be obtained by iteration.

3 Load Coverage Capability

The specific ground target to be observed is R, and its position coordinates under ECEF frame are (x_r, y_r, z_r), and the geodetic coordinates are (lon_r, lat_r, h_r). This paper only discussed that the satellite load beam is simple conical, and the cone angle is 2η, that is, the beam angle is 2η. Due to the influence of the satellite lateral swing angle and the earth curvature, the projection of the load beam on the earth surface is approximately an irregular elliptical geometry figure. In the direction of the satellite lateral swing, it is a semi-ellipse with the semi-major axis and semi-minor axis are a_1 and b, and in the opposite direction of the lateral swing, it is a semi-ellipse with the semi-major axis and semi-minor axis are a_2 ($a_1 > a_2$) and b. The two semi-ellipses form an irregular ellipse with a semi-minor axis b as shown in Fig. 2. The center of the ellipse is a projection point E of a satellite load beam center on the ground. Vector GE and ER composed of point E, sub-satellite point G and ground target R. The vector angle of the vector ***GE*** and ***ER*** is ∠GER. Thus, ∠GER can be denoted as:

$$\angle GER = \arccos\left(\frac{GR \cdot ER}{\|GR\|\,\|ER\|}\right) \tag{6}$$

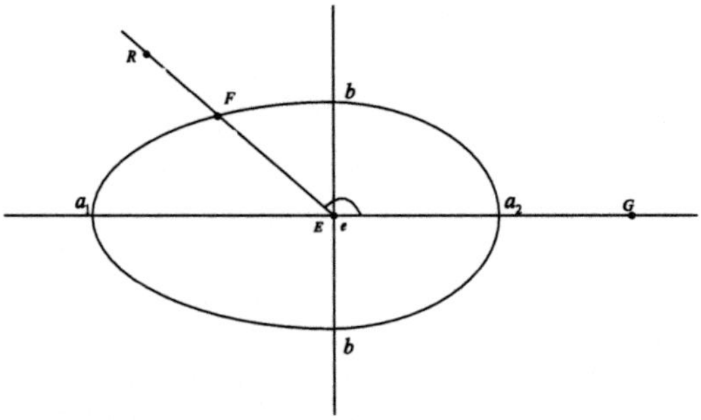

Fig. 2. Schematic diagram of satellite load projection

The angle between the satellite load beam boundary and the beam center is shown in Fig. 3. The angle between $(\beta - \eta, \beta + \eta)$. The vector angle ∂ between the beam center and the geocentric vector SO.

$$\partial = \arccos\left(\frac{ES \cdot OS}{|ES||OS|}\right) \tag{7}$$

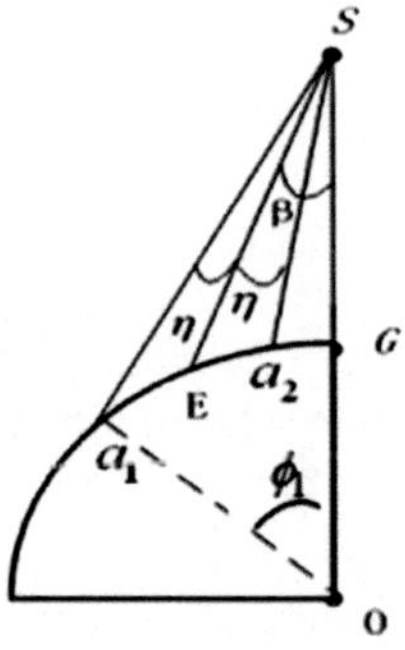

Fig. 3. Schematic diagram of satellite load lateral swing coverage area

The nadir angles ∂' of the load beam center point E and the projection ellipses a_1, a_2 and b of the beam on the ground are ∂'_0, $\partial'_1 = \partial - \eta$, $\partial'_2 = \partial + \eta$ and $\partial'_3 = \eta$, respectively ϕ_i $(i = 0,1,2,3)$ is the corresponding geocentric angle [6, 10, 13–15].

$$\phi_i = \arcsin\left(\frac{\|so\| \sin \partial'}{\mathrm{R_e}}\right) - \partial' \tag{8}$$

Then the radius arc lengths corresponding to points a_1, a_2 and b in the projection ellipse region are:

$$\begin{cases} d_1 = \dfrac{1}{2}(\phi_1 - \phi_0)\mathrm{R_e} \\ d_2 = \dfrac{1}{2}(\phi_2 - \phi_0)\mathrm{R_e} \\ d_3 = \dfrac{1}{2}\phi_3\mathrm{R_e} \end{cases} \tag{9}$$

A reference coordinate system is established with beam center E as the origin of the coordinate system, the direction of EG is the forward direction of x axis, and the operation direction of satellite is orthogonal to x-axis and is the forward direction of y axis. If the vector angle ∠GER is located in the second and third quadrants of the coordinate system plane, the coverage capability FE of the satellite load in the ground target direction is :

$$FE = \sqrt{\frac{d_1^2 d_3^2}{d_3^2 \cos^2 \angle GER + d_1^2 \sin^2 \angle GER}} \tag{10}$$

Otherwise, ∠GER is located in the first and fourth quadrants, and the coverage capability of FE is:

$$FE = \sqrt{\frac{d_2^2 d_3^2}{d_3^2 \cos^2 \angle GER + d_2^2 \sin^2 \angle GER}} \tag{11}$$

Because the satellite orbit is multiple circles and each circle has a vertical arc length, the vertical arc length is the shortest distance from the observation target to the satellite orbit. The cover sequence FE_n (n = 1, 2, …, ∞) of load can be obtained from the above equation. Define Δ is $\Delta = (FE_{\mathrm{n-1}} - FE_{\mathrm{n}})(FE_{\mathrm{n}} - FE_{\mathrm{n+1}})$. The shortest arc length sequence FE_j (j = 1, 2, ..., m) of the satellite circle passing through the intersection period lat_{r} can be obtained when the sign changes of Δ in per T interval.T is the orbital period of the satellite, $\mathrm{T_d}$ is the time period from the orbit control time to the time when the payload covers the observation target. m [16–18] is:

$$\mathrm{m} = \left\lceil \frac{\mathrm{T_d}}{\mathrm{T}} \right\rceil$$

Then the coverage longitude of the shortest arc length FE_{j} on the ground is:

$$FE_{\mathrm{lonj}} = FE_{\mathrm{j}} / lon_{\mathrm{eq}} \tag{12}$$

where lon_{eq} is the arc length corresponding to the latitude circle lat_r of the ground target observation point R, e_E is the eccentricity of the earth meridian circle,

$$lon_{eq} = \frac{2\pi a_e \cos lat_r}{360\sqrt{1 - e_E^2 \sin^2 lat_r}}.$$

4 Observation Design and Orbit Adjustment

Due to the longitude difference between the load beam and the ground target, orbit adjustment is needed to eliminate the longitude difference, then the load beam will effectively cover the ground target when it meets the ground target. If the longitude difference is eliminated, there will be changes of the nodal period in satellite orbit, and the increment of the nodal period is used to describe the satellite orbit control variables, so as to achieve the goal of covering at the specified time.

If the control time of the satellite orbit is T_c, the joint observation of the satellite requires that the time of the target should be completed within T_d period after the control time T_c. T_{sj} is the intersection period satellite passing through the target at the same latitude lat_r per cycle.

$$T_{sj} = 2\pi\left(\frac{a}{GM}\right)^{3/2} \tag{13}$$

where GM is the Earth's centric gravitation constant, $GM = 398600.5 \times 10^9 m^3/s^2$, a is the semi-major axis of orbit.The time difference ΔT_{sc} between T_{sj} and the orbit adjustment time T_c is: $\Delta T_{scj} = T_{sj} - T_c$. If the satellite observes the target at T_{sj}, Δn is the orbit adjustment operation circle starting from the orbit adjustment time T_c: $\Delta n_j = \frac{\Delta T_{scj}}{T}$. Longitude difference Δlon between satellite payload beam and ground target at the same latitude of the target can be denoted as:

$$\Delta lon_j = lon_{sj} - lon_r - FE_{lonj} \tag{14}$$

where, lon_{sj} is the geodetic coordinate longitude of sub-satellite point G at T_{sj}.

Δlon_j describes the longitude difference between the satellite payload beam and the ground target when the satellite payload beam passes through the same latitude circle of the ground target, and ΔT [16, 17]is to eliminate Δlon_j increment of the satellite orbit period.

$$\Delta T_j = \frac{\Delta lon_j}{\Delta n_j \omega_e} \tag{15}$$

Then, during the T_d period, the semi-major axis adjustment required by the satellite coverage target is Δa.

$$\Delta a_j = a - a_j' = \sqrt[3]{\frac{(T - \Delta T_j)^2 GM}{4\pi^2}} \tag{16}$$

The sequence Δa_j of satellite orbit adjustment control variables in $(T_c, T_c + T_d)$ is calculated, and the minimum control variables Δa are selected from Δa_j. Then these control variables are the minimum control variables required to complete the observation in the period from the starting time T_c to $T_c + T_d$ of the observation requirements.

$$\Delta a = \min\{\Delta a_j\} \tag{17}$$

5 Simulation Analysis

Supposing that the maximum lateral swing angle of satellite A and satellite B in the rolling direction is 30°, and orbital elements are shown in Table 1. If satellite A and satellite B are temporarily networked, the ground target R is explored, and the geodetic coordinates of the observation target R are (15.12, 117.85, 0).

Table 1. Kepler orbital elements of satellites A and B

Time (UTC)	a/[km]	e/[°]	i/[°]	Ω/[°]	ω/[°]	M_0/[°]
****-07-09T10:20:00	6867.94	0.0019	97.288	133.265	34.82	186.21
****-07-09T10:20:00	6867.94	0.0019	97.288	223.260	34.82	186.21

Taking ****-07-10 00 : 00 : 00 as the starting time, the visible time of the ground observation target are shown in Table 2.

Table 2. Visible period of target observation of satellites A and B

Visible start time (UTC)	Visible end time (UTC)	Visible time (s)	Coverage gap (h)
****-07-10 23:50:20	****-07-10 23:51:40	41.70	23.8
****-07-11 05:29:00	****-07-11 05:29:59	41.89	5.6
****-07-12 11:22:49	****-07-12 11:23:38	72.17	29.8
****-07-14 17:31:37	****-07-14 17:32:45	77.21	54.1
****-07-15 05:27:37	****-07-15 05:28:32	23.01	11.9
****-07-15 23:25:25	****-07-15 23:26:25	53.53	17.9
****-07-16 11:21:17	****-07-16 11:22:20	68.58	11.9
****-07-18 17:30:19	****-07-18 17:31:13	78.54	54.1
****-07-19 23:23:58	****-07-19 23:25:02	61.79	29.8

As can be seen from Table 2, satellites A and B observed the target , the visible duration was 518s and the total observation gap was about 238 hours. There was no target covered on ****-07-13, and the large coverage gap was composed of two sections of about 54 hours, which were as follows: ****-07-12 11:23:38 to ****-07-14 17:31:37, ****-07-16 11:22:20 to ****-07-18 17:30:19.

Choose to adjust the orbit of satellite A, and the control time is completed within 12 h after ****-07-10 02:00:00. Then the orbit control time and control variables are shown in Table 3.

Table 3. Orbit control time and control variables of satellites A

Control time (UTC)	The longitude difference at the same latitude of the load beam and the target	Orbit semi-major axis control variable (km)
****-07-10 01:11:01	−68.14	3178.652
****-07-10 01:50:15	117.79	−3948.306
****-07-10 02:45:31	−44.48	1546.355
****-07-10 03:24:45	141.41	−3960.966
****-07-10 04:20:01	−20.86	549.093
****-07-10 04:59:15	165.04	−3970.003
****-07-10 05:54:31	2.76	−4.350
****-07-10 06:33:45	−171.33	5315.505
****-07-10 07:29:02	26.39	−527.154
****-07-10 08:08:16	−147.71	3668.012
****-07-10 09:03:32	50.02	−927.838
****-07-10 09:42:46	−124.07	2620.469
****-07-10 10:38:02	73.65	−1245.210
****-07-10 11:17:16	−100.45	1852.786
****-07-10 12:12:32	97.28	−1503.050
****-07-10 12:51:46	−76.82	1255.891

It can be seen from the table that the optimal control variable is 4.35km of orbit reduction, and the orbit adjustment time is ****-07-10 05:54:31. After the orbit adjustment of satellite A, the observation time of the same ground target of satellite A and satellite B is shown in Table 4.

It can be seen from Table 4 that after the adjustment of satellite A, the visible duration of the observation of the target by satellite A and B reaches 788s. During the observation period from ****-07-10 to ****-07-19, there were at least once a day covered targets.The target is covered twice on ****-07-10 and****-07-19, in which the two large coverage gaps are shortened to 41 hours and 30 hours respectively, shortening the maximum revisit interval and achieving the effective coverage of dual-satellite to ground targets.

Table 4. Visible period of target observation of satellites A and B

Visible start time (UTC)	Visible end time (UTC)	Visible time (s)	Coverage gap (h)
****-07-10 05:39:41	****-07-10 05:40:58	77.34	5.66
****-07-10 23:50:20	****-07-10 23:51:40	80.56	18.1
****-07-11 11:31:09	****-07-11 11:32:21	72.17	11.3
****-07-12 11:22:49	****-07-12 11:23:38	49.08	23.2
****-07-13 17:37:39	****-07-13 17:38:55	75.95	30.2
****-07-14 05:33:10	****-07-14 05:33:50	39.85	11.9
****-07-15 23:25:25	****-07-15 23:26:25	59.58	41.8
****-07-16 11:21:17	****-07-16 11:22:20	63.08	11.9
****-07-17 17:30:52	****-07-17 17:32:04	71.82	30.1
****-07-18 14:56:00	****-07-18 14:57:01	61.08	20.4
****-07-19 05:01:04	****-07-19 05:02:18	73.36	14.0
****-07-19 23:23:58	****-07-19 23:25:02	64.31	18.3

6 Conclusion

When the joint observation of ground targets is carried out by satellite networking, the lateral swing can only be carried out in a limited range through satellite attitude adjustment and load lateral swing. Some satellite payloads do not have the ability of lateral swing. At the same time, due to the constraints of ground resolution and other factors, the maneuvering coverage ability of lateral swing adjustment for the ground target is limited. When the lateral swing adjustment still can not achieve the coverage of specific targets or the coverage does not meet the operational accuracy requirements, this method can be used to achieve effective coverage of the ground target within a specific period of time. It is also a preferred method of orbit adjustment for short-term observation of specific targets by satellite temporary networking.

References

1. Shaoyuan, C., Li, Z., Weijun, G. et al.: Geometric parameters analysis for side-swing imaging of a large field of view space camera. J. Infrared Laser Eng. **44**(6), 1872–1877 (2015)
2. Ningjuan, R., Xuxia, Z., Tuotuo, L. et al.: End to end simulation and Analysis of space optical remote sensing system. J. Spacecr. Recovery Remote Sens. **34**(6), 36–43 (2013)
3. Zhijun, J., Min, Y., Yang, S. et al.: Swinging strategy on earth observing satellites. J. Sichuan Ordnance **35**(7), 128–130 (2014)
4. Hangyu, C., Xiaojin, S., Chaoran, Z.: Misson planning and emergency imaging of domestic land observation satellite. J. Satell. Appl. **10**, 31–37 (2019)
5. Kai, R.: Target positioning technology of reconnaissance satellite based on distributed SAR system. J. Comput. Measur. Control **28**(9), 218–222 (2020)
6. Jianghan, Z., Xi, L., Chilong, M. et al.: Elevations selection approach of area target observation task using satellites resource. J. Geomatics Inf. Sci. Wuhan Univ. **31**(10), 868–870 (2006)

7. Yongjiao, X., Demin, W., Chundong, Q. et al.: Algorithm to determine the satellite side beam center. J. Appl. Electron. Tech. **41**(3), 105–115 (2015)
8. Xiaomin, G., Hongqun, Z., Jiawei, H., et al.: The research on computing the satellite observation area on swinging mode. J. Microcomput. Inf. **27**(11), 71–73 (2011)
9. Wei, Z., Youran, G.: A Calculation method for satellite coverage area. J. Mod. Radar **42**(7), 5–10 (2020)
10. Jiwei, Y., Wei, F., Li, H. et al.: Regional target planning algorithm of satellite imaging based on global grid. J. Spacecr. Eng. **30**(1), 31–37 (2021)
11. Hongyan, H., Chongde, W., Xiaoyong, W. et al.: Study of influence of swinging on the systemic parameters of the satellite and CCD camara. J. Spacecr. Recovery Remote Sens. **24**(4), 14–18 (2003)
12. Renwei, Z., Qinshi, T.: Satellite Orbit Attitude Dynamics and Control, pp. 3–17. Beihang University Press, Beijing (1998)
13. Yuhua, C., Shuxing, F., Qingbo, G. et al.: Modeling and Simulation of Spacecraft Military Application, pp. 54–61, 93–101. National Defense Industry Press, Beijing (2010)
14. Weichi, T., Jingang, H.: Spacecraft System Engineering, pp. 73–77. China Science and Technology Press, Beijing (2009)
15. Yan, H.: TT&C Network of Space, pp. 181–184. National Defense Industry Press, Beijing (2004)
16. Lin, L., Songjie, H., Jianfeng, C. et al.: Theory and Application of Spacecraft Orbit Determination, pp. 40–43. Publishing House of Electronics industry, Beijing (2015)
17. Lin, L., Jingshi, T.: Theory and Application of Spacecraft Orbit, pp. 50–53. Publishing House of Electronics Industry Press, Beijing (2015)
18. Jisheng, L.: Precise Orbit Determination of Artificial Satellite, pp. 77–83. The People's Liberation Army Press, Beijing (1995)

临近空间系统与应用技术

Thermal Characteristics Analysis of Airship Based on Multi-node Model

Yao Xuan(✉), Geng Baogang, Lu Du, Ye Hu, and Li Yunfei

North Institute of Nuclear Technology, Xi'an 710024, China
yaoxuan0601@163.com

Abstract. One-dimensional thermal model of stratospheric airship based on multi-nodes division is established in this paper. The variation law of temperatures at nodes which include solar cells, the upper skin covered by solar cells, the upper uncovered skin, the lower skin and internal helium in various time are analyzed by this algorithm. Furthermore, the key influencing factors of airship thermal characteristics, containing helium leakage rate, airship structure shape, seasonal changes, cloud cover and geographical environment, are researched in diverse condition. The results show that because of the low solar radiation absorption rate and high infrared absorption rate of the skin material, the temperature of lower skin is always higher than upper uncovered skin. Therefore, the helium temperature is somewhere in between. Compared with others element, the thermal characteristics are rarely affected by leakage of helium. Specifically, the amount of helium under the minimum take-off condition at night is unequal affected by the ground temperature; The increased cloud cover alters the amount of ground radiation and causes a downward trend in temperature of skin surface and internal gas; Considering the ground reflection coefficient and infrared emissivity of different geographical environments, the space station should be selected as far as possible over the ocean or grassland, which can control the maximum 'superthermal' amount of helium while increasing the temperature at night.

Keywords: Thermal characteristics analysis · Multi-node model · Convective heat transfer · Helium leak · Test environment

1 Introduction

Stratospheric airship is an efficient information platform with many advantages and has become a research hotpot in the past decade. The characters of long hang time, flexible deployment and large load capacity are gathered in the airship. In the course of airship's flight, the temperature of internal gas changes greatly due to the solar radiation and convection, which brings a certain threat to the airship's flight.

The thermal characteristics of airship was explored by many researchers in recent years. Stenfan believed that the airship had hot and cold surfaces [1], and divided its structure into upper and lower parts. Two-node model is established in his method. Fang Xiande calculated the temperature difference between day and night of the floating

L. Wang et al. (Eds.): CHREOC 2022, LNEE 969, pp. 59–71, 2023.
https://doi.org/10.1007/978-981-19-8202-6_6

gas inside stratospheric airship by using the two-node model [3]. Based on the three-dimensional model, Liu Tingting calculated the temperature of skin and internal gas inside airship [4]. Then, the effects of flight conditions and skin material parameters on the thermal characteristics of airship were analyzed.

Inspired by the above modules, one-dimensional model of stratospheric airship based on multi-nodes division is established in this paper. The stratospheric airship model consists of solar cells, upper covered skin, upper uncovered skin, lower skin and internal helium. Then, the response of the thermal characteristics to its own structure and external environment are analyzed more comprehensively in this work.

2 Thermodynamic Model of Stratospheric Airship

As shown in Fig. 1, the thermal environment of stratospheric airship is complex and mainly affected by solar radiation, long-wave radiation and convection. In this section, the specific heat exchange process is analyzed and the thermodynamic model of airship is established.

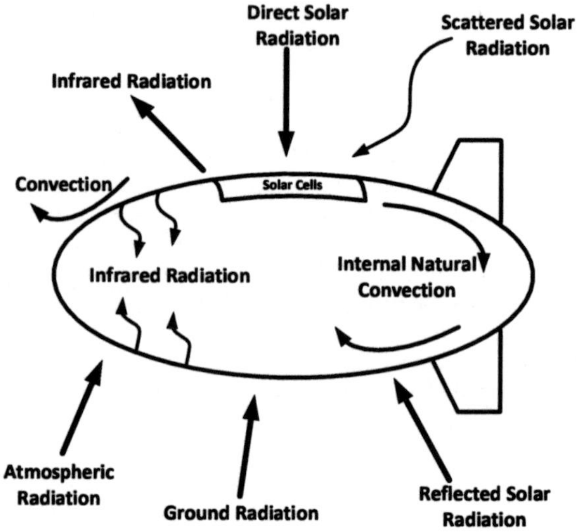

Fig. 1. Heat exchange process of stratospheric airship

2.1 Solar Radiation Model

Heat transfer by solar radiation dominates the heat source of stratospheric airship, mainly includes direct radiation, scattered radiation and reflected radiation. The intensity of solar radiation is mainly related to solar altitude angle and atmospheric transparency. According to the existing thermal environment model, the radiation definition of direct radiation I_{DN}, scattered radiation I_{suns} and reflected radiation I_{abd} received by airship are as follows.

$$I_{DN} = \tau_{atm} I_{sun} = \tau_{atm}[1 + 0.033\cos\left(\frac{2\pi n}{365}\right)] \cdot I_0 \tag{1}$$

$$I_{suns} = \frac{I_{sun}\sin ELV}{2} \cdot \frac{m(1 - \tau_{atm})}{m - 1.4\ln \tau_{atm}} \tag{2}$$

$$I_{abd} = C_{abd}(I_{DN}\sin ELV + I_{suss}) \tag{3}$$

where τ_{atm} and ELV denote atmospheric transmissivity and the solar altitude angle, m is the atmospheric optical quality and represents the degree to solar radiation affected by the atmosphere. $I_0 = 1367W/m^2$ and C_{abd} indicate solar constant and solar reflection coefficient.

2.2 Long-Wave Radiation Model

The long-wave radiation in the heat exchange process of airship mainly consists of ground radiation $I_{IR,ground}$, atmospheric radiation $I_{IR,sky}$, radiation from skin to outside $I_{IR,Ext}$ and radiation between inner surface of skin $I_{IR,In}$. The intensity of long-wave radiation is related to the surface temperature and the infrared emissivity, which can be expressed as $I_{IR} = \sigma\varepsilon_{IR}T^4$. Where σ is Stephen-Boltzmann constant and equals $5.67 * 10^{-3}$, ε_{IR} and T are the infrared emissivity and surface temperature of emitting object. For example, atmospheric radiation is bound up with atmospheric temperature and infrared emissivity which can be described as a function of atmospheric humidity and pressure.

Since the stratospheric airship works in the vicinity of 20 km, most of ground radiation received by the skin is absorbed by water vapor and carbon dioxide. Moreover, ground radiation is attenuated as the altitude increases. Therefore, the ground radiation received by the stratospheric airship is not only connected to the ground temperature and infrared emissivity, but also to the attenuation effect caused by the altitude. Attenuation coefficient is denoted by the atmospheric transmissivity τ_{atm} [13].

$$\tau_{IR,atm} = 1.716 - 0.5\exp\left(-0.65\frac{P_{atm}}{P_0}\right) + 0.5\exp\left(-0.095\frac{P_{atm}}{P_0}\right) \tag{4}$$

where P_0 and P_{atm} are the atmospheric pressure at ground and airship station.

2.3 Convection Model

The heat transfer of convection is determined by the contact area, temperature difference between exchanged substances and heat transfer coefficient, which can be expressed as follows.

$$Q_{Con} = ANu\frac{\lambda}{L}\Delta T \tag{5}$$

where A is effective area of convection, Nu denotes Nusselt number, λ represents coefficient of thermal conductivity and is determined by gas physical parameters, L expresses characteristic length. ΔT illustrates temperature difference of exchanged substances.

The calculation of Nusselt number is the key step to solving convection.. The Nusselt number of the flat body and sphere have been researched in the past decades [6], as shown in Table 1. In Table 1, Re, Pr and Gr are Reynolds number, Prandtl number and Grashof number [6]. When $Gr/\mathrm{Re}^2 \leq 0.1$, the influence of natural convection can be ignored. When $Gr/\mathrm{Re}^2 \geq 10$, the influence of forced convection can be ignored. When $0.1 < Gr/\mathrm{Re}^2 < 10$, both effects of natural convection and forced convection should be considered.

The main research object of this work is the stratospheric airship with spheroid structure. When calculating its Nusselt number, spheroid can be regarded as the combination of a flat body and a sphere. Nusselt number are expressed as $Nu = \beta Nu_L + (1 - \beta)Nu_D$. Where β is form factor of airship, which ranges from 0 to 1. Nu_L and Nu_D is the Nusselt number of a flat body and sphere. When the shape of the airship is close to a flat body, the value should be close to 1 [4].

2.4 Thermal Balance Equation

In order to meet the power demand of platform, the solar cells is symmetrically laid on the top of airship to receive part of solar radiation and convert into electricity. In addition, other solar radiation is absorbed as heat and causes an increase in temperature of solar cells. Moreover, solar cells can also receive infrared radiation and convect with the external environment. The thermodynamic differential equation of solar cells is shown as follows.

$$m_s c_s \frac{dT_s}{dt} = Q_{DN} + Q_{suns} + Q_{IR,sky} - Q_{IR,Ext} + Q_{Con,Ext} - Q_{en} - Q_{power} \tag{6}$$

$$Q_{en} = \frac{A(T_s - T_{f,c})}{\delta_s/\lambda_s + \delta_d/\lambda_d + \delta_{f,c}/\lambda_{f,c}} \tag{7}$$

where m_s and c_s are the mass and the specific heat capacity of solar cells. Q_{DN}, Q_{suns} and $Q_{IR,sky}$ represent the heat transfer of direct solar radiation, scattered solar radiation and atmospheric radiation received by solar cells respectively. $Q_{IR,Ext}$ and Q_{Con} indicate infrared radiation and convection with external environment. Q_{power} denotes the electricity converted by solar cells and Q_{en} is the heat transfer with the covered skin. δ_s, δ_d and δ_f express the thickness of solar cells, isolation layer and skin [6]. λ_s, λ_d and λ_f are the thermal conductivity of the three.

Considering the influence of solar cells, the airship skin is divided into three nodes: the upper covered part, the upper uncovered part and the lower part. Set the idealized precondition for uniform temperature at each node. The thermodynamic differential equation is shown as follows.

$$m_{f,c} c_{f,c} \frac{dT_{f,c}}{dt} = Q_{en} + Q_{IR,In} + Q_{Con,In} \tag{8}$$

Table 1. Nusselt numbers of convection

	Nusselt of forced convection	Nusselt of natural convection
Sphere body	$Nu_{F,D} = \begin{cases} 2 + 0.47\mathrm{Re}^{1/2}\,\mathrm{Pr}^{1/3} \ (\mathrm{Re} < 5 \times 10^5) \\ (0.0262\mathrm{Re}^{4/5} - 615)\,\mathrm{Pr}^{1/3} \ (5 \times 10^5 \le \mathrm{Re} \le 10^8) \\ 0.0262\mathrm{Re}^{4/5}\,\mathrm{Pr}^{1/3} \ (\mathrm{Re} \ge 10^8) \end{cases}$	$Nu_{N,D,out} = 2 + \frac{0.589(Gr \cdot \mathrm{Pr})^{0.25}}{(1+(0.469/\mathrm{Pr})^{9/16})^{4/9}}$ $Nu_{N,D,In} = \begin{cases} 0.59(GrPr)^{1/4} \ GrPr < 10^9 \\ 0.13(GrPr)^{1/3} \ GrPr > 10^9 \end{cases}$
Flat body	$Nu_{F,L} = \begin{cases} 0.664\mathrm{Re}^{1/2}\,\mathrm{Pr}^{1/3} \ (\mathrm{Re} < 5 \times 10^5) \\ (0.037\mathrm{Re}^{4/5} - 871)\,\mathrm{Pr}^{1/3} \ (5 \times 10^5 \le \mathrm{Re} \le 10^8) \\ 0.037\mathrm{Re}^{4/5}\,\mathrm{Pr}^{1/3} \ (\mathrm{Re} \ge 10^8) \end{cases}$	$Nu_{N,L} = \left\{0.6 + 0.387\left[\frac{Gr \cdot \mathrm{Pr}}{(1+(0.559/\mathrm{Pr})^{9/16})^{16/9}}\right]^{1/6}\right\}^2$

$$m_{f,un}c_{f,uc}\frac{dT_{f,uc}}{dt} = Q_{DN} + Q_{suns} + Q_{IR,sky} - Q_{IR,Ext} + Q_{IR,In} + Q_{Con,Ext} + Q_{Con,In} \quad (9)$$

$$m_{f,l}c_{f,l}\frac{dT_{f,l}}{dt} = Q_{suns} + Q_{abd} + Q_{IR,sky} + Q_{IR,ground} - Q_{IR,Ext} + Q_{IR,In} + Q_{Con} \quad (10)$$

$$m_{He}c_{He}\frac{dT_{He}}{dt} = Q_{Con,In} - P_{He} \cdot \frac{dV_t}{dt} + RT_{He} \cdot \frac{dm_{He}}{dt} \quad (11)$$

where $Q_{IR,In}$ is infrared radiation between skins. Q_{abd} and $Q_{IR,ground}$ are reflected solar radiation and ground radiation.

3 One-Dimensional Thermal Model Simulation of Stratospheric Airship Based on Multi-nodes Division

According to the above model, one-dimensional thermal model of stratospheric airship based on multi-nodes division is built by simulation in this section. Then, the thermal characteristics are analyzed and the temperature changes of solar cells, helium and skins are calculated by multi-node thermal model. Furthermore, the influences of structure shape, airspeed and gas leakage on the thermal characteristics of airship are researched through calculation.

3.1 Simulation Parameter Settings

See Table 2.

Table 2. Setting of simulation parameters

Parameter	Value	Parameter	Value
Longitude and latitude	N41.8, E87.4	Area of solar cells	90 m^2
Station height	19.5 km	Photon-to-electron conversion efficiency	16%
Volume of airship	36000 m^3	Solar radiation absorption rate	0.2
Long/maximum diameter	110 m/28 m	Infrared radiation absorption rate	0.75
Mass of helium	640 kg	Specific heat capacity of skin	2000
Surface area of skin	6600 m^2	Specific heat capacity of solar cells	1500

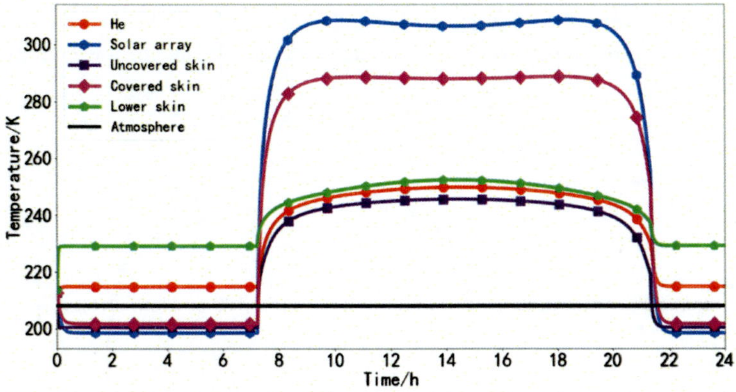

Fig. 2. The temperature of each node during the day and night

3.2 Simulation Results of Multi-node Temperature

The airship reaches thermal equilibrium under solar radiation, infrared radiation and convection. Based on the above parameters, the multi-node thermal model of stratospheric airship is solved. The temperature simulation results of each node are shown in Fig. 2.

The absorption rate of solar cells is between 60% and 90%, while the photon-to-electron conversion efficiency is only about 20%. During the day, the disparity between absorption and conversion leads to a significant heating effect of the solar cells. The highest temperature of solar cells can reach about 308 K. Affected by heat conduction, the temperature of the covered skin is significantly higher than uncovered skin. The maximum temperature of covered skin can reach up to 292 K. With the increase of the solar elevation angle, the productivity of solar array increases. Thus, the temperature decreases slightly at noon.

As shown in Fig. 2, while helium and lower skin maintain at a 'superthermal' state, the upper skin stay in a 'ultracold' state at night. Under the influence of ground radiation at night, the temperature of lower skin is significantly higher than upper skin and the temperature of helium is between the two. During the day, the temperature difference decreases gradually with the increase of solar radiation. However, because of the lower solar radiation absorption rate and higher infrared radiation absorption rate of the skin material, the lower skin temperature is still higher than the upper one.

3.3 Effect of Convection

Based on the thermodynamic model of airship, there are two main factors which affect the progress of convection. On the one hand, the shape of the airship results in the disparity of convection between flat body and sphere. On the other hand, various airspeeds cause the difference of forced convection.

As shown in Fig. 3, the helium temperature under different shape are simulated. With the increase of the shape coefficient, the temperature of helium tends to decrease. At noon, the temperature gap between the flat and sphere body is about 10 K. Therefore, flatter structures bring about stronger convective heat transfer and the lower temperature of helium.

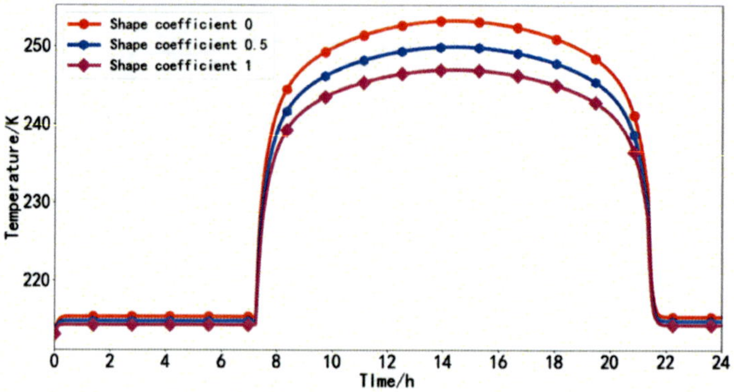

Fig. 3. Effect of shape factor on helium temperature

Figure 4 shows the simulation results of the maximum temperature and nighttime temperature of each node at different airspeed. During the day, the solar cells and skin are all in the 'superthermal' state. With the increase of airspeed, the effect of forced convection is enhanced and the node temperature gradually approaches to the atmosphere temperature. In addition, the temperature of helium gradually decrease under the influence of internal natural convection.

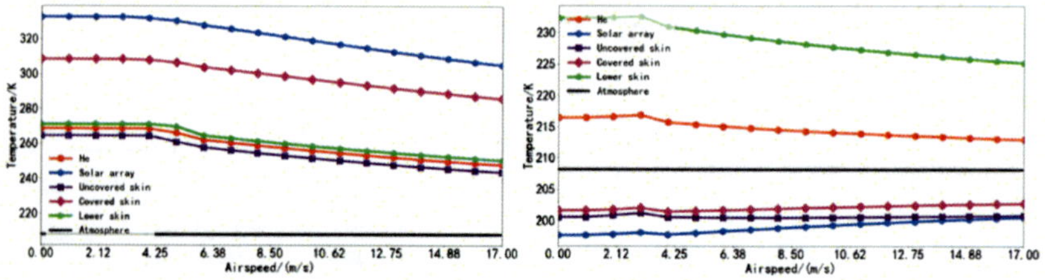

(a) Maximum temperature of each node (b)Nighttime temperature of each node

Fig. 4. Effect of airspeed on node temperature

At night, while the upper skin is in a 'ultracold' state, the lower skin stay in a 'superthermal' state. Since the temperature of 'superthermal' is higher than the 'ultracold', the temperature change law of the inner helium is consistent with the lower skin and shows a gradually decreasing trend with the increase of airspeed. Therefore, in order to reduce the risk of overpressure rupture and low-pressure instability, the airspeed should be increased during the day and reduced at night.

3.4 Effect of Gas Leakage

Helium is a small molecule gas and can not be completely isolated by any skin material. Thus, the problem of diffusion and penetration of helium must exist during the flight. In practical applications, the skin material is designed as laminated material to control the leakage of helium within a small range. The loss of helium leads to a change in the energy of internal fluid. As shown in Fig. 5, the temperature of helium under different leakage rate are calculated. When the leakage rate is lower than 10 kg/h, the temperature difference of helium is less than 1 K. Therefore, the energy loss caused by gas leakage is weak and the disparity of thermal characteristics are negligible.

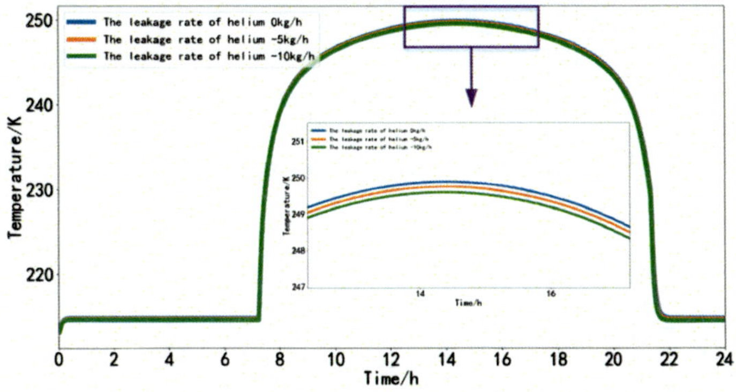

Fig. 5. Effect of gas leakage on helium temperature

4 Influence of Test Environment on Thermal Characteristics

The external factors that affect stratospheric airship include the intensity of solar radiation and infrared radiation which is determined by seasons, ground temperature, cloud cover and so on. The thermal characteristics effected by test environment of airship is calculated and analyzed in this section.

4.1 Effect of Season

Solar radiation varies in different seasons and affect the heat exchanges process of airship. Considering the disparity of ground temperature in different seasons, the results of helium temperature in 'spring equinox', 'summer solstice', 'autumn equinox' and 'winter solstice' are simulated, as shown in Fig. 6. Due to different subsolar points in different seasons, the high temperature period of helium is quite various. The high temperature period of helium on summer solstice is the longest, which is nearly 3 h longer than the 'spring equinox' and 'autumn equinox'. Besides, affected by ground temperature in winter, the ground radiation at night is weakened. Under this simulation condition, the airship can not complete take-off task. However, overfilling helium can be leveraged to solve this problem. As shown in Fig. 7, based on the design parameters such as airship structural mass and volume, the minimum take-off helium amount at different ground temperature is calculated. With the increase of ground temperature, the minimum take-off helium amount of airship shows a nonlinear attenuation trend.

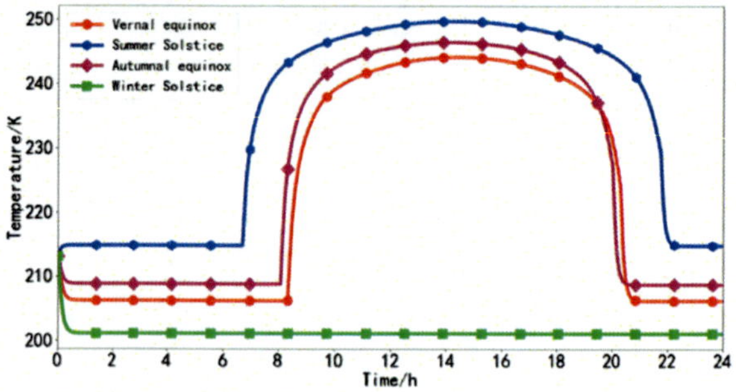

Fig. 6. Effect of seasons on helium temperature

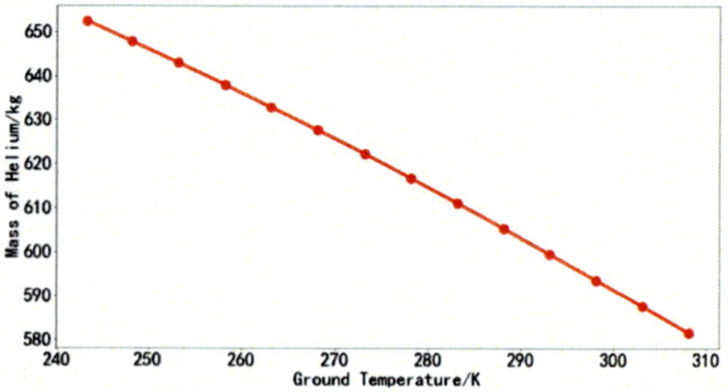

Fig. 7. Minimum take-off helium mass at different ground temperatures

4.2 Effect of Cloud Cover

Since the airship works in the space of 20 km, the clouds in the atmosphere will attenuate ground radiation. Ground and satellite data were employed to analyze the relationship between cloud radiative forcing and cloud amount: $CRF = 61.6c^{1.712}$. where c is the cloud amount. The ground radiation shows a nonlinear attenuation trend with the increase of cloud amount.

As shown in Fig. 8, the simulation results of the maximum temperature and nighttime temperature of each node at different cloud amount is calculated. Because the cloud cover directly changes the amount of ground radiation received by the lower skin, the lower skin temperature has the most dramatic change in value with the increase of cloud cover. Further, the natural convection of the internal helium and the upper skin is altered. At night, the temperature difference of lower skin under the condition of less cloudy and

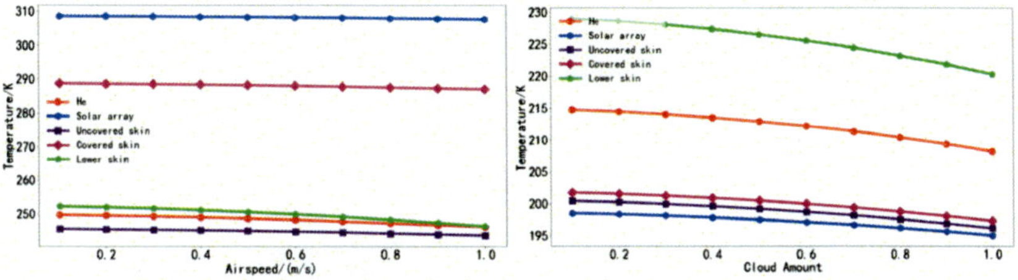

(a) Maximum temperature of each node (b) Nighttime temperature of each node

Fig. 8. Effect of cloud cover on node temperature

cloudy can reach 9 K and the temperature difference of the internal helium can reach 6 K. During the day, with the influence of solar radiation, the disparity of cloud cover on the skin and helium is reduced, the difference temperature of helium is only about 3 K in the case of less cloudy and cloudy conditions, as shown in Fig. 8.

4.3 Effect of Geographical Environment

The ground radiation and reflected solar radiation are influenced by the geographical environment where the stratospheric airship is stationed obviously. The ground infrared emissivity and solar reflection coefficient under different geographical environments are shown in Table 3 and the simulation results of helium temperature are shown in Fig. 9. Compared with the infrared emissivity in the ocean, grassland and snowfield which are resemble, the infrared emissivity in desert is the lowest. Therefore, the helium temperature of airship when it is stationed above the desert is about −8 K different from the other geographical environments at night.

Table 3. Thermal parameters of different geographical environments

Geographical environment	Reflection coefficient	Infrared emissivity
Desert	0.25	0.82
Ocean	0.15	0.99
Grassland	0.2	0.98
Snowfield	0.8	0.95

During the day, although the infrared emissivity of snowfield is slightly lower than that of ocean and grassland, the helium temperature of snowfield is about 10 K higher than ocean and grassland with the higher ground reflection coefficient. In this situation, the influence of solar reflected radiation on airship thermal characteristics is significantly higher than ground radiation. According to the above analysis, in order to effectively avoid overpressure rupture during the day and pressure loss at night, the station position of airship should be selected over the ocean or grassland. So, the maximum temperature of skin and helium during the day and the minimum temperature during the night are controlled by the suitable station position.

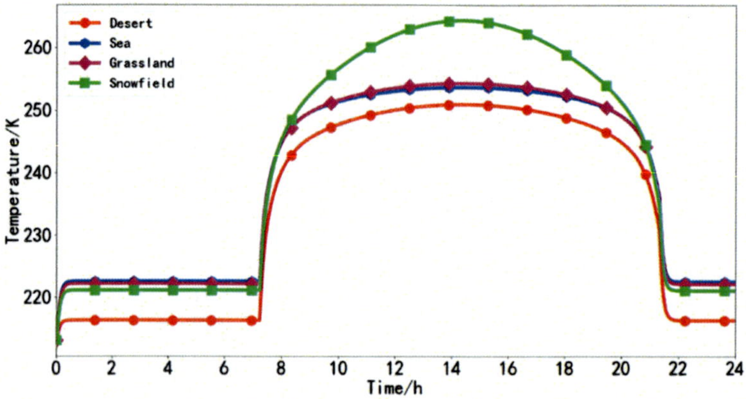

Fig. 9. Effect of geographical environment on helium temperature

5 Summary

The thermal characteristics of stratospheric airship is analyzed based on the multi-node model. Furthermore, the key influencing factors of airship thermal characteristics, containing helium leakage rate, airship structure shape, seasonal changes, cloud cover and geographical environment, are researched in diverse condition. The results show that the temperature of lower skin is always higher than upper uncovered skin because of the low solar radiation absorption rate and high infrared absorption rate of the skin material. Therefore, the helium temperature is somewhere in between. Compared with others element, the thermal characteristics are rarely affected by leakage of helium. Specifically, the flatter structural shape result in stronger convection and less 'superthermal' of skin and helium; The amount of helium under the minimum take-off condition at night is unequal affected by the ground temperature; The increased cloud cover alters the amount of ground radiation and causes a downward trend in temperature of skin surface and internal gas; Considering the ground reflection coefficient and infrared emissivity of different geographical environments, the space station should be selected as far as possible over the ocean or grassland, which can control the maximum 'superthermal' amount of helium while increasing the temperature at night.

References

1. Stefan, K.: Thermal effects on a high altitude airship. In: Lighter-Than Air Conference (1983)
2. Xinlin, X., Defu, L., Xiaochuan, Y.: Thermal characteristics of stratospheric aerostats and their research. Acta Aeronautica ET Astronautica Sinica **30**(4), 7 (2009). https://doi.org/10.1007/978-90-313-7627-8_7

3. Xiande, F., Weizhi, W., Xiaojian, L.: A study of thermal simulation of stratospheric airships Spacecr. Recovery Remote Sens. **02**, 9–13 (2007)
4. Tingting, L.: Thermal charcteristics of stratospheric airships at long duration floating conditions. Graduate School of National University of Defense Technology
5. Xiaojian, L.: Numerical simulation of thermal-structure coupling for near space airship. Nanjing University of Aeronautics and Astronautics (2013)
6. Shiming, Y., Wenquan, T.: Heat Transfer, 4th edn. Higher Education Press (2006)
7. LA Carlson WJ Horn 1981 New thermal and trajectory model for high-altitude balloons Aiaa Paper 20 6 500 507
8. X Xianghua C Xuetao L Xingang 2009 Thermal analysis of a stratospheric airship Tsinghua Unic. Sci. Tech. 11 4
9. Xiquan, D., Baike et al.: A climatology of midlatitude continental clouds from the ARM SGP central facility. Part II: Cloud Fraction and Surface Radiative Forcing. J. Clim. (2006)
10. Shuchen, Y., Taichang, G., Xiaoyong, S.: Research on cloud surface long-wave radiative forcing by using SBDART. J. Atmos. Environ. Opt. **7**(4) (2012)
11. D Xiaolong M Zhenyu Y Xixiang 2020 Thermal characteristics analysis of a stratospheric aerostat based on multi-layer node model J. Shanghai Jiao Tong Univ. 54 7 6
12. Xiaohui, Q.: Research on the thermal performance and environment control system for the near-space airship. Nanjing University of Aeronautics and Astronautics
13. Keith, F., Kreider, J.F.: Numerical Prediction of the Performance of High Altitude Balloons (Revised February 1974)
14. C Chen W Xiaoliang 2021 Thermal dynamic model and thermal characteristics of airships considering skin transmittance J. ShangHai Jiao Tong Univ. 55 7 10
15. Helei, Z., Xiande, F., Qiumin, D.: Investigation on internal natural convection of stratospheric airship. J. Astronaut. **37**(7), 8 (2016)
16. C Chen W Xiaoliang 2020 Thermal sensitivity factors analysis of stratospheric airships J. Zhejiang Univ. Eng. Sci. 54 1 11
17. Xiaojian, L., Xiande, F., Dapeng, L.: Research on thermal characteristics of solar cells of stratospheric airship. In: The 3 China High Resolution Earth Observation Conference (2014)
18. Xiaoliang, W., Weicheng, X.: Thermal characteristics analysis of stratospheric airships envelope. Equip. Environ. Eng. **17**(1), 7 (2020)
19. Daoming, X.: Numerical Study on the Thermal Characteristic and Thermal-Mechanical Effect of Stratospheric Aerostat. Nanjing University of Science & Technology (2018)
20. Tingting, L., Zhenyu, M., Xixiang, Y.: Influence of solar cells on thermal characteristics of stratospheric airship (1) (2018)

Analysis of Space Contamination Effect on the Detecting Data from "XY-1" Satellite

Yuan Xiao-xue(✉), Yang Dong-sheng, Zang Wei-guo, and Y. U. Qian

Beijing Institute of Spacecraft Environment Engineering, Beijing 100094, China
xilingxue2012@126.com

Abstract. The detecting data from the space contamination effect detector onboard "XY-1" satellite is analyzed in this article. Summaries are made via analyzing the detecting data of the space contamination effect detector in more than four years, via analyzing the relationship between the contamination mass and the capability of $GaInP_2$ solar cells. It can be drawn that the measures for the screening of space materials are effective and satisfy the contamination control demands for the whole process of spacecraft. The current results can be used for the contamination control design of solar cells, and also, in checking out and correcting the contamination control techniques on the ground for spacecrafts.

Keywords: Space contamination · In-orbit detection · Solar cell · Non-metallic materials · Outgassing

1 Introduction

Non-metallic materials of spacecraft released a series of organic gas molecules, among which phthalates and silicones are the common substances for in-flight spacecrafts [1, 2]. Those gas molecules may deposit on sensitive surfaces [3] such as optical elements, thermal control coating and solar cell, by direct transportation, reflection and atmospheric scattered transportation, which can result in the decrease of the property or even the suspension of work [4].

The red-blue photoconductive detector is installed on the XY-1 satellite [5], made by Beijing Institute of Spacecraft Environment Engineering. It was launched successfully on December 19th, 2012. One of the test goals is detecting the contamination effects on it. It may serve as a reference for the protection of solar cell and compare to the analyzed simulation data and take effective measures therefore (Figs. 1 and 2).

2 Detecting Principle

The space contamination effect is obtained via analyzing the relationship between the contamination mass and the performance of $GaInP_2$ solar cells used for the space contamination effect detector onboard XY-1 [6].

The contaminant components of non-metallic materials outgassing are usually phthalate and siloxane based on the test results of spacecraft thermal vacuum test [7–10].

The comparison of the transmittance between the cleaned and the contaminated optical glasses because of the phthalate deposition (1×10^{-6} g/cm^2) is shown in Fig. 3.

L. Wang et al. (Eds.): CHREOC 2022, LNEE 969, pp. 72–78, 2023.
https://doi.org/10.1007/978-981-19-8202-6_7

Fig. 1. Space environment effect detector with red-blue optical sensitivity

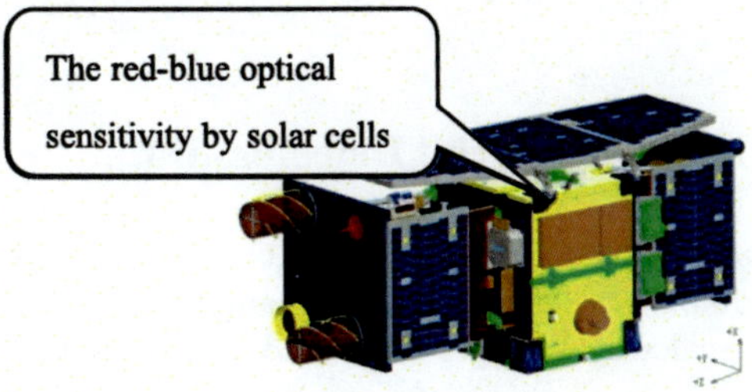

Fig. 2. The location of space environment effect detector with red-blue optical sensitivity

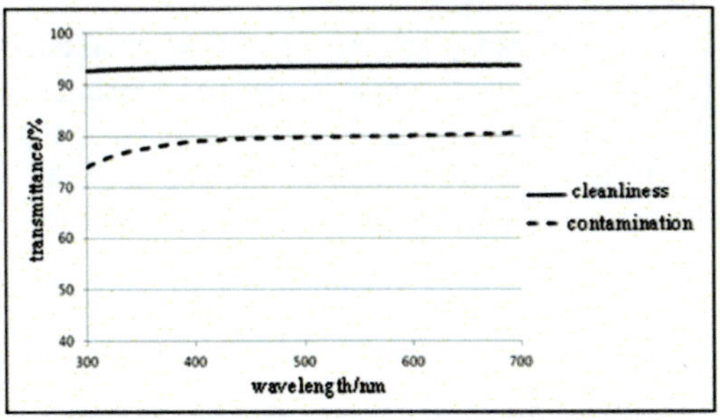

Fig. 3. Comparison graph of transmittance between cleaned and contaminated optical glasses

The 300-700nm transmittance of the optical glasses deposited by different contamination masses is shown in Table 1.

The transmittance of the optical specimens decreased by 13.6%, 18.0% and 38.6%, respectively, under different simulated contamination masses of 1×10^{-6} g/cm^2, 1×10^{-5} g/cm^2 and 1×10^{-4} g/cm^2.

According to the $GaInP_2$ solar cell volt-ampere characteristics we chose four pieces of 2 cm × 2 cm solar cells in series as the contamination effect detector, to ensure the enough duty ratio, and at the same time on both ends of the cells access 40Ω sampling resistor. The short circuit current of $GaInP_2$ solar cells is obtained by sampling the voltage values at both ends of the resistance.

Table 1. The transmittance of the optical glasses deposited different contamination masses

Contamination/(g/cm^2)	1×10^{-6}	1×10^{-6}	1×10^{-5}	1×10^{-5}	1×10^{-5}	1×10^{-4}	1×10^{-4}
Number	1	2	3	4	5	6	7
Cleanliness transmittance/%	93.4	93.4	93.6	93.6	93.6	93.3	93.2
Contamination transmittance/%	80.5	79.2	77.5	73.4	76.0	53.9	55.4
Transmittance loss/%	12.9	14.2	16.1	20.2	17.6	39.4	37.8

3 Analysis of Original In-Orbit Detection Data

The original data from the space contamination effect detector includes the output voltage, the temperature, the rolling angle and the yaw angle of XY-1. The rolling angle and yaw angle are used to calculate the light angle according to the following formula.

$$\alpha = 90 - \arctan\left(\frac{1}{\sqrt{tg^2\beta + \mathrm{tg}^2\gamma}}\right) \tag{1}$$

In the above formula, α refers to the light angle, β refers to the rolling angle, and γ refers to the yaw angle. The data transmitted by the space contamination effect detector from December 19th, 2012 through June 28th, 2017 are shown in Figs. 4 and 7. Figures 4 and 5 show the output voltage and the temperature curve of the $GaInP_2$ solar cells. Figures 6 and 7 show the satellite's rolling angle and yaw angle.

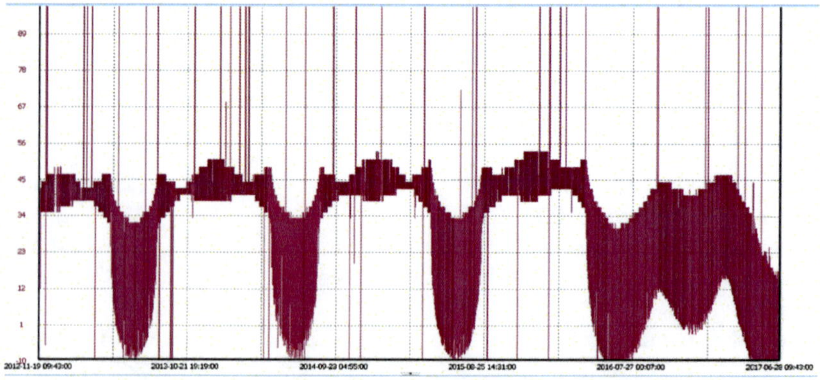

Fig. 4. The graph of output temperature

The following conclusions can be drawn from Figs. 4, 5, 6 and 7.

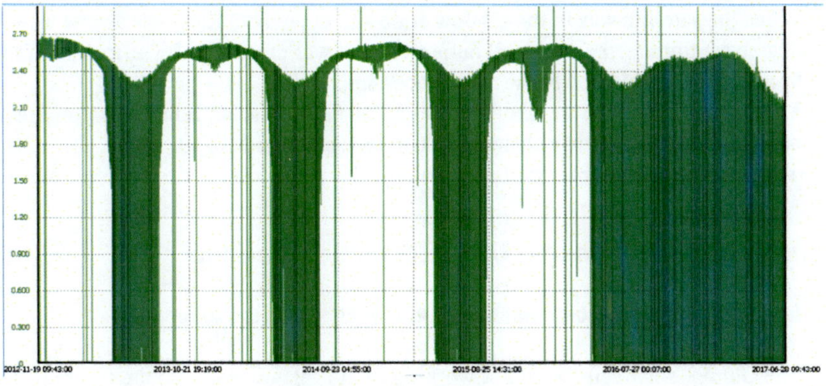

Fig. 5. The graph of output voltage

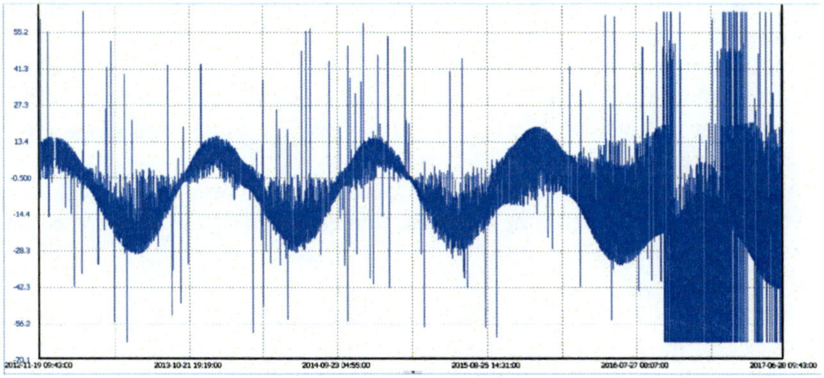

Fig. 6. The graph of rolling angle

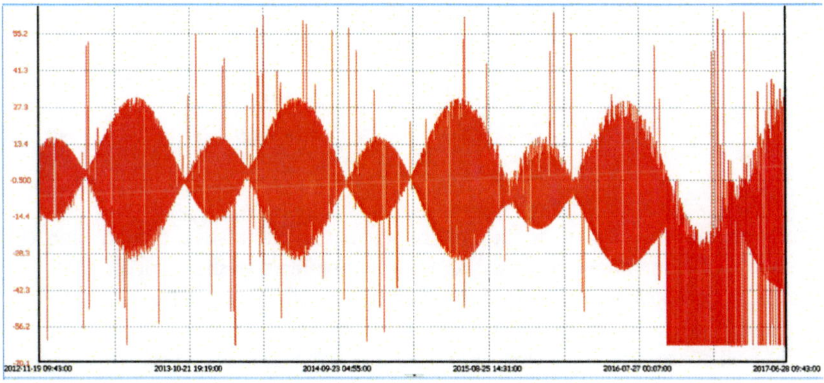

Fig. 7. The graph of yaw angle

(1) When the satellite was in the shadow zone and the temperature was as low as −10 degree centigrade, the solar cell output voltage was close to 0V. The impact of this time is to be considered in the data processing.
(2) The voltage, temperature and light angle of $GaInP_2$ solar cells were consistent, indicating that the voltage changes with the temperature and light angle. The effects of temperature and light angle on the voltage should be corrected in the data analysis.

4 In-Orbit Detection and Data Correction Analysis

The detection data of in-orbit contamination is analyzed in the following steps.

(1) The $GaInP_2$ solar cells were not working in the shadow zone, so the test points in the shadow area were removed.
(2) According to the temperature coefficient of $GaInP_2$ solar cells measured on the ground, the effect of temperature on the output voltage of the solar cells was excluded.
(3) The light angle of the satellite was calculated, and then the effect of the light Angle on the output voltage was eliminated by the interpolation method according to the relationship between the light angle and the voltage obtained in the ground calibration experiment.
(4) According to the relationship between the contamination deposition and the output voltage of the solar cells, the contamination environment and effect of on-orbit satellite were analyzed.

After processing the above data according to the relationship between the temperature coefficient and the light Angle of $GaInP_2$ solar cells, the deviation between ground calibration test results and in-orbit detection data was analyzed.

Calibration test results based on the ground, the relationship between the $GaInP_2$ solar cells power output rate and the shading rate is shown in formula (1), the relationship between the shading rate and the deposition quantity is shown in formula (2), and then the quantity of in-orbit contamination deposition was calculated.

$$\delta = 0.348 \cdot 1\mu^3 + 0.584\mu^2 + 0.067 \cdot 9\mu \tag{2}$$

$$\rho = -\frac{\ln(1-\mu)}{5000} \tag{3}$$

Through the analysis of contamination data, the cumulative amount and daily increase rate of contamination in each stage were obtained, as shown in Table 2.

Through the above analysis, the following summaries can be made:

(1) The first phase, from Dec 19th to Dec 30th, 2012, was the contamination reducing stage, when the contaminants of the $GaInP_2$ solar cells reduced out of the desorption or stripping effect in the vacuum. The second phase, from Dec 30th, 2012 to Apr 18th, 2013, was the contamination increasing stage. During the period, the cumulative amount of contamination increased with a growing incremental speed.

Table 2. The result of contamination cumulating mass and increase rate per day

Date	Total contamination mass g/cm^2	Average daily contamination mass $g/(cm^2 \cdot d)$
2012/11/19	9.893×10^{-6}	
2012/11/30	0.000	-8.99×10^{-7}
2012/12/30	5.175×10^{-6}	1.73×10^{-7}
2013/05/19	2.641×10^{-5}	1.25×10^{-7}
2013/10/30	3.228×10^{-5}	3.65×10^{-8}
2014/03/30	4.013×10^{-5}	5.23×10^{-8}
2014/09/30	5.532×10^{-5}	8.44×10^{-8}
2015/03/30	6.119×10^{-5}	3.26×10^{-8}
2015/09/30	6.997×10^{-5}	4.88×10^{-8}
2016/03/30	7.228×10^{-5}	1.28×10^{-8}
2016/09/30	7.879×10^{-5}	3.61×10^{-8}
2017/03/30	8.012×10^{-5}	7.38×10^{-9}
2017/06/28	8.254×10^{-5}	1.34×10^{-8}

The third phase, from Apr 18th to Aug 30th, 2013, was the contamination increasing stage. In the period the cumulative amount of contamination increased with a slow-down incremental rate. The fourth phase, from Aug 30th to Dec 19th, 2013, was the contamination descending stage. In the period the cumulative amount of contamination descended tardily because the light enhanced.

(2) The daily increase in contamination was between 7.38×10^{-9} g/cm^2 and 1.73×10^{-7} g/cm^2, and the annual average was 4.56×10^{-8} g/cm^2. The value was consistent with the requirement that the contamination in the total process of spacecraft development should be controlled within 1×10^{-7} g/cm^2.

(3) The overall contractive change trend of the cumulative contamination data showed that volatile phthalate acid aster reduced gradually, while the non-volatile siloxane became the main body to render a lasting impact of contaminants The deposition of phthalate decreased gradually from 6×10^{-7} g/cm^2 to 3×10^{-7} g/cm^2. The deposition of siloxane decreased 2.5×10^{-7} g/cm^2 in the first ten days, and increased 7 $\times 10^{-7}$ g/cm^2 consistently, with the average daily increment of 1.2×10^{-8} g/cm^2.

5 Conclusion

The space contamination effect detector carried by XY-1 satellite is to detect the space contamination environment and effects by evaluating the influence of contaminants released by non-metallic materials on the performance of $GaInP_2$ solar cells. Based on the results of the ground calibration test and the data acquired from the satellite, this paper analyzes the deposition of the in-orbit non-metallic materials. The results show that the screening and control of spacecraft materials are effective to satisfy the contamination control demands for the whole process of spacecraft development. Meanwhile, there is still space for improvement in the control of siloxane contaminants.

References

1. Yuan, X.X., Pang, H.W., Zhou, C.L.: Study on the attenuation of the solar cell power output caused by molecular contamination. J. Astronaut. **28**(1), 118–122 (2007)
2. Yuan, X.X., Zhou, C.L., Yang, D.S.: The contamination control of aerospace optical system. Spacecr. Environ. Eng. **29**(2), 168–172 (2012)
3. Sun, W., Yan, S.G., Tian, D.B., et al.: Numerical simulation of a ZY satellite orbit-control thruster plume. Spacecr. Environ. Eng. **25**(3), 276–281 (2008)
4. Sun, W., Tian, D.B., Yan, S.G.: Analysis of plume contamination effects on the thermal control coatings of a satellite. Spacecr. Environ. Eng. **25**(6), 538–541 (2008)
5. Yuan, X.X., Yi, Z., Yang, D.S., et al.: The analysis of detecting data for the space environment effect detector with red-blue optical sensitivity by solar cells. Spacecr. Environ. Eng. **31**(5), 476–482 (2014)
6. Yuan, X.X., Yi, Z., Yang, D.S., et al.: The study of a new technology for space environment effects monitor and application. Space Electr. Technol. **10**(1), 30–32 (2013)
7. Jiao, Z.L., Pang, H.W., Yi, Z., et al.: Composition analysis of contaminant in thermal vacuum test for spacecraft. Spacecr. Environ. Eng. **27**(6), 711–714 (2010)
8. Li, N., Yang, D.S., Yu, Q., et al.: Detection and analysis of contamination inside the spacecraft in vacuum thermal test. Vacuum **50**(3), 17–19 (2013)
9. Jiao, Z.L., Pang, H.W., Yi, Z., et al.: Identification of molecular contaminant in thermal vacuum test by using GC/MS. Chin. Space Sci. Technol. **28**(6), 64–67 (2008)
10. Yang, D.S., Zang, W.G., Yu, Q., et al.: Contamination measurement and control of Chinese spacecraft. In: 64th International Astronautical Congress, Beijing, September 23–27 (2013)
11. Wood, B.E., Bertrand, W.T., Lesho, J.C., et al.: Midcourse space experiment (MSX) satellite measurements of contaminant films using QCMs—5 years in space. AIAA 2001-2956
12. Wood, B.E., Hall, D.F., Lesho, J.C., et al.: On-orbit midcourse space experiment (MSX) satellite environment flight experiments. AIAA 99-0252

Thermosphere Neutral Densities at Dusk/Dawn Derived from Space-Borne Atmospheric Density Detectors

Yegui Wang[1](✉), Xiaofang Zhang[1](✉), Qifa Cai[1], and Guangming Chen[2,3]

[1] Mailbox 5111, Beijing 100094, China
wangyg0110@163.com, zhangxiaofang1979@aliyun.com
[2] Beijing Aerospace Control Center, Beijing 100094, China
[3] Science and Technology on Aerospace Flight Dynamics Laboratory, Beijing 100094, China

Abstract. Based on the in-situ observed mass densities derived from the space-borne atmospheric density detectors on the Atmospheric density detection and Precise Orbit Determination (APOD) and another Chinese satellite (CHN-sat), from 2008–2018, the neutral density variations of the dawn/dusk thermosphere mass density at 460 km, 630 km altitudes are presented in this paper. Our results reveal: (1) The densities observed by APOD show consistent variations with that in empirical model results, and the latitudinal- seasonal structure shows that the largest density maxima appear near equinoxes in the northern high latitudes with the higher maximum near March than that in October at solar minimum. (2) Equatorial observations of CHN-sat are in good agreement with model results, while significant differences between observations and MSISE00 densities occur at high latitudes under the very low solar activity conditions. In solstices, the measured densities by CHN-sat at high latitudes in the summer hemisphere are distinctly higher than those in the winter hemisphere, which indicates that the relative importance of the conjunct contribution from oxygen atom and temperature, and the winter helium bulge maybe should be adjusted in the empirical model. (3) The weak geomagnetic disturbance can strengthen the high-latitude density at dawn and the low-latitude density maxima at dusk at CHN-sat altitudes.

Keywords: Upper thermosphere · Mass densities from in-situ observations · APOD · Space-borne atmospheric density detector · MSISE00 · JB2008

1 Introduction

The mass density of thermosphere above 90 km altitude is a crucial parameter for Low Earth Orbit (LEO) prediction due to the atmospheric drag on satellites in this region. As the primary driver, fluctuations in the level of solar radiation occur over a wide range of time scales from solar rotations up through the solar cycle, producing thermosphere neutral density and composition spanning these same time scales. Additionally, the thermosphere-ionosphere (T-I) system responses to the electrical energy and energetic particles from the solar wind and magnetosphere, through nonlinear interactions of

L. Wang et al. (Eds.): CHREOC 2022, LNEE 969, pp. 79–92, 2023.
https://doi.org/10.1007/978-981-19-8202-6_8

dynamical, electrodynamical, and chemical processes on different temporal and spatial scales (e.g., [8, 19]). In addition to the vertical, exponential solar activity, geomagnetic dependences, thermosphere density varies with local time, season, latitude, and longitude (e.g., [5, 6]). Although the existence of repeatable and distinct variations in T-I system has been recognized for a long time, the dynamic process and the underlying mechanisms are not fully explained.

Thermosphere mass density data resources mainly include drag-derived, accelerometer-derived, neutral mass spectrometer, ultraviolet remote sensing, and so on [6]. Due to the coarse spatial and temporal resolution, the daily globally averaged density derived from satellite drag data provides little information about the local time and latitudinal dependence of the intra-annual oscillations. The high-resolution density observations deduced from accelerometer measurements on the satellites such as the Gravity field and steady-state Ocean Circulation Explorer (GOCE), the Challenging Minisatellite Payload (CHAMP), Gravity Recovery and Climate Experiment (GRACE), can cover almost the pole-to-pole latitudes given their high-inclination satellite orbits (e.g., [7, 9, 16]). Since these satellites process 24-h local solar time (LST) approximately every ~130 days, the estimated densities deduced from these satellites are usually combinations of varying season, height, and local time, besides the effects of geomagnetic and solar activity. It is not always easy or possible to separate these variations which are deduced from different effecting sources. Since the sun-synchronous satellite is fixed at a nearly constant dawn and dawn dusk LST, it can provide neutral density observations which minimize or avoid mixes of the solar activity and seasonal variations with influences of the changing local time.

Little research has been made in the high altitude thermosphere density; especially the global seasonal-latitude structure above 500 km is not known, mainly due to lack of observations. The neutral mass density estimates from in-situ observations along two sun-synchronous satellite orbits, which worked near 630 km and 460 km orbit altitudes, offer a good opportunity to analyze the characteristic of oscillations in the upper thermosphere neutral density. The focus of this paper is the mass density variations in the high-altitude thermosphere at dusk and dawn on different temporal and spatial scales, including the density variations with season, latitude, solar flux level and geomagnetic activity.

2 Data and Methodology

2.1 Densities Derived from Space-Borne Atmospheric Density Detectors

The in-situ measurements by atmospheric density detectors are derived from two satellite missions. Figure 1 demonstrates examples of the nature of satellite observations along orbits. The daily satellite altitude and the corresponding solar flux and geomagnetic activity indexes are shown in Fig. 2. It can be seen that the APOD satellite orbits from January 2016 to December 2018, are fixed at dusk (18:20–20:00 LST) and dawn (06:20–8:00 LST) with the daily average altitude of 450–470km. Another Chinese satellite (CHN-sat) mission operated at dusk (16:55–17:40 LST) and dawn (04:55–05:40 LST) sectors near 625–640 km altitudes from March 2008 to December 2014 (Fig. 2a). The solar minimum of the 24th solar cycle occurred from 2008–2009. As shown in Fig. 2c, the

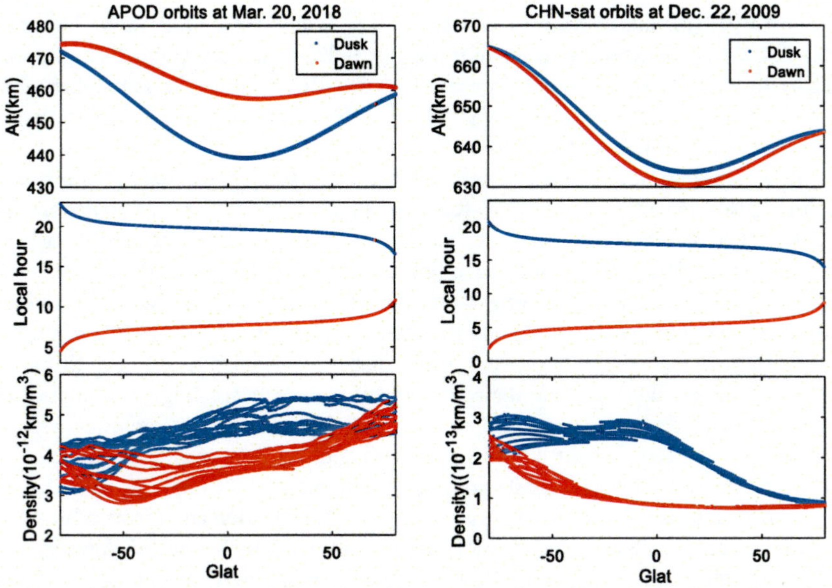

Fig. 1. Examples of the nature of in-situ satellite observations in a day, are shown along the orbital paths of APOD (left), and CHN-sat (right) satellites. From top to bottom, the panels show altitudes (top), local hours (middle), and density estimates (bottom) versus latitudes.

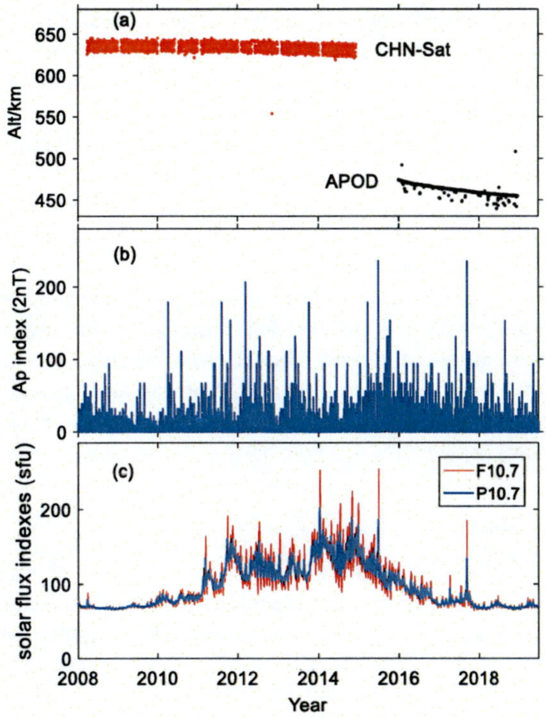

Fig. 2. **a** Daily satellites height, **b** geomagnetic Ap index, and **c** the solar activity indexes $P_{10.7}$ (blue), $F_{10.7}$ (red).

solar flux $F_{10.7}$ (in units of sfu, where 1 sfu = 10^{-22} Wm^{-2} Hz^{-1}) was below 70 during 2008–2009, while it ascended to above 160 in 2013–2014, and then gradually descended back to 70 in 2018. Severe geomagnetic storms usually occur at solar maximum when the extreme Ap value can even exceed 100 (Fig. 2b), and the geomagnetic disturbance at solar minimum is distinctly lower.

The in-situ thermosphere mass densities were detected by the space-borne atmospheric density detectors which were designed by the National Space Science Center (NSSC) of China. The in-situ temperature and pressure estimates are detected along orbit through voltage detections by the space-borne sensor [11, 14]. Then the neutral mass density (ρ) can be inferred according to the equation: $\rho = N_a . m_2$. N_a here indicates the atmosphere number density in regards to pressure and temperature detections, and m_2 denotes the corresponding atmospheric molecular mass average [3, 11, 14]. Since the satellites payloads don't include the spectrometer, the corresponding gas composition can't be detected directly. The molecular mass averages derived from empirical models are alternatively used to deduce the neutral mass density. Tang et al. [14] and Li et al. [11] have provided the detailed data process procedure.

APOD is a project aiming at neutral density in-situ detection and precis orbit determination using CubeSat, and some scientific instruments including the Atmospheric Density Detector are boarded. APOD was launched into a near circular polar orbit at ~520km altitude on September 20, 2015, and descended into the mission orbit altitude of ~460km on October 27, 2015, with an inclination of about 97.4°. The accuracies of the estimated APOD density have been testified by Tang et al. [14] and Li et al. [11]. Their investigations have shown the accuracy of the calibrated APOD density is comparable to the representative empirical models, e. g., JB2008 and MSISE00 model with a ~10% standard deviation of the background density. Calabia et al. [5] used APOD densities to assess a thermospheric neutral density model which was based on the GRACE densities of 13-years.

The CHN-sat mission provides a long-term continuous atmosphere mass density data which are also derived from space-borne atmospheric Density Detector. The CHN-sat densities are estimated through the similar data process procedure to that of APOD measurements and the detailed instrument performance specification was presented by Li et al. [12]. The CHN-sat operated at an altitude range of 620–660km with a quasi-circular orbit. The detector was recalibrated regularly through the ground monitor and calibration system. The atmosphere density is the level 3 data products which are processed through the data process system on the ground. The CHN-sat observations of the atmosphere neutral density have been investigated, such as the comparisons between the observation and model during geomagnetic storms [10], or the thermosphere density variations in solstice and equinox [12].

2.2 Empirical Models

Although an empirical model cannot elucidate the underlying physical mechanisms, it can provide the mean thermosphere climatology based on the data constructing the model. Most empirical models (e.g., [1, 13]), can describe the global density variations.

In order to check the uncertainties of density measurements, calculations from empirical models are compared with the measured densities. The empirical model calculates the density at every data point with the same input parameters as that of the measurement condition, and differences between models predictions and measurements are analyzed. Since an empirical model also has errors, two empirical models, MSISE00 and JB2008, are both used to compare with observations and further verify those results.

The MSISE00 model [13] is an empirical model of the atmosphere covering the altitude from the ground to exosphere space. The word 'MSIS' denotes two primary data sources: Mass Spectrometer and Incoherent Scatter Radar. The total mass density data derived from satellite drag are also taken in the model. The inputs of the model include the solar flux $F_{10.7}$ index, and the geomagnetic Ap index. Outputs of the model are total mass density, neutral temperature and compositions.

The JB2008 model [1] is constructed using a combination of solar and geomagnetic indices to reconstruct exospheric temperature and the total mass density. The variations in the semiannual density which result from EUV heating are presented by new semiannual density equations. And the Dst index is used to model Geomagnetic storm effects as the driver of global density changes. The model was validated through comparisons with daily density drag data previously derived from numerous satellites at altitudes from 175 to 1000 km.

2.3 Methodology

The thermosphere neutral density varies on different spatial and temporal scales which origin from several different drivers, so it is important to consider the thermospheric density variation under the similar conditions with other effects are eliminated or minimized. First, the altitude normalization is made for each satellite. By using an empirical model computation, the density measurements are normalized to an averaged orbit altitude according to the following equation : $\rho(h, t) = \rho(z, t)\frac{\rho_{M(h,t)}}{\rho_{M(z,t)}}$. The term ρ(z,t) is the observed density on the satellite orbit altitude (z) at the observing time (t), and ρ(h,t) is the normalized neutral density to the constant averaged height (h), while ρ_M denotes the corresponding neutral mass density computed from empirical model, such as JB2008 or MSISE00. The altitude variation of each satellite is not large, which is less than 50 km either for APOD or for CHN-sat satellite, then the error induced by this altitude normalization is considered to be small.

Second, the density data are grouped according to local time, latitude, season, and the solar flux condition. Four season periods are categorized: March equinox (March-May), September equinox (September-November), June solstice (June-August), and December solstice (December-February). The solar composite index $P_{10.7}$ defined as $P_{10.7} = (F_{10.7} + F_{10.7A})/2$, is used to describe the solar flux levels. The term of $F_{10.7A}$ here denotes the 81-day average of the $F_{10.7}$ index. The $P_{10.7}$ index is considered as a more appropriate representation index of solar radiation flux [7, 18]. Third, the density observations along satellite orbits are averaged and binned over the geographic latitudes (Lat) and times with a 1° (Lat) × 1 h (hour) grid. The binned data sets are then reprocessed and categorized into seven latitude regions: equator (−5°N–5°S), low-lat. (5–25°), mid-lat. (25–50°) and high-lat. (50–75°) regions in the southern and northern hemispheres respectively.

The geomagnetic conditions are described by geomagnetic Kp and ap indexes. The ap index (in unit of 2nT) is 3-h resolution and the daily average is denoted by Ap. Geomagnetic quiet-time are determined by Kp < 3 for both the preceding 6 h and the current 3-h. When the seasonal variation characteristic is analyzed, only the quiet-time data are used. Since the thermospheric neutral mass density is sensitive to solar radiation input during geomagnetic quite-time, the geographic coordinate is used.

In order to verify the observation results, the Space Environment Technologies (SET) HASDM (U.S. Air Force High Accuracy Satellite Drag Model) density dataset is also used for comparisons. The SET HASDM gridded data is 1 h (local hour) × 10° latitude with a 3-h time resolution from 2000–2019, which is considered suitable for evaluating the global model densities [15].

3 Results and Discussion

The daily varying densities with time are compared at latitudes from equator to high latitudes. Results for equator, middle latitude 30°S and high latitude 70°S are shown in Fig. 3. In regards to thermospheric neutral mass densities at APOD orbits, except at solar minimum, the varying measured densities show consistent variations with that in HASDM database and the empirical model results, whether the seasonal characteristic of two maxima in equinoxes or the density growth with the increasing solar flux are agreeable. Meanwhile, the MSISE00 densities are significantly higher than observations under the low solar activity conditions, for example the densities at APOD orbits in 2018 (Fig. 3, left column). The similar larger MSISE00 density than observations at solar minimum has been denoted by several researches (e.g., [16]).

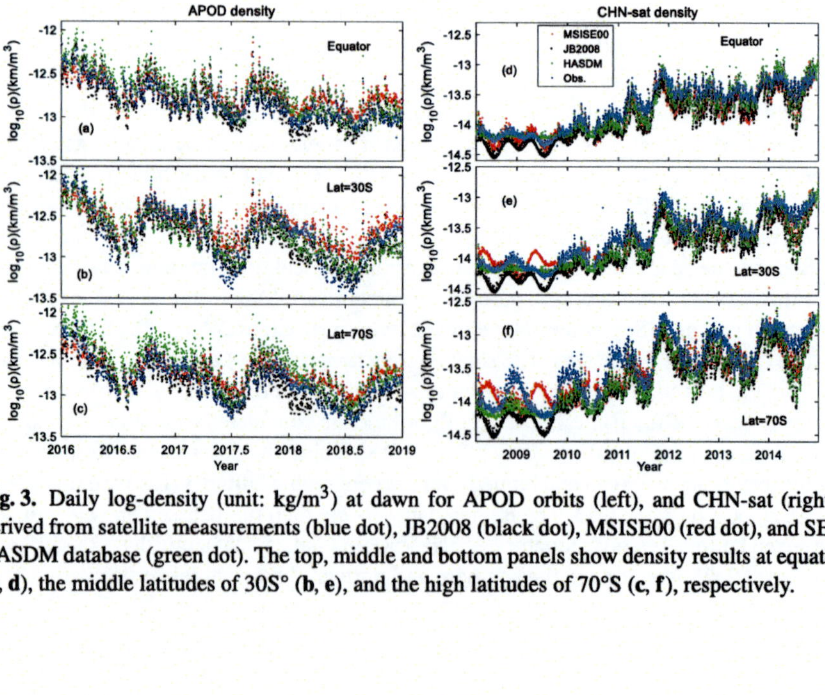

Fig. 3. Daily log-density (unit: kg/m^3) at dawn for APOD orbits (left), and CHN-sat (right), derived from satellite measurements (blue dot), JB2008 (black dot), MSISE00 (red dot), and SET HASDM database (green dot). The top, middle and bottom panels show density results at equator (**a**, **d**), the middle latitudes of 30S° (**b**, **e**), and the high latitudes of 70°S (**c**, **f**), respectively.

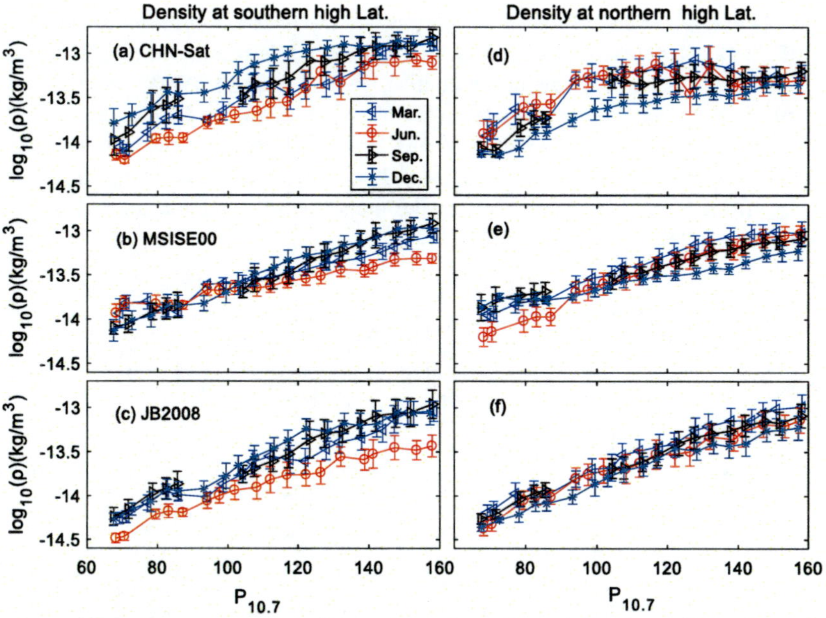

Fig. 4. Statistical distribution of log-density (unit: kg/m^3) at dawn for four seasons in the southern (left column) and northern (right column) high latitude regions (50–75°) versus the binned $P_{10.7}$. The neutral densities are from CHN-sat measurements (**a**, **d**), MSISE00 (**b**, **e**) and JB2008 (**c**, **f**) predictions. The $P_{10.7}$ index is 5 sfu binned and the bar denotes the standard deviation of each bin.

For the total mass density at CHN-sat orbit altitudes, contributions from anomalous oxygen [13] are included in MSISE00 predictions. Obviously, the CHN-sat observations are in best agreement with the SET HASDM densities (Fig. 3, the right column). On the one hand, equatorial observations of CHN-sat from 2008–2014, and measurements at high latitudes during 2011–2014 (Fig. 3d) are in good agreement with model results. On the other hand, significant differences between observations and MSISE00 densities occur at middle-high latitudes under the very low solar activity conditions during 2008–2010 (Fig. 3e, f). In contrast to the distinct annual and semiannual variations shown in MSISE00 results, neutral densities derived from CHN-sat show only annual variation at high latitudes. Meanwhile, results from MSISE00 and Jb2008 seem to be quite different from each other with opposite phases. Although CHN-sat densities of 70°S, especially the density maxima in summer near January, are higher than results of both models, the trend of density variation seems consistent with that of JB2008 densities. In fact, the thermospheric neutral densities on the very low solar radiation condition, both from the observations and model predictions have shown abnormal characteristics (e.g., [2, 17]).

Statistical results for different latitudes under different solar fluxes are conjunctly analyzed to assure the statistically significant results. Distributions of density at high latitudes in Fig. 4 exhibit distinct differences between observations and model results under low solar flux conditions. The mass densities at high latitudes in the Northern

Hemisphere (NH) from CHN-sat observations are the lowest in the winter, December, regardless of solar radiation level (Fig. 4d–f). Meanwhile results from JB2008 agree well with CHN-sat measurements. In contrast, results from MSISE00 show the mass density minima at the northern high latitudes in June under $P_{10.7}$ <85. Concerning to the neutral density in the SH (Fig. 4a–c), both the CHN-sat observations and computations of JB2008 show minima in June and maxima in December regardless of solar flux levels, different from MSISE00 outputs in that the model density minima under $P_{10.7}$ <80 appearing in December and March.

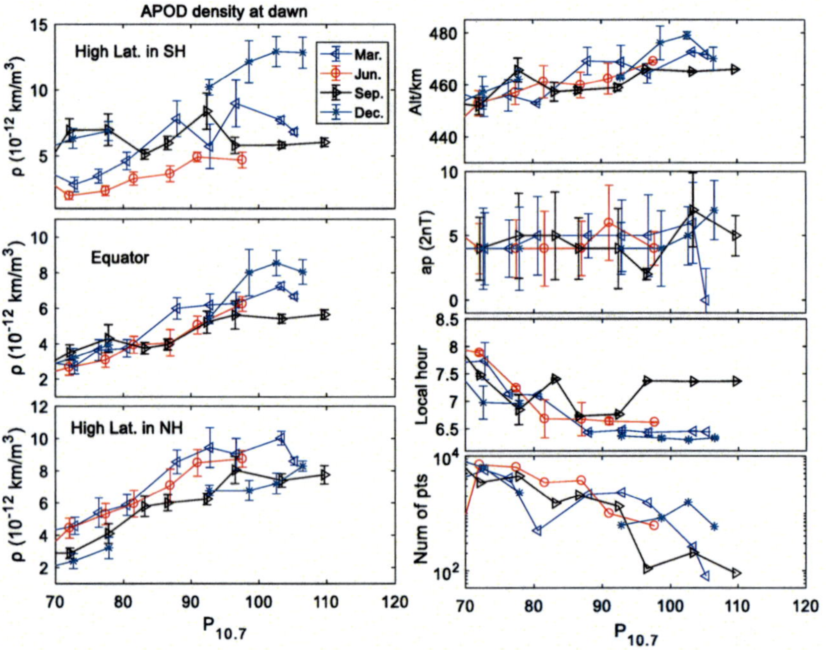

Fig. 5. Statistical APOD density averages (left) versus the binned $P_{10.7}$ in three latitude regions, equator (5°S–5°N), the northern high latitudes (50–75°N), and southern high latitudes (50–75°S). The right column shows the corresponding median altitude, ap index, solar local hour, and the number of data samples in each bin.

Figure 5 shows statistical median densities derived from APOD measurements in three latitudes. It can be seen that the APOD density observations under $P_{10.7} > 90$ are invariable with the solar radiation increase, which seems unreasonable. Meanwhile, the corresponding data spots number (Fig. 5, the right bottom panel) is small, which indicates it statistically insignificant. In fact, only under the low solar activity condition when the $P_{10.7}$ range is ~70–80, there are the simultaneous APOD observations in the four seasons for statistical analysis. Thus, in order to reduce the error caused by other effecting factors, the intra-annual variations in the APOD densities are analyzed based on the data under the low solar flux condition of $P_{10.7}$ ~70–80.

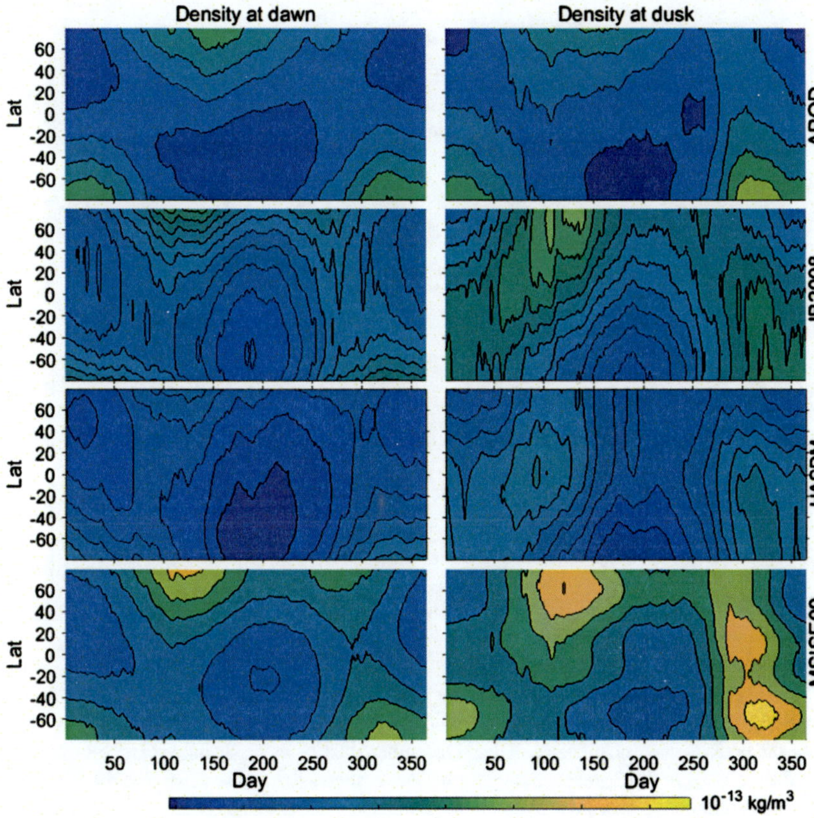

Fig. 6. Latitudinal-seasonal distribution of the total mass densities (unit: 10^{-13}kg/m^3) at APOD orbits. Panels from the top to bottom are APOD satellite measurements, JB2008, HASDM, and MSISE00 densities at dawn (left) and dusk (right). APOD and HASDM densities are performed on the low solar flux condition of $P_{10.7}$: ~70–80, and model densities are computed under $P_{10.7} = 75$.

Latitudinal structures of seasonal variation in neutral mass density at 460 km altitudes are shown in Fig. 6, in which the mass density data sets are smoothed by a 27-day running average to remove short-term variabilities. The largest density maxima in the densities appear in the local summer at southern high latitudes, while near equinoxes in the northern poles. The neutral density minima occur at high latitudes in local winter, with the lowest one in the Southern Hemisphere (SH) near June and the second minimum in the NH near December. Of the two maxima in the NH, the maximum in March is distinctly stronger than that in October. The similar characteristic with stronger density maximum in March equinox at the northern high latitude was also found in GOCE measurements results [16], and from GRACE observations [4], in which the mass density variations showed two maxima around June in the NH and only one in December in the SH.

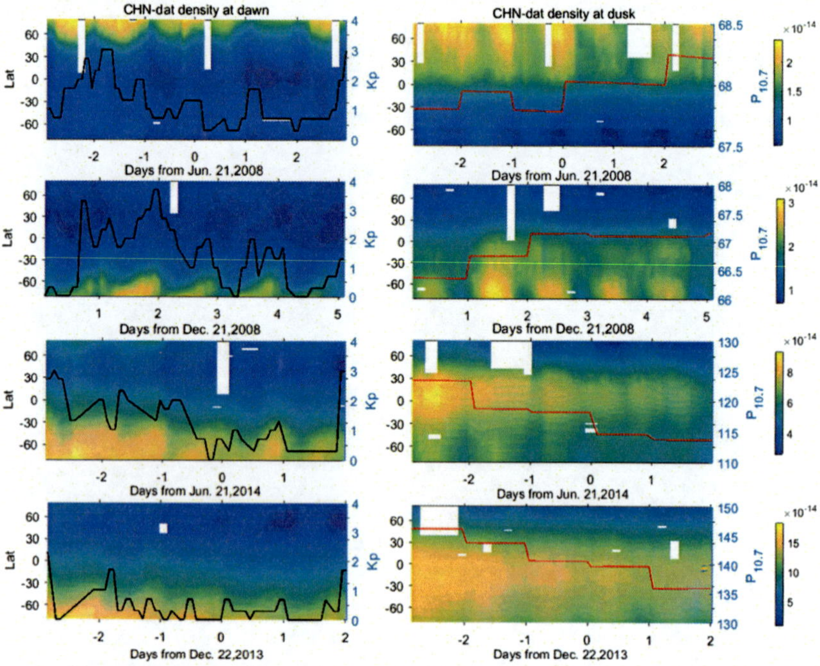

Fig. 7. Hourly CHN-sat densities versus time and geographic latitude in solstices at solar minimum (2008–2009) and solar maximum (2013–2014). Panels from the top to bottom show densities for the periods centers at June solstice, December solstice in 2008, June solstice in 2014 and December solstice in 2013, respectively. Left column shows density at dawn and the black lines denote the geomagnetic index Kp given on the right axis. Densities at dusk and the solar flux index $P_{10.7}$ (red lines) are shown in the right column.

The latitude structure of the mass density variations calculated by MSISE00 at 630 km altitude under the very low solar radiation is quite different from that observed by CHN-sat. Figure 7 shows densities at CHN-sat orbits during 2–3 days around the June and December solstices at solar minimum and solar maximum. The solar flux level is low and stable for the two periods in 2008 shown in the top two panels, with the nearly constant $P_{10.7}$ ~67, when the geomagnetic activity is almost quiet with the highest Kp ~ 3^+. For the periods of 2013–2014, the solar flux level is distinctly higher with the $P_{10.7}$ range ~115–145. The corresponding predictions of MSISE00 and JB2008 are shown in Figs. 8 and 9, in which the modeled densities are computed from the exactly same conditions as that of measurements.

In solstices at solar minimum, the observed densities (Fig. 7, the top two panels) from middle-high latitudes are obviously higher in the summer hemisphere than in the winter hemisphere. The JB2008 predictions (Fig. 9) with the completely similar conditions also show higher density in the summer high latitudes at dawn, however the densities are distinctly weaker than CHN-sat observations. Contrastively, the hemisphere asymmetry of the densities at high latitudes in solstices shown in MSISE00 computations

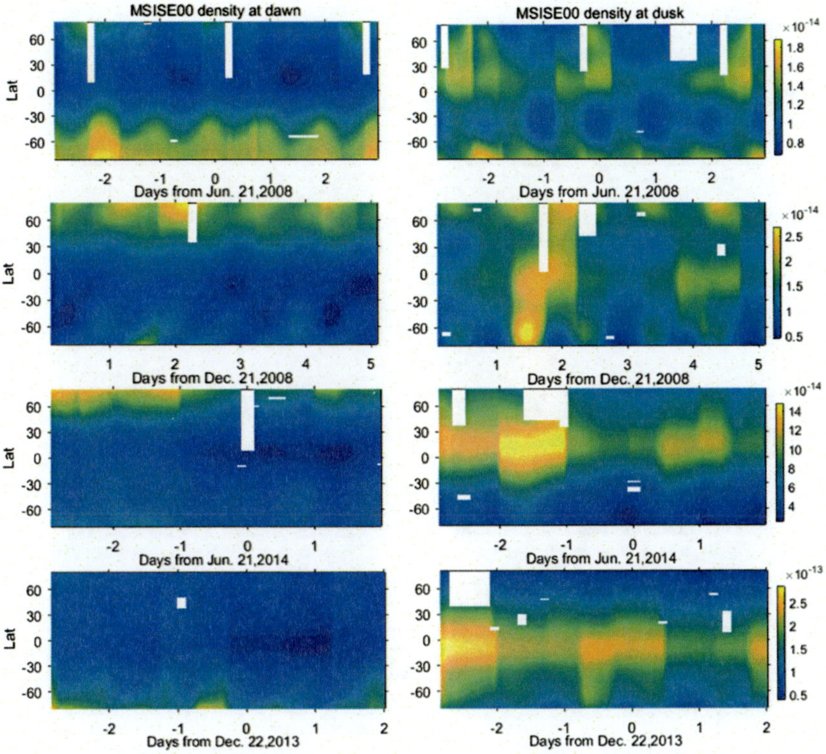

Fig. 8. Same as Fig. 7, but densities from MSISE00 predictions.

seems in reverse phase with that of CHN-sat observations. That is, MSISE00 densities at high latitudes in the winter hemisphere are larger than those in the winter hemisphere. This phenomenon may indicate that the relative importance of the conjunct contribution from temperature and oxygen atom, and the winter helium bulge maybe should be redistributed.

In regards to the density comparisons at solar maximum in solstices, on the one hand, the model densities from MSISE00 and JB2008 are in good agreements with the CHN-sat observations in December, that is, the high-latitude density in the summer (southern) hemisphere is significantly stronger than that in the winter hemisphere. On the other hand, the CHN-sat results in June solstice seem different from MSISE00 and JB2008 predictions. The CHN-sat densities at dawn in the northern hemisphere are lower than that in the southern hemisphere on June 2014, which seems contrary to the model results (Figs. 7, 8 and 9). The reasons for this phenomenon are not clear and may be worth further research and discussion. In addition, the high-altitude thermosphere densities seem sensitively variable on the geomagnetic disturbance condition. Figure 10 shows some examples of the CHN-sat densities at solar minimum when the geomagnetic condition is weakly disturbed. It can be seen that the low-latitude density maxima at

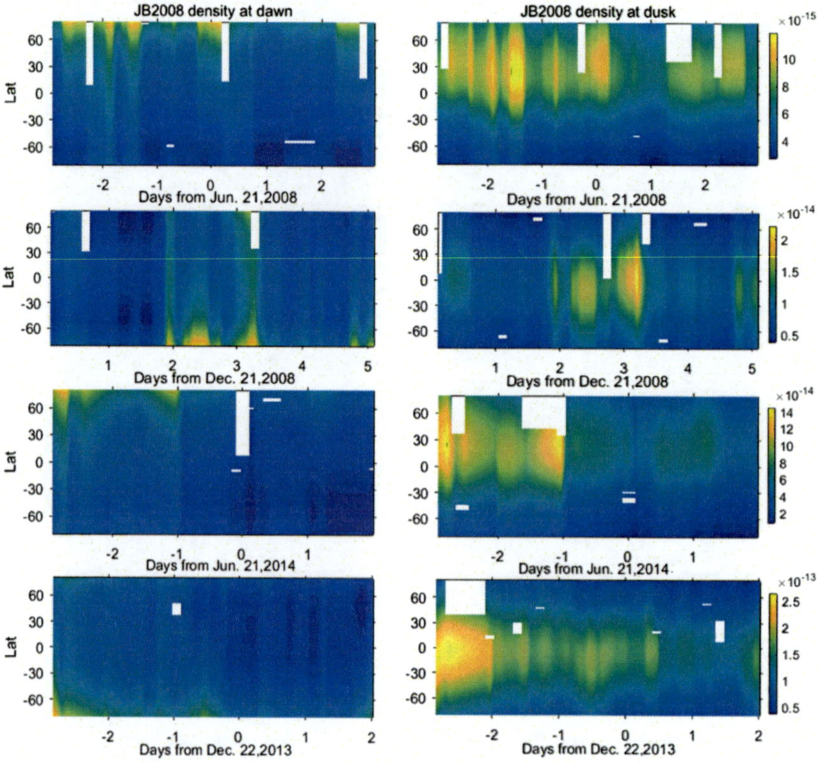

Fig. 9. Same as Fig. 7, but densities from JB2008 predictions.

dusk (Fig. 10, the right bottom panel) and the high-latitude density at dawn are distinctly strengthened.

4 Summary

The thermosphere undergoes distinct variations, which are characterized by significant spatial-temporal structure, dependent on altitude, latitude, solar flux and geomagnetic activity. Based on the in-situ mass densities detected by the space-borne atmospheric density detectors on two sun-synchronous satellites, this paper presents the characteristic of oscillations in upper thermosphere near 460 km and 630 km altitudes. Some new or more detailed characteristics of thermosphere density variations are summarized as follows:

The densities observed by APOD show consistent variations with that in empirical model results, meanwhile the MSISE00 densities are higher than observations at solar minimum (Fig. 3). Latitudinal- seasonal structures show that the largest density maxima in the densities appear in the local summer at southern high latitudes, while near equinoxes in the northern poles. Of the two maxima in the NH, the maximum in March is distinctly stronger than that in October under low solar flux conditions (Fig. 6).

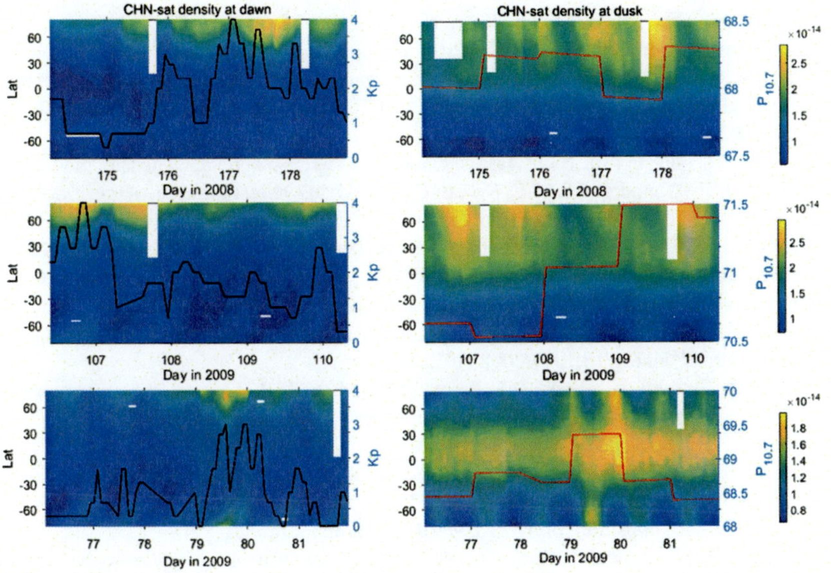

Fig. 10. Similar with Fig. 7, but examples of CHN-sat densities at solar minimum during weak geomagnetic disturbances.

Equatorial observations of CHN-sat are in good agreement with model results, while significant differences between observations and MSISE00 densities occur at high latitudes under the very low solar activity conditions (Figs. 3, 4, 7 and 8). In solstices, the measured densities from middle-high latitudes in the summer hemisphere are distinctly higher than those in the winter hemisphere. Inversely, MSISE00 densities at high latitudes are larger in the winter hemisphere, which indicates that the relative importance of the conjunct contribution from oxygen atom and temperature, and the winter helium bulge should be adjusted.

The high-altitude thermosphere density variation is sensitive to the geomagnetic activity, and the weak geomagnetic disturbance can clearly strengthen the low-latitude density maxima at dusk and the high-latitude density at dawn at CHN-sat altitudes (Fig. 10).

References

1. Bowman, B.R., Tobiska, W.K., Marcos, F.A., Huang, C.Y., Lin, C.S., Burke, W.J.: A new empirical thermospheric density model JB2008 using new solar and geomagnetic indices. In: AIAA/AAS Astrodynamics Specialist Conference, 18–21 August, Honolulu, Hawaii, paper AIAA 2008–6438 (2008)
2. Bruinsma, S.L., Forbes, J.M.: Anomalous behavior of the thermosphere during solar minimum observed by CHAMP and GRACE. J. Geophys Res. **115**, A11323 (2010). https://doi.org/10.1029/2010JA015605

3. Clemmons, J.H., Friesen, L.M., Katz, N., Ben-Ami, M., Dotan, Y., Bishop, R.L.: The ionization gauge investigation for the streak mission. Space Sci. Rev. **145**, 263–283 (2009)
4. Calabia, A., Jin, S.: New modes and mechanisms of thermospheric mass density variations from GRACE accelerometers. J. Geophys. Res. **121**(A10), 11 (2016). https://doi.org/10.1002/2016JA022594
5. Calabia, A., Tang, G., Jin, S.G.: Assessment of new thermospheric mass density model using NRLMSISE-00 model, GRACE, Swarm-C, and APOD observations. J. Atmos. Sol.-Terr. Phys. **199**, 105207 (2020). https://doi.org/10.1016/j.jastp.2020.105207
6. Emmert J.T.: Thermospheric mass density: a review. Adv. Space Res. **56**(5), 113 773–824 (2015). https://doi.org/10.1016/j.asr.2015.05.038
7. Guo, J., Wan, W., Forbes, J.M., Sutton, E., Nerem, R.S., Bruinsma, S.: Interannual and latitudinal variability of the thermosphere density annual harmonics. J. Geophys. Res. **113**, A08301 (2008). https://doi.org/10.1029/2008JA013056
8. Jones, M., Emmert, J.T., Drob, D.P., Picone, J.M., Meier, R.R.: Origins of the thermosphere-ionosphere semiannual oscillation: reformulating the "thermospheric spoon" mechanism. J. Geophys. Res. Space Phys. **123**(11–12) (2018)
9. Lei, J., Matsuo, T., Dou, X., Sutton, E., Luan, X.: Annual and semiannual variations of thermospheric density: EOF analysis of CHAMP and GRACE data. J. Geophys. Res. **117**, A01310 (2012). https://doi.org/10.1029/2011JA017324
10. Yongping, L., Guangwu, Z., Guotai, Q., et al.: Significant differences of thermosphere density between the model and the obvervation values during different altitudes and geomagnetic disturbances. Chin. J. Geophys. **57**(11), 3703–3714 (2014). (in Chinese)
11. Li Xie, X., Jiyao, T.G., et al.: Processing and calibrating of in-situ atmospheric densities for APOD. Chin. J. Geophys. **61**(9), 3567–3576 (2018). (in Chinese)
12. Yongping, L., Yueqiang, S., Xinyue, W., Zhenyu, F.: Variations of thermospheric density during equinox and solstice. Sci. Technol. Rev. **37**(6), 104–113 (2019). (in Chinese)
13. Picone J.M., Hedin A.E., Drob D.P., Aikin A.C.: Nrlmsise-00 empirical model of the atmosphere: statistical comparisons and scientific issues. J. Geophys. Res. Space Phys. **107**(A12), SIA 15-1–SIA 15-16 (2002)
14. Tang, G., Li, X., Cao, J., Liu, S., Chen, G., Man, H., Zhang, X., Shi, S., Sun, J., Li, Y., Calabia, A.: APOD mission status and preliminary results. Sci. China Earth Sci. **63**, 257–266 (2020). https://doi.org/10.1007/s11430-018-9362-6
15. Tobiska, W.K., Bowman, B.R., Bouwer, S.D., Cruz, A., Wahl, K., Pilinski, M.D., et al.: The SET HASDM density database. Space Weather **19**, e2020SW002682 (2021). https://doi.org/10.1029/2020SW002682
16. Libin, W., Jiuhou, L., Eelco, D., et al.: Seasonal variations of thermospheric mass density at dawn/dusk from GOCE observations. Ann. Geophys. **36**(2), 489–496 (2018)
17. Solomon, S.C., Woods, T.N., Didkovsky, L.V., Emmert, J.T., Qian, L.: Anomalously low solar extremeultraviolet irradiance and thermospheric density during solar minimum. Geophys. Res. Lett. **37**, L16103 (2010). https://doi.org/10.1029/2010GL044468
18. Xiong, C., Lühr, H., Schmidt, M., Bloßfeld, M., Rudenko, S.: An empirical model of the thermospheric mass density derived from champ satellite. Ann. Geophys. **36**(4), 1141–1152 (2018)
19. Zhang, X.F., Liu, L., Liu, S.T., Wu, Y.P.: A statistical study on the response of thermospheric total mass density to geomagnetic storms. Chin. J. Geophys. **58**(9), 3023–3037 (2015). https://doi.org/10.6038/cjg20150901. (in Chinese)

对地精确感知技术

Semi-automatic Annotation Method for Semantic Segmentation of Synthetic Aperture Radar Images

Nai-Rong Zheng[1], Zi-An Yang[1], Hong Yang[2,3], Yue Sun[2], and Feng Wang[1](✉)

[1] Key Laboratory for Information Science of Electromagnetic Waves (Ministry of Education), School of Information Science and Technology, Fudan University, Shanghai 200433, China
fengwang@fudan.edu.cn

[2] Aerospace Information Research Institute, Chinese Academy of Sciences, Beijing 100094, China

[3] School of Electronics and Information, Northwestern Polytechnical University, Xi'an 710129, China

Abstract. With the availability of public high-resolution remote sensing datasets, it has become a trend to realize land cover classification using deep learning methods. However, supervised semantic segmentation requires an accurate label at the pixel level as the ground truth of image. This is a challenge to remote sensing data with big data characteristics, and manual annotation is challenged with a heavy workload and low efficiency. In this paper, a semi-automatic annotation method is proposed and verified on synthetic aperture radar (SAR) images obtained by Gaofen-3 (GF-3) and the Chinese Aeronautic Remote Sensing System (CARSS). The method uses semantic segmentation to classify the homologous images initially, then iterates the coarse classification results as ground truth. The coarse classification results of multiple iterations will be corrected and added to the base dataset. The whole system is updated positively, and the model is constantly optimized to improve the performance of the auxiliary semi-automatic annotation model. Experimental results demonstrate that the pixel accuracy of the final classification results can reach higher than 80%.

Keywords: Semantic segmentation · Semi-automatic annotation · Synthetic aperture radar · Gaofen-3 · Chinese aeronautic remote sensing system

1 Introduction

Remote sensing is a comprehensive and interdisciplinary subject that emerged in the 1960s. The data sources of remote sensing include optical, synthetic aperture radar (SAR), and hyperspectral. The application of remote sensing data for classifying land cover types has played a more critical role in the fields of resource survey [1], urban planting [2], and sea ice monitoring [3]. With the development of deep learning in recent years, it has gradually become a trend to adopt the semantic segmentation method in deep learning to classify land cover. And more and more public datasets have been available.

L. Wang et al. (Eds.): CHREOC 2022, LNEE 969, pp. 95–101, 2023.
https://doi.org/10.1007/978-981-19-8202-6_9

For example, Tong et al. released the GID dataset [4], which contains optical images of more than 60 different cities in China with a total area of 50000 square kilometers. Wang et al. Released the AIR-PolSAR-Seg dataset [5], which contains 500 fully polarized SAR images of 512 * 512 with a resolution of 8 meters. Zhong et al. released the WHU-Hi dataset [6], which includes Hyperspectral images of multiple locations in Wuhan. These publicly available datasets drive the development of deep learning in the land cover classification of remote sensing data. However, the land cover classification using supervised semantic segmentation requires accurate labels at the pixel level as the ground truth of images. And it is impossible to manually generate ground truth in a short time, especially in the field of big data remote sensing. Although the semi-automatic labeling method proposed by Shi et al. [7] can reduce part of the labeling workload, their method is complicated, and the categories are still limited.

This paper proposes an auxiliary semi-automatic annotation method for labeling SAR images for semantic segmentation. This method combines deep learning and adopts negative feedback to iterate continuously, which significantly improves the accuracy of model prediction to reduce the workload of correcting the ground truth.

2 Data and Methodology

Aiming at the difficulty of label generating in supervised learning of land cover classification from large-scale remote sensing images, a semi-automatic auxiliary annotation method is proposed in this paper. By using deep learning, experiments are carried out on the single-polarization SAR images of the GF-3 satellite and the full-polarization SAR images of CARSS.

2.1 The Experimental Data of GF-3 and CARSS

The data used for experimental verification in this paper are from GF-3 and CARSS. The basic information is shown in Table 1.

Table 1. The basic information of experimental data.

Data source	No	Region	Polarization mode	Resolution (m)	Image size (w, h)
GF-3	1	Pearl harbor	Single-polarization (HH)	1	15296 * 12303
	2	Coronado	Single-polarization (HH)	1	14680 * 11210
CARSS	3	Wanning, Hainan	Full-polarization	0.5	10454 * 9322
	4	Wanning, Hainan	Full-polarization	0.5	14747 * 11113

Each image has a corresponding label, which is classified according to the features of its land cover. The GF-3 data divides the types of land cover into 9 categories: water, bare soil, roads, industry, residence, vegetation, woodland, plantation, and human-built. The CARSS data divides land cover types into 9 categories: water, bare soil, road, industry, vegetation, residence, plantation, other, and farms. The GF-3 and CARSS data are shown in Fig. 1.

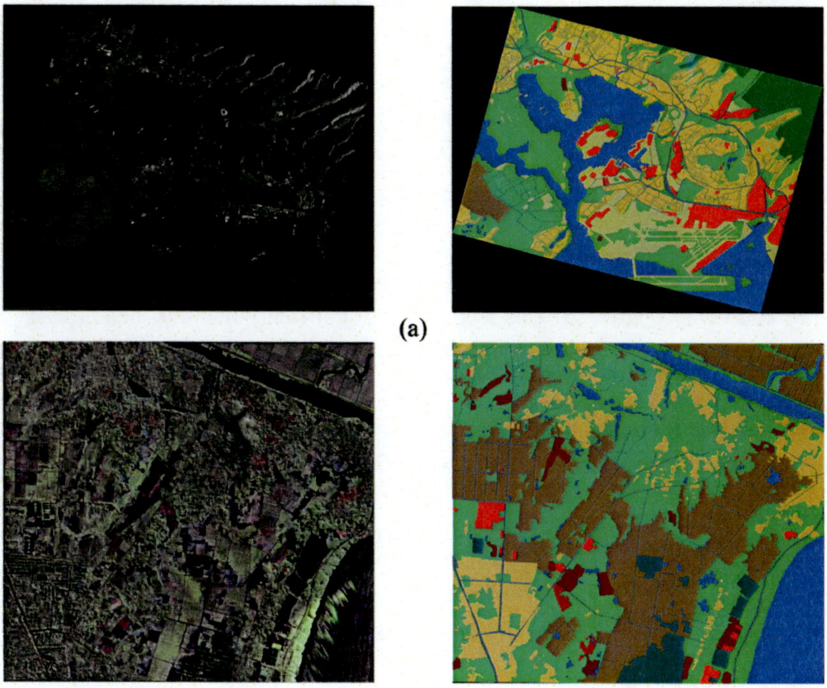

(a)

(b)

Fig. 1. The experimental SAR images and their ground truth. **a** Pearl Harbor; **b** Wanning, Hainan

2.2 Auxiliary Semi-automatic Annotation Method

The efficiency and accuracy advantages of deep learning for processing big data are employed to improve the efficiency of label generation. The process of the auxiliary semi-automatic annotation method proposed in this paper is shown in Fig. 2.

As shown in Fig. 2, the base dataset is trained by a neural network to obtain a basic semantic segmentation model, which can be used to predict subsequent new images. When new images are obtained, the model can be used for testing to get coarse classification results. The newly added images and coarse classification results are added to the training set as augmented data for iterative training, which means the model is adjusted and optimized through a negative feedback method. After many iterations, the

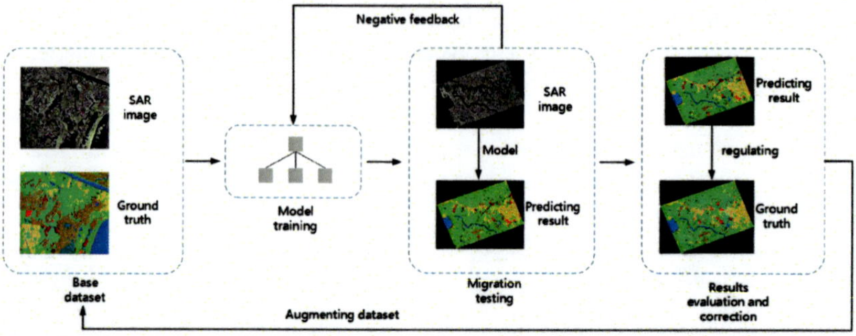

Fig. 2. The realization of proposed auxiliary semi-automatic annotation method.

model fully learns the simple information of the newly added data and generates preliminary classification results. Finally, the preliminary generated classification results are corrected to obtain the ground truth of SAR images. Whenever an image is added, the above steps can be repeated to continuously expand the base dataset so that the model can be constantly optimized. This method can significantly improve the efficiency of remote sensing data semantic segmentation label creation in the case of big data.

2.3 The Evaluation Indicators

After the labels are semi-automatically generated, their classification accuracy is evaluated to simply assess the reduction in ground truth production workload. The evaluation metrics used in this paper are Intersection over Union (IoU) and Pixel Accuracy (PA).

IoU is the ratio of the intersection and union between the predicted results and the actual results. The formula is as follows:

$$IoU = \frac{target \cap prediction}{target \cup prediction} \tag{1}$$

PA is the ratio of the correct pixels to the total pixels, and the formula is as follows:

$$PA = \frac{\sum_{i=1}^{n} p_{ii}}{\sum_{i=1}^{n}\sum_{j=1}^{n} p_{ij}} \tag{2}$$

where n is the number of classes, and p_{ij} is the number of pixels that belong to class i and are predicted to be class j.

3 Experiment and Discussion

Section 2 describes the data and methods of the experiments in this paper. This section will introduce the parameter settings, experimental process, and results. Since the scene of SAR images is large, which is needed to be preprocessed before training. For GF-3 images, we use N times the mean of non-zero data to truncate and stretch the original SAR data (NMT) [7] and maps the image pixel value to 0–255. The NMT is realized as follow:

$$I = uint8\left[\frac{\mathrm{I}(\mathrm{I} \leq \mathrm{N}\cdot\mathrm{mean}(\mathrm{I}))}{\mathrm{N}\cdot\mathrm{mean}(\mathrm{I})} \times 255\right] \in [0, 255] \tag{3}$$

where, *I* represents the pixel value of each channel, *mean(I)* represents the mean of all values of each channel, and *N* represents a multiple. Here, *N* is set as 3.

For the images obtained by CARSS, NMT is adopted to process the fully-polarization images. Then HH, HV, and VV polarized images are used as R, G, and B channels respectively to synthesize color images. In the GF-3 data, the image numbered one is selected as the initial base dataset, and the image numbered three is selected as the initial base dataset for the CARSS image. Image No. 2 and No. 4 were respectively used as their new data for iterative auxiliary semi-automatic annotation.

After image preprocessing, images and labels are simultaneously sampled by the sliding window. The window size is set to 512, and the step size is set to 256. The HRNet [8] is used for training. All the batch sizes and epochs of this paper are set to 16 and 50, respectively. The experimental environment of this paper is implemented on a Linux system with two NVIDIA GeForce RTX 3090 Ti GPUs. The experimental results on GF-3 and CARSS are presented in Tables 2 and 3.

Table 2. The experimental results on the data of CARSS.

Data	IoU									PA
	Water	Bare soil	Road	Industry	Vegetation	Residence	Plantation	Others	Farms	
Base data	0.9798	0.049	0.2524	0.0635	0.7409	0.564	0.304	0	0.5934	0.8631
Iter1	0.9829	0.0645	0.2691	0.0645	0.7482	0.5704	0.3179	0	0.6327	0.8682
Iter2	0.9837	0.0668	0.2945	0.0624	0.753	0.5871	0.3224	0	0.6568	0.8716

Table 3. The experimental results on the data of GF-3.

Data	IoU									PA
	Water	Bare soil	Road	Industry	Residence	Vegetation	Woodland	Plantation	Human-built	
Base data	0.968	0	0.2441	0.0762	0.5731	0.4809	0	0	0.4183	0.8041
Iter1	0.9691	0	0.2506	0.0711	0.5761	0.4865	0	0	0.4209	0.8074

From Tables 2 and 3, the auxiliary semi-automatic annotation method proposed in this paper can achieve the expected results whether the data of GF-3 or CARSS. And as the number of iterations increases, the accuracy improves. In terms of base data, the accuracy rate can reach more than 80%, which dramatically improves the efficiency of label production. The visualization of the experimental results is shown in Fig. 3.

As shown in Fig. 3, the classification results of base data have been able to classify most of the categories roughly. But there is a certain amount of noise, which is also a big problem for the subsequent auxiliary semi-automatic annotation. However, in iter2, the noise is significantly reduced (such as the results in the red box), which dramatically improves the efficiency of subsequent label correction.

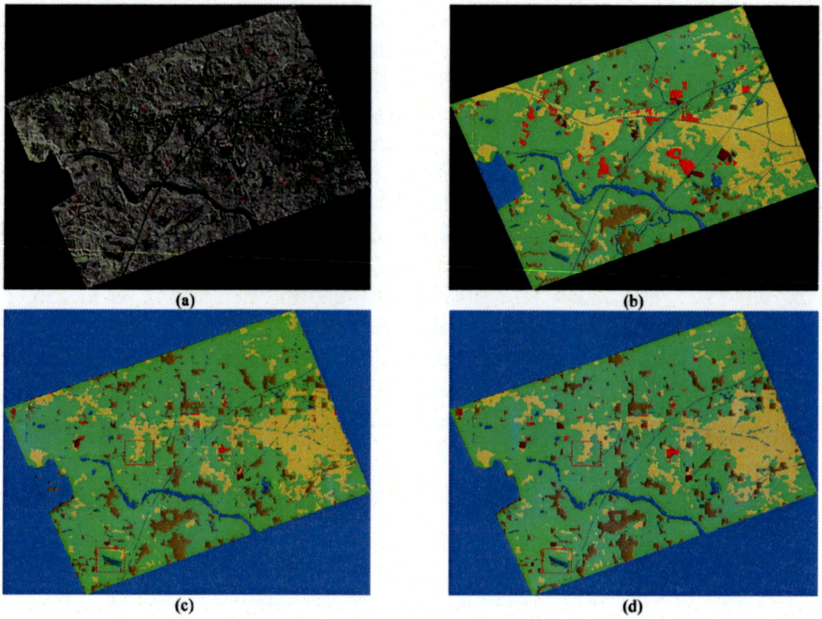

Fig. 3. The visualization of the experimental results on CARSS data. **a** SAR image; **b** ground truth; **c** the result of base data; **d** the result of iter2

4 Conclusion

This paper proposes a semi-automatic annotation method using neural network assistance for the large-scale remote sensing images in semantic segmentation. Based on iteration, the pixel classification of new images is performed and then corrected when the accuracy is over 80%, which significantly reduces the workload of manually labeling. After the completion of label correction, the new ground truth is generated to expand the base dataset and improve the generalization ability of the basic model. Under the circumstance that the amount of data increases, the auxiliary effect is further enhanced, and the efficiency is improved.

Acknowledgment. This work was supported in part by the China High-resolution Earth Observation System (CHEOS)-Aerial Observation System Project (30-H30C01-9004-19/21).

References

1. Chen, Q., Kuang, G., Li, J., et al.: Unsupervised land cover/land use classification using PolSAR imagery based on scattering similarity. IEEE Trans. Geosci. Remote Sens. **51**(3), 1817–1825 (2013)
2. Lv, Q., Dou, Y., Niu, X., et al.: Urban land use and land cover classification using remotely sensed SAR Data through deep belief networks. J. Sensors, 538063 (2015)

3. Dierking, W., Busche, T.: Sea ice monitoring by L-band SAR: an assessment based on literature and comparisons of JERS-1 and ERS-1 imagery. IEEE Trans. Geosci. Remote Sens. **44**(4), 957–970 (2006)
4. Tong, X.-Y., Xia, G.-S., Lu, Q., et al.: Land-cover classification with high-resolution remote sensing images using transferable deep models. Remote Sens. Environ. **237**, 111322 (2020)
5. Wang, Z.Z., Xuan, Yan, Z., Kang, J., Sun, X.: AIR-PolSAR-Seg: a large-scale data set for terrain segmentation in complex-scene PolSAR images. IEEE J. Select. Topics Appl. Earth Observ. Remote Sens. (2021)
6. Zhong, Y., Hu, X., Luo, C., et al.: WHU-Hi: UAV-borne hyperspectral with high spatial resolution (H2) benchmark datasets and classifier for precise crop identification based on deep convolutional neural network with CRF. Remote Sens. Environ. **250**, 112012 (2020)
7. Shi, X., Fu, S., Chen, J., et al.: Object-level semantic segmentation on the high-resolution Gaofen-3 FUSAR-map dataset. IEEE J. Select. Topics Appl. Earth Observ. Remote Sens. **14**, 3107–3119 (2021)
8. Wang, J., Sun, K., Cheng, T., et al.: Deep high-resolution representation learning for visual recognition. IEEE Trans. Pattern Anal. Mach. Intell. **43**(10), 3349–3364 (2021)

Dominate Tree Species Classification on Large-Scale Mountainous Areas Using Voting Strategy-Based Ensemble Learning Method

Pengfei Zheng[1], Panfei Fang[1], Peiwei Liu[1], Weiheng Xu[2,3], Guojun Wang[4], and Leiguang Wang[2,3](✉)

[1] Faculty of Forestry, Southwest Forestry University, Kunming, China

[2] College of Big Data and Intelligent Engineering, Southwest Forestry University, Kunming, China

wlgbain@126.com

[3] Key Laboratory of National Forestry and Grassland Administration on Forestry and Ecological Big Data, Southwest Forestry University, Kunming, China

[4] Yunnan Tobacco Company Qujing Company, Qujing, China

Abstract. Remote sensing technology provides an economical and efficient means for obtaining the distribution of dominant tree species. However, remote sensing mapping of dominant tree species have remained an ongoing challenge, such as massive data and low accuracy of a single machine learning method, especially over large-scale mountainous areas. In this study, the multi-source variables and multi-classifier decision fusion was used to separate the dominant tree species on the Google Earth Engine cloud computing platform. We combined with multi-source data such as Sentinel-2 satellite imagery, bioclimate, topography, and forest inventory data and constructed time series for NDVI and REP to evaluate different variables combinations. At the same time, tree species classification was carried out following two approaches: non-stratified and stratified. Four different Machine learning algorithms, Random Forest (RF), Support Vector Machine (SVM), XGBoost, and Maximum Entropy (MaxEnt) were used as component classifiers to construct two decision fusion models of serial integration and Bayesian parallel integration respectively and completed the spatial distribution of 10 main dominant tree species mapping in the mountainous areas of northwest Yunnan. The nine tree species were classified with an overall accuracy of 74.44% using serial integration of MaxEnt and RF. This study confirmed that combination of multi-source data multi-source variables and the decision fusion method can be used to provide a reference for the classification of dominant tree species in large-scale mountainous areas.

Keywords: Dominant tree species · Machine learning · Time series · Decision fusion · Google earth engine

L. Wang et al. (Eds.): CHREOC 2022, LNEE 969, pp. 102–115, 2023.
https://doi.org/10.1007/978-981-19-8202-6_10

1 Introduction

Forest is an important natural resource that plays an important role in balancing the earth's ecosystem and is also an important part of national resource [1]. Forest surveys systematically collect information about the composition of forests, covering various spatial scales, from stand level to woodland level, from regional to national and even global levels [2]. Among them, the classification of tree species is a basic task of forest investigation, which is of great significance to the sustainable management of forests. Due to the needs of forest management and ecological protection, tree species information is widely used in biodiversity assessment and monitoring, community dynamics and species contributions to ecosystem functions and services, mapping of wildlife habitats, and monitoring of pests and diseases in forests [3–6]. However, for the acquisition of large-scale forest survey information, the traditional forest survey method mainly adopts the method of ground survey, which requires a lot of manpower and time cost, and is difficult to meet the actual needs due to problems such as long period, low efficiency, and unstable accuracy application requirements [7]. With the development of remote sensing technology and the increasing availability of remote sensing data, the use of remote sensing images for forest resource surveys and high-precision tree species mapping have become focus.

Multispectral, hyperspectral, and high spatial resolution optical satellite remote sensing data have been extensively used for forest types and tree species classification at various scales. In addition, the development of LiDAR and SAR active remote sensing has enabled the acquisition of forest spatial structure information and stand scale, even single-tree-scale forest parameters provide the possibility [8, 9]. But these explorations used high-resolution data, hyper-spectral images, or these sets of data in combination with other ancillary data such as LiDAR have obtained acceptable classification accuracy of tree species, the greatest limitation of these data is their high cost and low availability [10, 11]. And these studies mapping tree species only for relatively small areas of several hundred to several thousand square kilometers. The vegetation types in these areas are single, the distribution of tree species is relatively concentrated, the samples are easy to obtain, and the samples obtained in the area are generally highly representative. However, there are relatively few studies on tree species identification in some atypical, large-area mountainous areas. In these areas, the harsh climatic conditions and complex terrain environment in most mountainous areas hinder obtaining low-cost, high-accuracy ground survey data, which brings great challenges to the production of classified samples and the selection of high-quality tree species samples. In addition, mountainous areas have diverse vegetation types, complex forest structures, and relatively discrete distribution of tree species compared with typical areas [12]. It is limited to obtain accurate tree species information only by using spectral features.

For a wide range of classification, forest type and tree species in the medium-resolution sensors such as Landsat and Sentinel-2 data map forest scene have become the most widely and frequently data. Landsat and Sentinel data open access multispectral data are very useful for classification mapping of forest types, tree species and spatial and temporal analysis of distribution changes due to their high temporal and spatial resolution optical satellite images. There are differences in the phenological changes of different tree species. Many studies have shown that using multi-temporal or time-series data can capture phenological differences between tree species and improve classification accuracy such as Kollert use more than one year of Sentinel—2 images of scene seasonal cloud-free images, the classification accuracy of five different tree species groups has reached about 84.4%, and the results show that multi-temporal images are helpful to improve the classification accuracy [13]. Grabska et al. used Sentinel-2 images and environmental data to map 11 tree species in the Polish Carpathian Mountains, with an accuracy of 85%, demonstrating the potential of Sentinel-2 image data and topographic information in mapping forest species in the mountain region [14]. Cheng et al., used Sentinel-2 time series data to extract temporal, spatial and spectral features to map forest types in four typical regions of China based on the GEE platform and machine learning algorithm, with an accuracy of over 85% [15].

Due to the special geographical environment of the region, there are obvious differences in the classification accuracy of different classification algorithms for different dominant tree species. The multi-classifier ensemble learning algorithm converts the traditional classification results based on single-dominant classifiers into comprehensive classification results that comprehensively consider multiple classifiers, which can reduce the accidental errors caused by a single classification algorithm [16]. Vasilakos et al., applied six traditional data classifiers and nine ensemble methods on the multi-temporal Sentinel-2 image dataset to identify olive groves, oaks, black groves in a heterogeneous Mediterranean landscape on the island of Lesvos, Greece Pine, hardwood, and other cover types. The results show that the voting method outperforms the base classifier [17].

The advent of a wide variety of remote sensing cloud computing technology services facilitated the geospatial data processing on a large-scale. Google Earth Engine has been widely used in vegetation monitoring by remote sensing in large and even global regions by its powerful computing power in data mining, processing and analysis [18]. This study based on Google Earth Engine platform combined with Sentinel-2 remote sensing images, terrain data, and environmental factor data such as bioclimate, integrates the maximum entropy niche model and machine learning algorithms such as Random Forest (RF), Support Vector Machine (SVM), and XGBoost to map tree species in the mountainous areas of northwest Yunnan.

2 Materials and Methods

2.1 Study Area

The study area is located in northwestern of Yunnan Province, at junction of Yunnan, Tibet and Sichuan provinces, including Diqing Prefecture, Dali Bai Autonomous Prefecture and Koksang (Fig. 1), covering 77229 km^2.

The topography of study area is complex and diverse and high in north and low in south, the three mountains are Meili Snow Mountain, Yun ling Snow Mountain, and Zhong dian Snow Mountain, which are arranged in sequence from west to east. The highest elevation is 6740 meters from Kawagebo Peak of Meili Snow Mountain, and the lowest elevation is 730 meters in Yunlong County. The three-dimensional climate is obvious, and the dry and wet seasons are distinct.

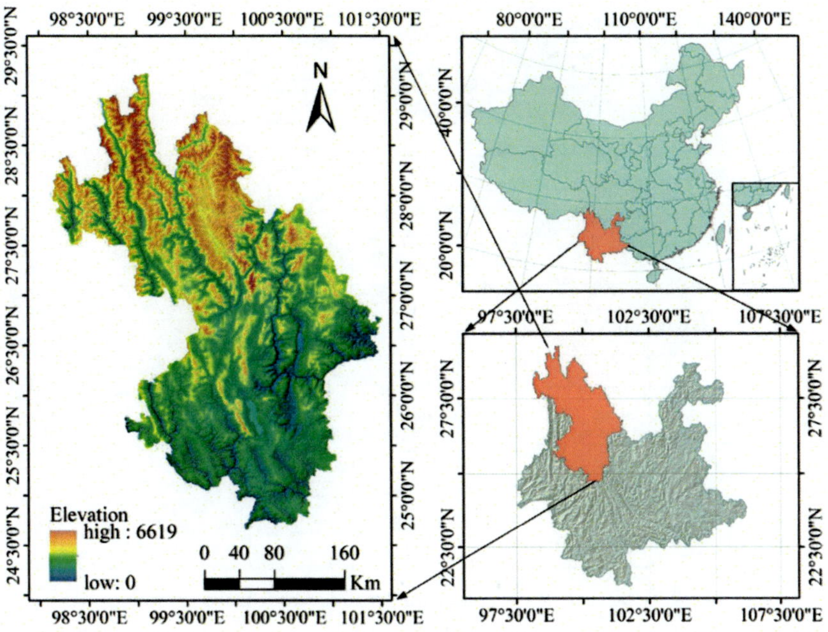

Fig. 1. Study area

The heterogeneity of topographic and climatic conditions leads to the formation of various vegetation communities along suitable environmental conditions. The abundant forest resources and diversity tree species are deciduous dominated by *Pinus yunnanensis*, *Pinus densata*, *Abies fabri*, *Quercus*. Figure 2 shows the flowchart of the proposed classification method.

2.2 Dataset

2.2.1 Ground Reference

In this study, we constructed the ground reference data using The Third National Forest Resources Inventory in 2016 for ground sample generation. To select high-confidence data as the ground truth, we counted the standard deviation of the NDVI bands of the annual composited images of the forest and non-forest patches, further screened the forest and non-forest according to the NDVI standard deviation, then obtained the sample points from the selected patches transition points. For the filtering of tree species samples, we also carry out the same process.

2.2.2 Satellite Data and Auxiliary Data

The Sentinel-2 (S2) satellite has been available since 2015 (S2A) and the second sensor (S2B) was launched in 2017. The Sentinel-2 has 13 spectral bands spanning from the visible (VIS) and near-infrared (NIR) bands to short-wave infrared (SWIR) bands and

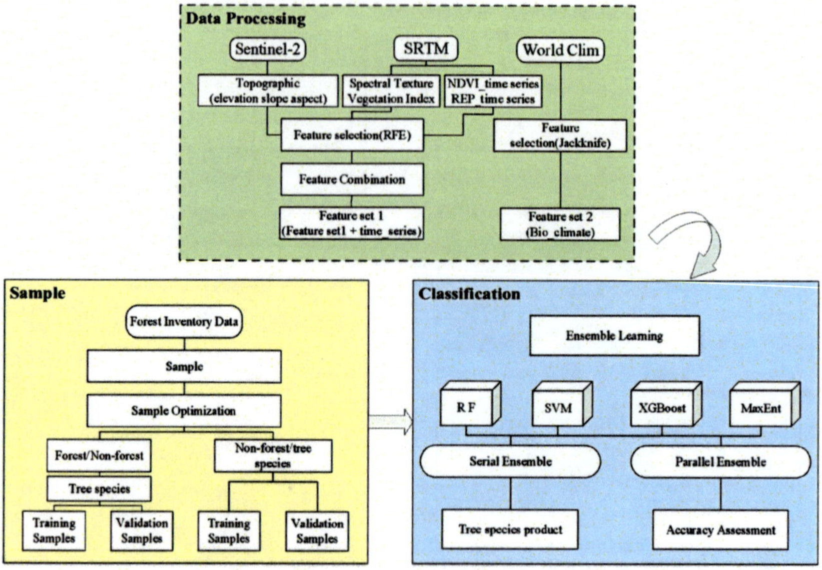

Fig. 2. Flowchart of the proposed method for tree species mapping.

allows for an ideal revisit time of five days (Table 1). The ten spectral bands of Sentinel-2 images ranging from the visible and near-infrared (NIR) to the shortwave infrared (SWIR) were considered in this study. The spectral information of red-edge bands in Sentinel-2 data shows a potential performance application in evaluating the classification of vegetation.

All sentinel-2 images for the period 2016–2021 are composited by median to calculate texture metrics and vegetation index. The Google Earth Engine cloud platform provides the 'glcmTexture' function to calculate GLCM. After GLCM calculation, a series of indicators reflecting texture characteristics are generated. In this study, six texture parameters including average, correlation, difference, variance, contrast and cluster shadow were selected. The vegetation index is calculated from multi-spectral remote sensing image analysis and has a certain indication of vegetation growth status and helps to identify dominant tree species. In order to explore various effects on distinguishing dominant tree species, this paper calculates various vegetation indices, the calculation formulas of various vegetation indices are shown in the following table.

In addition, the topographic data SRTMGL1_003 with a resolution of 30m and the 19-dimensional bioclimatic data WorldClim provided by Google Earth Engine will be used to extract topographic and climate features. After integrating these multi-source spatial data into each tree species, we can obtain multi-dimensional environmental feature descriptions for each tree species.

2.3 Time Series Construction

In general, the growth pattern of a vegetated pixel is determined by the inherent cycle of the species within the pixel and is largely affected by the local environmental conditions

[19]. We can find the special rules representing the growth patterns of different tree species from the vegetation index time series products with continuous observations and high quality, and then extract the differences in the tree species time series to identify different tree species.

For nine tree species, two vegetation index (NDVI, REP) were collected at a 10-day interval in the entire 2016, resulting in a total of 36 temporal data records for each vegetation index. Due to the cloud and show contamination, time series may be missing some features in different time node. This paper performed linear interpolation on the reconstructed 10-day time series to obtain a complete time series with no missing features. Then the Savitzky-Golay (SG) algorithm was used to eliminate outliers and smooth time series features (Fig.3).

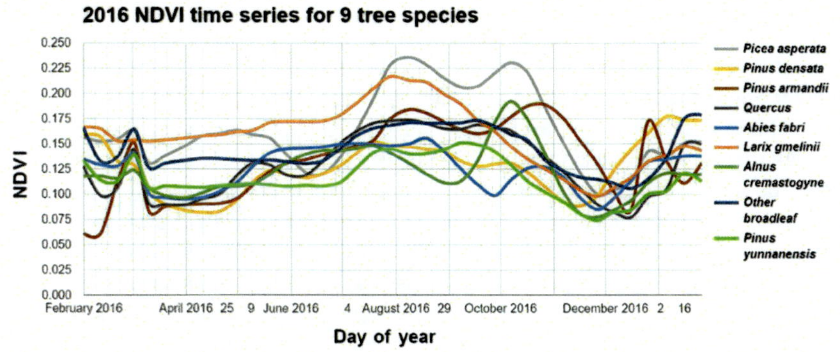

(a)

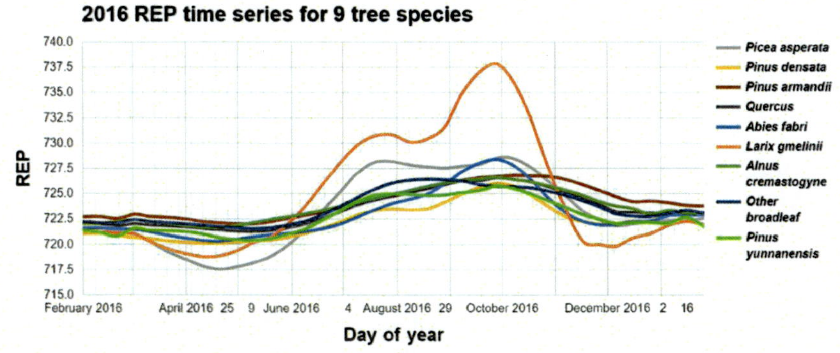

(b)

Fig. 3. Time-series spectral profiles in different vegetation index.

2.4 Classification

2.4.1 Hierarchical Classification Method

The hierarchical classification method divides the classification problem into different levels. For each level, the optimal features and samples can be filtered to train the classifier. In this paper, sentinel-2 satellite images are used to create and map the forest and non-forest with a resolution of 10 m in Yunnan Province, and then determine the spatial distribution of dominant tree species within the forest region. When classify tree species together with these non-forest cover types, we face a challenge from the interference of non-forest cover types. Classifiers trained for all categories do not ensure optimal results for forest type classification. To solve this problem, the hierarchical classification method divides the classification problem into different levels. For each layer, we can train the classifier with different features and samples to increase the degree of differentiation between classes [12]. Additionally, we also added non-forest and nine other tree species into the classification process to verify whether error transmission was more obvious during hierarchical classification.

2.4.2 Decision Fusion

In recent years, ensemble learning models are becoming increasingly popular among researchers in the field of regression and classification analysis [20]. The core of ensemble learning is to perform decision fusion on the initial classification results of multiple base classifiers to create an ensemble classifier to improve classification results. A successful multi-class ensemble system largely depends on the component classifiers. A highly diverse component classifier is the key to building an effective ensemble learning system. The more accurate and different the component classifiers are, the better the ensemble effect [21]. In this paper, four classifiers, SVM, RF, XGBoost and MaxEnt, which are quite different and often obtain higher accuracy in land use, forest scene classification and prediction of species distribution, were employed as the base classifiers to generate the initial classification. Since the voting strategy plays a critical role in ensemble learning models, different voting strategy leading to diverse classification results of ensemble learning.

Here, two paradigms of parallel integration and serial integration was selected to evaluate the ability of ensemble learning in tree species classification. The first paradigm is the parallel pattern. Three component classifiers are used for separate classification, and then decision fusion is carried out based on Bayesian decision making. The second integration paradigm is the serial pattern, which is chain executed in sequential mode, that is, the result of the previous classifier is the input of the next classifier in the chain. Here, we took the classification result of MaxEnt as the input of RF and SVM. MaxEnt algorithm uses a variety of ecological environment factors, RF and SVM use spectral and texture features to achieve the fusion of feature level and decision level (Fig. 4).

3 Results

3.1 Feature Importance Assessment

The 16-dimensional more important variables were used for the classification of non-forest and nine tree species in a non-hierarchical classification process. The variables are ranked by the level of importance. Among them, the elevation, REP time series and wetness scored the higher. The importance of input variables in the classification of the nine tree species following the stratified approach is presented in Fig. 5.

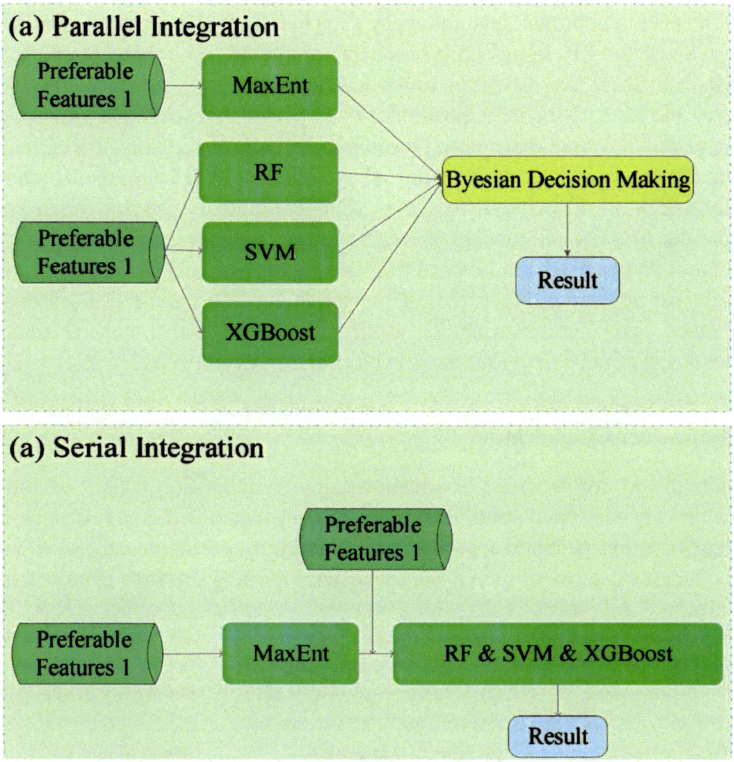

Fig. 4. Schematic diagram of decision fusion method.

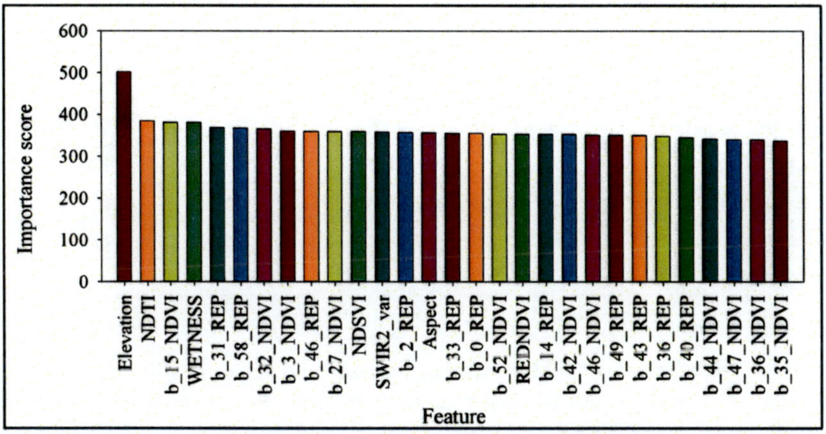

Fig. 5. Optimal feature importance analysis

29-dimensional optimal features were selected from 237 feature candidates for XGBoost, SVM and RF classification based on correlation hierarchical cluster screening and the built-in Jackknife algorithm of the Maximum Entropy software. The Elevation contributed the most to the classification of tree species that related to the topographic and climatic conditions of the region. The time series calculated from the REP and NDVI was more important for the classification of tree species. This suggests that changes in vegetation index are associated with seasonal phenological changes, which are spectral responses to seasonal changes in vegetation growth-senescence alternation [22]. The temporal and spatial variability of vegetation phenological changes help to distinguish different vegetation types [23]. Time-series synthetic and multi-seasonal composite methods can capture variability in different spectra, which has been extensively demonstrated in complex environments such as mountain forests [24, 25].

3.2 Tree Species Classification

The results of the tree species classification using Random Forest (RF) performed on non-stratified approach and stratified approach is presented in Table 1. Compared with the results of the non-stratified approach, where nine tree species were classified together inside the forest mask, we observed a higher accuracy using stratified approach for most tree species.

Table 1. Accuracy assessment for on non-stratified approach and stratified approach.

Tree species	Non-stratified approach		Stratified approach	
	UA	PA	UA	PA
Pinusyunnanensis	70.55	67.50	73	70
Quercus	58.02	66.23	58	67
Abies fabri	73.70	84.03	58	43
Pinus densata	67.93	85.01	59	13
Alnus cremastogyne	70.83	72.34	71	78
Picea asperata	56.11	38.75	76	87
Other broadleaf	64.99	0.39	76	63
Pinus armandii	75.15	64.14	68	88
Larix gmelinii	74.36	59.49	62	34

Table 2 shows the overall classification accuracy and Kappa coefficient of different methods in this area. (In this table, M represents MaxEnt, R represents RF, S represents SVM, X represents XGBoost, ALL represents all variables as input, OPT represents preferable variables) Among them, M classification accuracy is the lowest, OA is only 54.44%, and kappa is 0.47. When all remote sensing factors are used for classification, S-ALL classification accuracy is slightly higher, OA was 68.95%. The classification accuracy of the three classifiers X-ALL, S-ALL and S-ALL is similar when using the filtered features, but the classification accuracy is about 1% lower than that using all the features. However, the feature dimension is reduced from 237 dimensions to 29dimensions, which reduces the computational pressure and improves the classification efficiency. Among them, the MR serial method has the highest accuracy. OA was 74.44% and Kappa was 0.70.

Table 2. Classification accuracy of different methods.

	Method	OA (%)	Kappa
Component classifier	M	53.44	0.47
	X-ALL	67.76	0.63
	S-ALL	68.95	0.64
	R-ALL	68.57	0.64
	X-OPT	67.30	0.62
	S-OPT	67.58	0.65
	R-OPT	67.83	0.63
Integrated classifier	X-XSR	68.28	0.64
	M-X	73.76	0.70
	M-S	72.93	0.69
	M-R	74.44	0.70

Figure 6 is a heat map of the three accuracy indicators in this area, which shows the advantages of the ensemble method in improving the classification accuracy. As can be seen from the figure, the best UA, PA, and F1 for multiple categories all come from the ensemble method. The classifiers using all features also achieved better accuracy for some categories, demonstrating the performance of component classifiers and the ability of spatio-temporal-spectral features to characterize tree species. Among the nine dominant tree species, *Abies fabri*, *Pinus densata*, *Pinus yunnanensis* and *Pinus armandii* have better classification accuracy, and all three indicators are higher in each classifier algorithm, while the categories with lower accuracy are *Larix gmelinii,* other broadleaf and *Picea asperata.*

3.3 Mapping of Dominant Tree Species

Figure 7 shows the best classification results provided by the serial ensemble method M-R. It can be seen from the figure that the *Abies fabri* trees are distributed in areas with higher altitudes, such as *Pinus densata*, *Pinus armandii*, and *Alnus cremastogyne. Pinus yunnanensis* has a wide distribution and a large area and is distributed in areas outside the high-altitude areas of Shangri-La, Deqin, and Gongshan. *Abies fabri* and *Larix gmelinii* are abundant and occupy high-altitude areas within zone one. The spatial distribution of the nine tree species reflects the obvious heterogeneity of forest site conditions in this

Method	indicator	Pinus yunnanensis	Quercus	Abies fabri	Pinus densata	Alnus cremastogyne	Picea asperata	other broadleaf	Pinus armandii	Larix gmelinii
M	UA(%)	0.57	0.42	0.79	0.68	0.48	0.46	0.33	0.70	0.45
	PA(%)	0.48	0.49	0.68	0.67	0.31	0.40	0.56	0.66	0.64
	Fi(%)	0.52	0.45	0.73	0.68	0.38	0.43	0.42	0.68	0.53
X-ALL	UA(%)	0.70	0.60	0.74	0.68	0.71	0.56	0.63	0.76	0.75
	PA(%)	0.70	0.69	0.84	0.85	0.72	0.42	0.41	0.63	0.64
	Fi(%)	0.70	0.64	0.79	0.76	0.72	0.48	0.49	0.69	0.69
S-ALL	UA(%)	0.73	0.62	0.77	0.75	0.72	0.56	0.50	0.75	0.49
	PA(%)	0.73	0.65	0.84	0.81	0.71	0.50	0.44	0.72	0.24
	Fi(%)	0.73	0.63	0.80	0.78	0.72	0.53	0.47	0.73	0.32
R-ALL	UA(%)	0.73	0.58	0.58	0.59	0.71	0.76	0.76	0.68	0.62
	PA(%)	0.70	0.67	0.43	0.13	0.78	0.87	0.63	0.88	0.34
	Fi(%)	0.72	0.62	0.49	0.21	0.74	0.81	0.69	0.77	0.44
X-OPT	UA(%)	0.73	0.58	0.77	0.68	0.67	0.55	0.58	0.70	0.48
	PA(%)	0.70	0.63	0.86	0.84	0.72	0.41	0.37	0.67	0.29
	Fi(%)	0.72	0.60	0.81	0.75	0.70	0.47	0.45	0.68	0.36
S- OPT	UA(%)	0.72	0.58	0.78	0.68	0.69	0.57	0.57	0.68	0.55
	PA(%)	0.70	0.65	0.87	0.85	0.70	0.41	0.40	0.65	0.26
	Fi(%)	0.71	0.62	0.82	0.76	0.70	0.47	0.47	0.67	0.35
R- OPT	UA(%)	0.72	0.58	0.77	0.68	0.69	0.56	0.60	0.72	0.57
	PA(%)	0.70	0.63	0.87	0.88	0.73	0.39	0.38	0.65	0.25
	Fi(%)	0.71	0.60	0.82	0.77	0.71	0.46	0.47	0.68	0.35
M-XSR	UA(%)	0.70	0.60	0.77	0.70	0.70	0.58	0.61	0.72	0.72
	PA(%)	0.69	0.66	0.85	0.86	0.71	0.43	0.44	0.67	0.71
	Fi(%)	0.70	0.63	0.81	0.77	0.71	0.49	0.51	0.69	0.72
M-X	UA(%)	0.75	0.67	0.82	0.74	0.74	0.67	0.67	0.79	0.60
	PA(%)	0.75	0.65	0.86	0.89	0.73	0.60	0.54	0.80	0.41
	Fi(%)	0.75	0.66	0.84	0.80	0.73	0.63	0.60	0.79	0.49
M-S	UA(%)	0.76	0.66	0.81	0.72	0.73	0.68	0.65	0.77	0.54
	PA(%)	0.73	0.65	0.86	0.90	0.73	0.55	0.51	0.80	0.40
	Fi(%)	0.75	0.66	0.83	0.80	0.73	0.61	0.57	0.78	0.46
M-R	UA(%)	0.76	0.67	0.82	0.74	0.74	0.67	0.70	0.81	0.65
	PA(%)	0.74	0.67	0.88	0.90	0.74	0.61	0.53	0.79	0.31
	Fi(%)	0.75	0.67	0.85	0.81	0.74	0.64	0.60	0.80	0.42

Fig. 6. Heat map of accuracy index of each classification method.

area. Most tree species are concentrated and have obvious vertical zonality, which is beneficial to distinguish them by environmental factors and spectrum.

As shown in Fig. 8, the visual comparison of tree species classifications by the base classifiers (RF and MaxEnt) and the ensemble learning method (M-R) with multiple voting strategies on selected part of the study area.

4 Discussion and Conclusions

In this study, the remote sensing characteristics displayed by the dominant tree species and their corresponding environmental and geological conditions are considered, and the phenological information contained in the multi-temporal remote sensing images is combined with the environmental information such as terrain, temperature and rainfall. Force and various machine learning algorithms, Sentinel_2 images, SRTM and World-clim and other data as data sources, proposed a variety of ecological niche, machine learning model and decision fusion models to map nine dominant tree species in the northwestern mountainous area of Yunnan. Compared with the conventional machine learning methods, what makes Ensemble Learning different is that it further utilized advantage of base classifiers. In this study, the classifier based on the MaxEnt-RF serial ensemble has the best performance, with OA of 74.44% and Kappa of 0.70, which

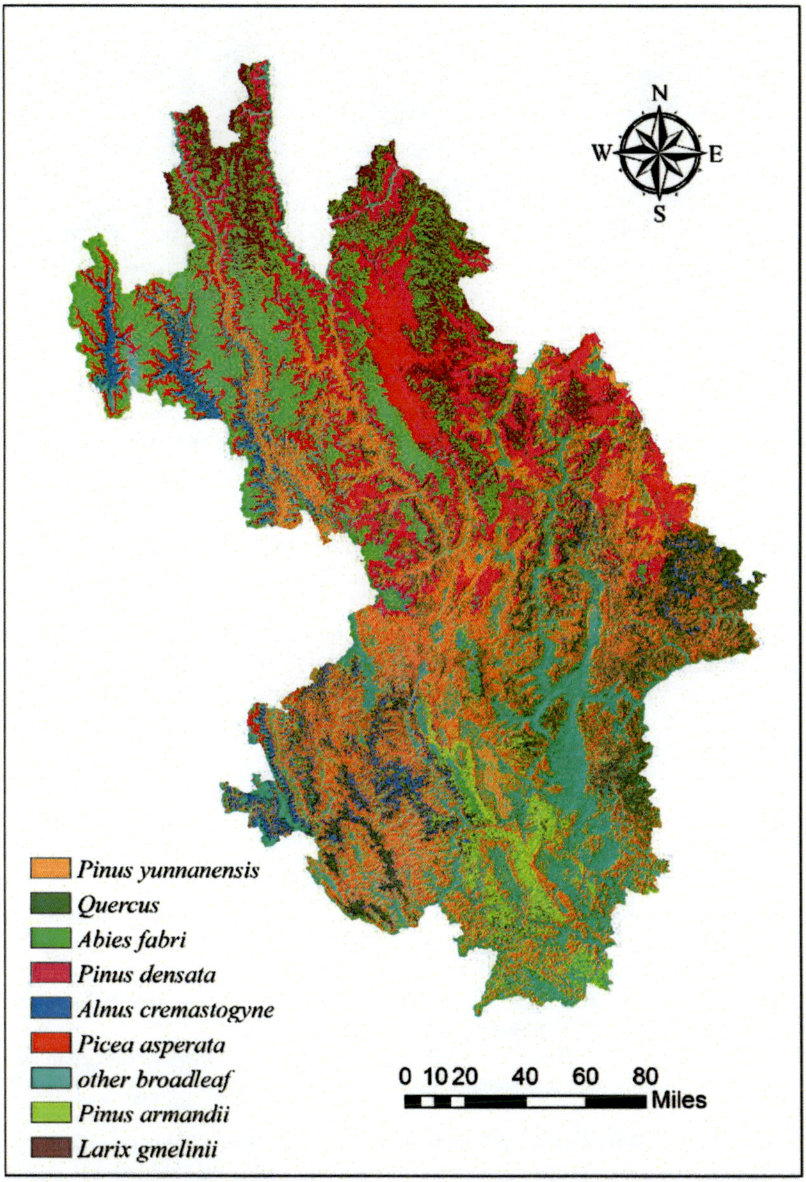

Fig. 7. Classification results of nine dominant tree species by the serial ensemble method M-R.

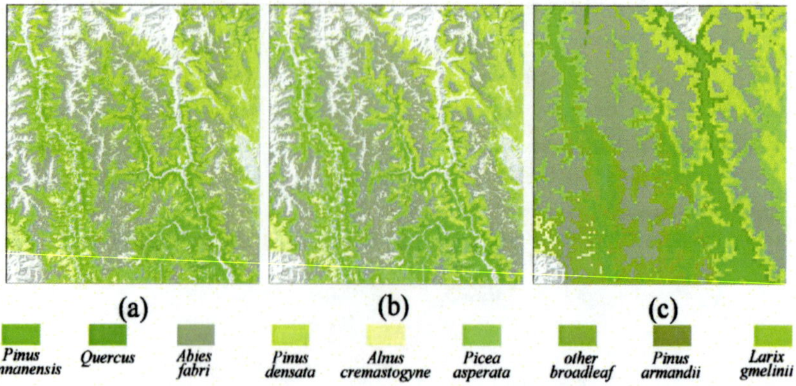

Fig. 8. Visual comparison of tree species classifications by the base classifiers and the ensemble learning method with multiple voting strategies on selected part of the study area. **a** M-R method. **b** RF. **c** MaxEnt.

proves the effectiveness of the decision fusion method in the classification of dominant tree species in large-scale mountainous areas.

Acknowledgements. This work was supported in part by the National Natural Science Foundation of China under Grants 32160369, 31860182, 41961053, and 41771375, in part by the Key Development and Promotion Project of Yunnan Province under Grant 202102AE090051; in part by the Fund of Reserve Talents for Young and Middle-Aged Academic and Technological Leaders of Yunnan Province under Grant 2018HB026.

References

1. Torahi, A.A., Chand Rai, S.: Land cover classification and forest change analysis, using satellite imagery-a case study in Dehdez area of Zagros Mountain in Iran. J. Geogr. Inf. Syst. **3**(1), 1 (2011)
2. White, J.C., et al.: Remote sensing technologies for enhancing forest inventories: a review. Can. J. Remote Sens. **42**(5), 619–641 (2016)
3. Chambers, D., et al.: Challenges in modelling the abundance of 105 tree species in eastern North America using climate, edaphic, and topographic variables. For. Ecol. Manag. **291**, 20–29 (2013)
4. Lehmann, J.R.K., et al.: Analysis of unmanned aerial system-based CIR images in forestry—A new perspective to monitor pest infestation levels. Forests **6**(3), 594–612 (2015)
5. Shang, X., Chisholm, L.A.: Classification of Australian native forest species using hyperspectral remote sensing and machine-learning classification algorithms. IEEE J. Select. Topics Appl. Earth Observ. Remote Sens. **7**(6), 2481–2489 (2013)
6. Hyde, P., et al.: Mapping forest structure for wildlife habitat analysis using multi-sensor (LiDAR, SAR/InSAR, ETM+, Quickbird) synergy. Remote Sens. Environ. **102**(1–2), 63–73 (2006)
7. Holmgren, P., Thuresson, T.: Satellite remote sensing for forestry planning—a review. Scand. J. For. Res. **13**(1–4), 90–110 (1998)

8. Ferro-Famil, L., Pottier, E., Lee, J.-S.: Unsupervised classification of multifrequency and fully polarimetric SAR images based on the H/A/Alpha-Wishart classifier. IEEE Trans. Geosci. Remote Sens. **39**(11), 2332–2342 (2001)
9. Korpela, I., et al.: Tree species classification using airborne LiDAR–effects of stand and tree parameters, downsizing of training set, intensity normalization, and sensor type. Silva Fennica **44**(2), 319–339 (2010)
10. Fricker, G.A., et al.: A convolutional neural network classifier identifies tree species in mixed-conifer forest from hyperspectral imagery. Remote Sens. **11**(19), 2326 (2019)
11. Hartling, S., et al.: Urban tree species classification using a WorldView-2/3 and LiDAR data fusion approach and deep learning. Sensors **19**(6), 1284 (2019)
12. Hościło, A., Lewandowska, A.: Mapping forest type and tree species on a regional scale using multi-temporal Sentinel-2 data. Remote Sens. **11**(8), 929 (2019)
13. Kollert, A., et al.: Exploring the potential of land surface phenology and seasonal cloud free composites of one year of Sentinel-2 imagery for tree species mapping in a mountainous region. Int. J. Appl. Earth Observ. Geoinf. **94**, 102208 (2021)
14. Grabska, E., Frantz, D., Ostapowicz, K.: Evaluation of machine learning algorithms for forest stand species mapping using Sentinel-2 imagery and environmental data in the polish carpathians. Remote Sens. Environ. **251**, 112103 (2020)
15. Cheng, K., Wang, J., Yan, X.: Mapping forest types in China with 10 m resolution based on spectral–spatial–temporal features. Remote Sens. **13**(5), 973 (2021)
16. Ponti Jr, M.P.: Combining classifiers: from the creation of ensembles to the decision fusion. In: 2011 24th SIBGRAPI Conference on Graphics, Patterns, and Images Tutorials. IEEE (2011)
17. Vasilakos, C., Kavroudakis, D., Georganta, A.: Machine learning classification ensemble of multitemporal Sentinel-2 images: the case of a mixed mediterranean ecosystem. Remote Sens. **12**(12), 2005 (2020)
18. Gorelick, N., et al.: Google earth engine: planetary-scale geospatial analysis for everyone. Remote Sens. Environ. **202**, 18–27 (2017)
19. Qiu, Y., et al.: Spatiotemporal fusion method to simultaneously generate full-length normalized difference vegetation index time series (SSFIT). Int. J. Appl. Earth Observ. Geoinf. **100**, 102333 (2021)
20. Gandhi, I., Pandey, M.: Hybrid ensemble of classifiers using voting. In: 2015 International Conference on Green Computing and Internet of Things (ICGCIoT). IEEE (2015)
21. Modzelewska, A., et al.: Multitemporal hyperspectral tree species classification in the Białowieża forest world heritage site. For.: Int. J. For. Res. **94**(3), 464–476 (2021)
22. Zhong, L., Gong, P., Biging, G.S.: Efficient corn and soybean mapping with temporal extendability: a multi-year experiment using Landsat imagery. Remote Sens. Environ. **140**, 1–13 (2014)
23. Zeng, L., et al.: A review of vegetation phenological metrics extraction using time-series, multispectral satellite data. Remote Sens. Environ. **237**, 111511 (2020)
24. Wang, C., et al.: Phenology-assisted classification of C3 and C4 grasses in the US Great Plains and their climate dependency with MODIS time series. Remote Sens. Environ. **138**, 90–101 (2013)
25. Li, R., et al.: Classifying forest types over a mountainous area in southwest China with landsat data composites and multiple environmental factors. Forests **13**(1), 135 (2022)

A Radar Clutter Model Selection Method Based on Multi-layer Perceptron

Shu-Qi Lei[1], Dong-Xiao Yue[2], and Feng Wang[1](✉)

[1] Key Laboratory for Information Science of Electromagnetic Waves (Ministry of Education), School of Information Science and Technology, Fudan University, Shanghai 200433, China
{sqlei20,fengwang}@fudan.edu.cn

[2] Department of Automation, Tsinghua University, Beijing 100084, China
dxyue15@fudan.edu.cn

Abstract. The radar clutter model is a vital characteristic of the environment. As the scene becomes more complicated, the closed-loop design of cognitive radar system can use feedback information to improve target detection performance. We propose a new model selection method based on the MLP(Multi-Layer-Perceptron) and simulated clutter curves of generalized models. The trained MLP is used to select the clutter model automatically for real-time clutter. Compared with traditional model selection methods, our method holds instantaneity that MLP abandons the comparation of fitting precision between parametric model and real non-parametric clutter model while using a neural network to extract a representation of clutter features. The experimental results on real-time clutter demonstrate that our method has a comparable fitting performance to traditional model selection methods.

Keywords: Clutter model selection · Generalized statistical model · Radar clutter · MLP

1 Introduction

The proposal of cognitive radar [1] technology aims at improving the intelligent perception and interaction of radar systems. From that, the model selection of environmental clutter and timely feedback is essential for improving target detection and tracking performance. There have been several mature model selection methods. One is the study of clutter classification [2, 3], and the other is based on the criteria of fitting goodness [4]. The clutter classification adopts a deep learning method to classify simulated clutter, and evaluating performance is described by classification accuracy. However, this method has uncertainty in complex scenarios of cognitive radar systems due to the limited simulation capability of clutter. The traditional model selection methods calculate the fitting goodness between the measured clutter and modeling results. Mean-square error (MSE), Kullback_Leibler divergence [5] (KLD), and Kolmogorov–Smirnov test (KS) are well-known quantitative evaluation criteria. For many statistical models, the fitting and evaluating process of candidate models adds complexity that is not applicable in the changing clutter environment to judge clutter.

L. Wang et al. (Eds.): CHREOC 2022, LNEE 969, pp. 116–123, 2023.
https://doi.org/10.1007/978-981-19-8202-6_11

How to determine the fine model from real-time clutter without clear labels is essential to analyzing the environment or design waveform. It's easy to acquire the non-parametric clutter curve, which provides amplitude fluctuation critical characteristics. At the same time, the non-parametric curve of real-time clutter also is the core description of clutter in the traditional selection method. Therefore, we proposed a method of optimal model selection based on model automatic classification. The MLP [6] is proposed and performed well in deep learning. Even though deep learning has evolved with various feature extractors, the importance of a fully-connect layer with an activation function is essential. For the diversity and complex fitting process of radar clutter, an MLP is first adopted to train simulated clutter curves and then used to select a model of real-time clutter. Before the training, the simulated samples were parametric clutter curves of generalized models, which hold much more flexibility than standard statistical models and abandon the potential diversity of actual clutter. In the test procedure, the testing sample is the non-parametric curve of real-time clutter, and then the trained MLP output optimal model.

This paper proposes a novel method for clutter model selection, which uses the core of traditional methods and the nonlinear mapping ability of the neural network. The method includes simulation, training, and testing processes. The remainder of this paper is organized as follows: Sect. 2 introduces the simulation method of generalized statistical models and the MLP network. Section 3 presents the simulation data and evaluation of different model selection results. Finally, Sect. 4 gives the summary.

2 Theory and Method

2.1 Simulation Method of Generalized Statistical Models

Generally, we divide statistical distributions into two categories: the exponential and non-exponential distribution families. We analyze classic statistical models from probability function and Log-Cumulants (LC) [7]. As shown in Fig. 1, the LC manifold diagram gives a more intuitive description of the relationship and practical range of LC statistics for different statistical models with different colors. From Fig. 1, we find that GΓD and Fisher distribution [8] are generalized models with a robust modeling capability.

The parametric clutter curves simulation method is based on MoLC [7]. We need expressions for the probability density distribution and theoretical expressions of the first three-order LCs $(\tilde{c}_{1GG}, \tilde{c}_{2GG}, \tilde{c}_{3GG})$ of the GΓD in [8], $(\tilde{c}_{1K}, \tilde{c}_{2K}, \tilde{c}_{3K})$ of the K distribution in [9] and $(\tilde{c}_{1F}, \tilde{c}_{2F}, \tilde{c}_{3F})$ of the Fisher distribution in [10], which are presented in the Table 1. The simulation procedure is presented in the Algorithm 1.

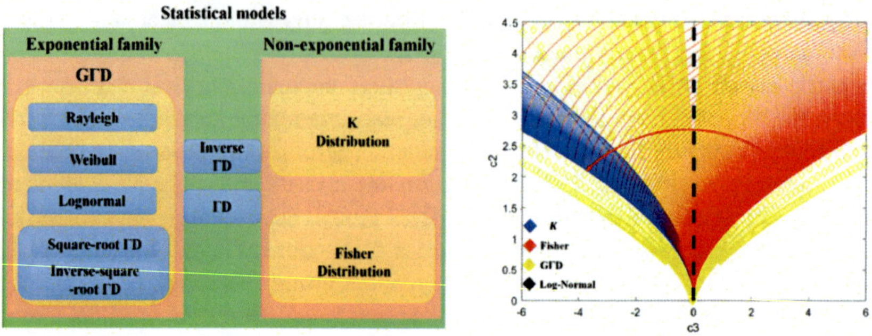

(a) Relationships among statistical models. (b) LC diagram.

Fig. 1. Relationship of generalized statistical models

Table 1. Probability distribution and LCs of generalized models

Models	Probability density distribution	Three-order LCs
GΓD	$p_{GG}(x) =$ $2\frac{\lvert\mu_{GG}\rvert s_{GG}{}^{s_{GG}}}{m_{GG}\Gamma(s_{GG})}\left(\frac{x}{m_{GG}}\right)^{2\mu_{GG}s_{GG}-1}$ $\cdot\exp\left\{-s_{GG}\left(\frac{x^2}{m_{GG}}\right)^{\mu_{GG}}\right\}$	$2\tilde{c}_{1GG} = \ln(m_{GG}) + \frac{\Psi(s_{GG})-\ln(s_{GG})}{\mu_{GG}}$ $4\tilde{c}_{2GG} = \frac{\Psi(1,s_{GG})}{\mu_{GG}{}^2}$ $8\tilde{c}_{3GG} = \frac{\Psi(2,s_{GG})}{\mu_{GG}{}^3}$
K	$p_K(x) =$ $\frac{4\frac{s_K m_K}{\mu_K}x}{\Gamma(s_K)\Gamma(m_K)}\left(\frac{s_K m_K}{\mu_K}x^2\right)^{\frac{s_K+m_K}{2}-1}$ $\cdot B_{s_K-m_K}\left(2x\sqrt{\frac{s_K m_K}{\mu_K}}\right)$	$2\tilde{c}_{1K} = ln\left(\frac{\mu_K}{m_K\cdot s_K}\right) + \Psi(s_K) + \Psi(m_K)$ $4\tilde{c}_{2K} = \Psi(1, s_K) + \Psi(1, m_K)$ $8\tilde{c}_{3K} = \Psi(2, s_K) + \Psi(2, m_K)$
Fisher	$p_F(x) =$ $\frac{2s_F}{m_F\mu_F}\frac{\Gamma(s_F+m_F)}{\Gamma(s_F)\Gamma(m_F)}\frac{\left(\frac{s_F}{m_F\mu_F}x\right)^{2s_F-1}}{\left(1+\frac{s_F}{m_F\mu_F}x^2\right)^{s_F+m_F}}$	$2\tilde{c}_{1F} = ln\left(\frac{\mu_F\cdot m_F}{s_F}\right) + \Psi(s_F) - \Psi(m_F)$ $4\tilde{c}_{2F} = \Psi(1, s_F) + \Psi(1, m_F)$ $8\tilde{c}_{3F} = \Psi(2, s_F) - \Psi(2, m_F)$

$\Gamma(\cdot)$ denotes the Gamma function, $\Psi(\cdot)$ denotes Digamma function and $\Psi(n,\cdot)$ represents the n th Polygamma function. B_n denotes the n th-order Bessel function

Algorithm1: Simulated parametric curves based on the LCs.

Input: The K or Fisher distribution parameters are given randomly, which includes the range of prior real-time clutter models.

Output : The discrete sequence of the parametric clutter curves on generalized models.

While true:

a set of (μ_K, s_K, m_K) and (μ_F, s_F, m_F)

Calculate the $(\tilde{c}_{1K}, \tilde{c}_{2K}, \tilde{c}_{3K})$ or $(\tilde{c}_{1F}, \tilde{c}_{2F}, \tilde{c}_{3F})$ by three-order LCs in Table 1

Set $(\tilde{c}_{1K}, \tilde{c}_{2K}, \tilde{c}_{3K})$ and $(\tilde{c}_{1F}, \tilde{c}_{2F}, \tilde{c}_{3F})$ equal $(\tilde{c}_{1GG}, \tilde{c}_{2GG}, \tilde{c}_{3GG})$

Estimate the $(m_{GG}, s_{GG}, \mu_{GG})$

For the normalized x values, the discrete $p_{GG}(x)$, $p_K(x)$, $p_F(x)$ are computed.

end

2.2 The Optimal Model Selection Method of MLP

The MLP includes the input, hidden, and output layers. Each neuron in the previous layer connects to all the neurons in the next layer by a weight matrix. The activation function is a nonlinearity operation to fit any distribution. Figure 2 presents its principle of classification. Figure 3 presents the procedure of MLP selection methods, which include three steps: simulation, training, and prediction.

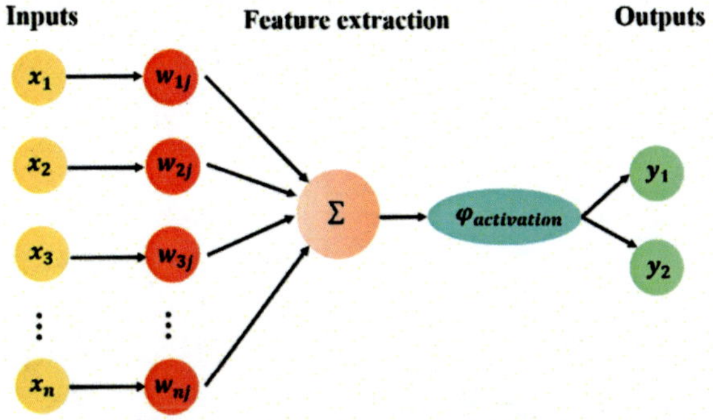

Fig. 2. Feedforward process of MLP

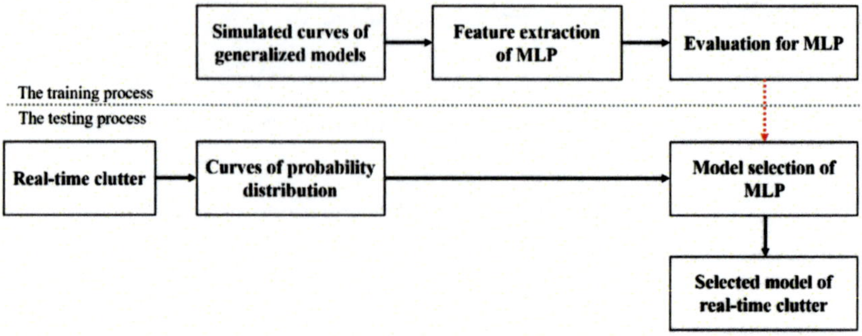

Fig. 3. The procedure of MLP model selection

3 Experiments on Model Selection

3.1 The Training Process of Simulated Parametric Curves

A. Simulated samples

The simulation data is based on the introduced method in 2.1. For the controversial LCs of clutter, we simulate all possible parametric clutter curves and discrete them as sequences. We simulate 1620 sets of $(\widehat{c_3}, \widehat{c_2})$ located in the region of K distribution and 3622 sets of $(\widehat{c_3}, \widehat{c_2})$ in the region of Fisher distribution. Figure 4 presents simulated LCs and parametric curves.

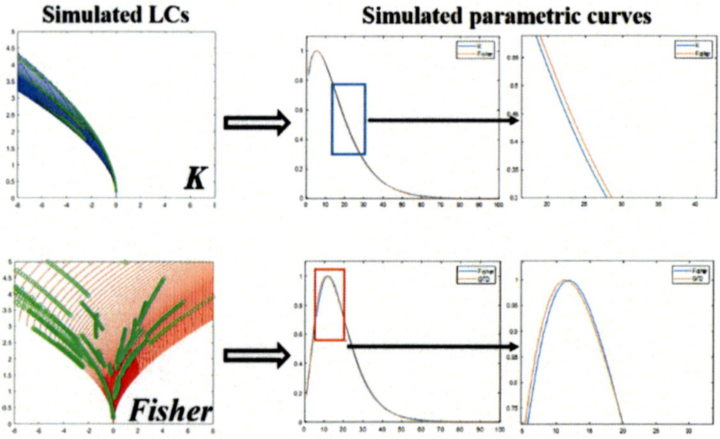

Fig. 4. Presentation for simulated clutter curves

B. Classification performance of the MLP

For all simulated curves, we compared their similarity in advance. Among them, 80% of the simulation data with a significant difference is used for training and other simulation data for testing. The MLP has three hidden layers with 150 neurons. We also give the evaluation of MLP on simulated curves, and the classification accuracy in Fig. 5 represents that the trained MLP can be used to predict the model of real-time clutter.

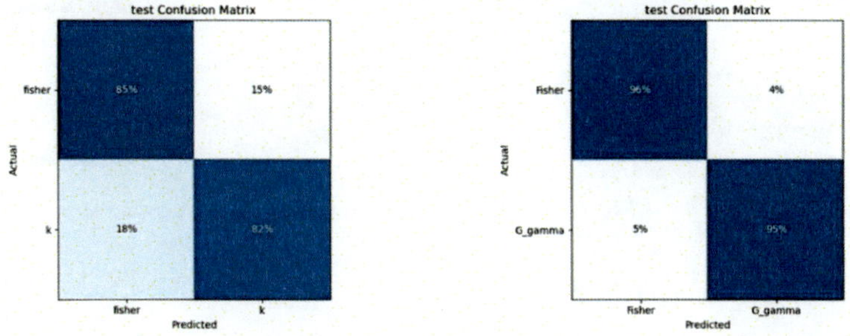

(a) Fisher and K distribution (b) Fisher and GΓ distribution

Fig. 5. Confusion matrix of model classification

3.2 Comparation of Model Selection Methods

For three model selection methods: MSE, KLD, and MLP-base, the same real-time clutter data are used to select the clutter model. Then, we use the KS test to evaluate model selection methods based on the cumulative curves F(·). MSE, KLD, and KS values are in Table 2.

Table 2. Calculation of goodness of fitting

MSE	KLD	KS
$MSE = \frac{1}{N}\sum_1^N \|\|p_A(x) - p_B(x)\|\|^2$	$KLD = \frac{1}{N}\sum_1^N \|p_A(x) \cdot ln\left(\frac{p_A(x)}{p_B(x)}\right)\|$	$KS = MAX\|F(p_A(x) - F(p_B(x)\|$

A. Real-time clutter of the airborne radar

The real-time clutter is the HH single-polarization data obtained by a UAV-borne MiniSAR system. The waveform is FMCW, and the down perspective is 60°. Figure 6 shows the real-time clutter data of five land covers, LCs of all samples, and presentation of clutter curves.

B. KS value of model selection results

Table 3 shows the model selection results on three methods, and Table 4 evaluates the results in Table 3. None of the real-time clutter samples fit the K distribution for each land cover.

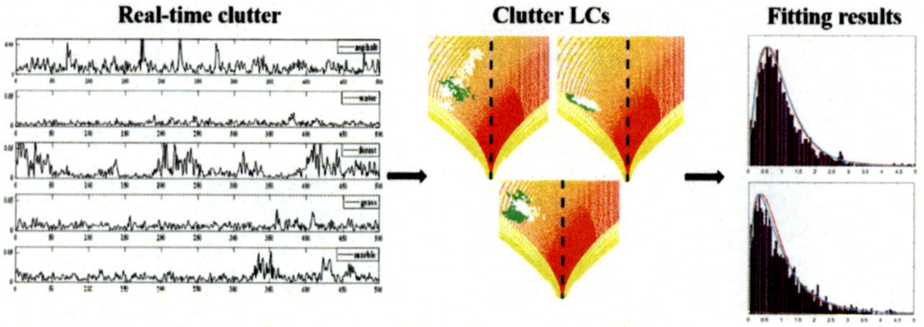

Fig. 6. Clutter data of five land covers

Table 3. Comparison of model selection results

Criterion	Model	Forest	Water	Marble	Grass	Asphalt
MSE	Fisher	18	174	292	242	292
	GΓD	340	84	32	87	8
KLD	Fisher	80	242	279	28	300
	GΓD	279	16	45	301	0
MLP	Fisher	39	252	304	130	300
	GΓD	320	6	20	199	0

Table 4. Evaluation of model selection methods

Criterion	Forest	Water	Marble	Grass	Asphalt
MSE	3.3249	3.0800	5.0174	6.0187	2.8330
KLD	3.6415	2.6253	4.7458	4.6783	2.8159
MLP	**3.1286**	**2.4158**	5.0234	5.0477	**2.8159**

From the results in Table 3, it can be found that the MLP selects the clutter model is feasible completely. Compared with traditional methods, the prediction of MLP and the model with better fitting goodness are comparable. At the same time, the MLP method is more accessible than traditional methods, so it's more suitable for the cognitive radar system to realize the interaction of environmental information and the target detector.

4 Summary

Adaptive feedback of real-time environment information such as clutter model selection in a cognitive radar system is a crucial problem. This paper proposed a novel model selection method integrating simulation, training, and testing. The key of our method is based on clutter curves, which synthesize the core of traditional model selection methods and make full use of the ability of a neural network to fit and learn nonlinear data. Compared with the traditional model selection methods, this method realized automatic model selection for real-time clutter, and its performance is evaluated based on abundant real-time clutter data. Experimental results show that the MLP model selection method has a similar performance as the traditional model selection method, and it's easier to be applied to practice.

References

1. Haykin, S., Xue, Y., Setoodeh, P.: Cognitive radar: step toward bridging the gap between neuroscience and engineering. Proc. IEEE **100**(11), 3102–3130 (2012)
2. Deng, C., Haykin, S.: A multi-layer neural network classifier for radar clutter. In: 1990 IJCNN International Joint Conference on Neural Networks, vol. 1, pp. 241–246 (1990)
3. Hailong, W., Shouyuan, D., Xu, W., Zhao, S.: Sea clutter recognition based on dual-polarization weather radar. In: 2019 International Conference on Meteorology Observations (ICMO), pp. 1–3 (2019)
4. Jafari, M., Maghsoudi, Y., Zoej, M.J.V.: A new method for land cover characterization and classification of polarimetric SAR data using polarimetric signatures. IEEE J. Select. Topics Appl. Earth Observ. Remote Sens. **8**(7), 3595–3607 (2015)
5. Qin, X., Zou, H., Zhou, S., Ji, K.: Region-based classification of SAR images using Kullback-Leibler distance between generalized gamma distributions. IEEE Geosci. Remote Sens. Lett. **12**(8), 1655–1659 (2015)
6. Pal, S.K., Mitra, S.: Multilayer perceptron, fuzzy sets, and classification. IEEE Trans. Neural Netw. **3**(5), 683–697 (1992)
7. Krylov, V.A., Moser, G., Serpico, S.B., Zerubia, J.: On the method of logarithmic cumulants for parametric probability density function estimation. IEEE Trans. Image Process. **22**(10), 3791–3806 (2013)
8. Yue, D.X., Xu, F., Frery, A.C., Jin, Y.Q.: Synthetic aperture radar image statistical modeling: part one-single-pixel statistical models. IEEE Geosci. Remote Sens. Magaz. **9**(1), 82–114 (2021)
9. Shankar, P.M.: A compound scattering pdf for the ultrasonic echo envelope and its relationship to K and Nakagami distributions. IEEE Trans. Ultrason. Ferroelectr. Freq. Control **50**(3), 339–343 (2003)
10. Bombrun, L., Beaulieu, J.: Fisher distribution for texture modeling of polarimetric SAR data. IEEE Geosci. Remote Sens. Lett. **5**(3), 512–516 (2008)

Urban Surface Solid Waste Detection Based on UAV Images

Yang Liu[1,2], Peng Gou[1,2](✉), Wei Nie[1,2], Nuo Xu[1,2,3], Tianyu Zhou[1,2], and Yalan Zheng[4]

[1] Big Data Technoloy Research Center of Nanhu Laboratory, Jiaxing, China
goupeng@nanhulab.ac.cn
[2] Beijing Big Data Advanced Technology Institute, Beijing, China
[3] Department of Geography, University of California, Los Angles, USA
[4] School of Geography, Nanjing Normal University, Nanjing, China

Abstract. The monitoring of illegal dumping of urban solid waste requires timeliness. Conventional methods are costly and have difficulty resolving the problem in remote regions timely. In this paper, a method based on semantic segmentation technology, which considers the multiscale characteristics and the disordered distribution of garbage, is proposed for automatic detection of surface solid waste in unmanned aerial vehicle (UAV) images. First, an asymmetrically decomposed atrous convolutional group was embedded to the center of the semantic segmentation network and the residual structure is used to enhance the learning ability of the spatial relationships. Then, a semantic flow alignment module was combined with attention mechanism to solve the disordered edges of information fusion. Experimental results demonstrate that the proposed method achieved better segmentation performance.

Keywords: Semantic segmentation · Semantic flow · Atrous convolution · Waste detection

1 Introduction

As the increasing global population and the accelerated urbanization, solid waste is increasing. The management and disposal of waste is a global challenge. Lured by illegal interests, dumping or stacking construction waste, domestic waste, and setting up disposal sites without authorization etc. occur from time to time, which increases the difficulty in supervision. A small number of supervision blind spots form solid waste storage sites, which have a huge impact on the environment.

The detection of garbage dumps is mainly carried out by urban management (Ministry of Housing and Urban-Rural Development), ecological environment, water conservancy and other departments at all levels. The traditional detection methods are based on reports from the masses or regular on-site inspections. These methods require a lot of manpower and resources, and the investigation and punishment usually lag behind. Moreover, due to the diverse distribution environment of garbage, there are inspection

L. Wang et al. (Eds.): CHREOC 2022, LNEE 969, pp. 124–134, 2023.
https://doi.org/10.1007/978-981-19-8202-6_12

blind angle that cannot be reached by staff in the area with complex environment, and it is easy to form a long-term accumulation status. Therefore, it is necessary to find a high-precision, high-efficiency, low-cost, and information-based informal garbage dumping detection method, which is an urgent problem to be solved by urban management and other departments, mainly including information acquisition and information processing.

Hema Begur [1] proposes an edge-based intelligent mobile service system for illegal dumping detection and monitoring in San Jose. Angelino [2] established a multi-feature detection algorithm based on the four main features of type, state, location, and activity on optical satellite images of Naples and Caserta provinces to automate the detection of illegal dumping. This kind of traditional extraction algorithm work mainly focuses on designing a suitable feature set for the target for expression and automatic identification. However, the manual design of features requires a lot of preliminary work to analyze and compare various features. Although the extraction effect is good in some samples, which cannot be applied to global automatic extraction on a large scale due to the structure and texture of garbage dumps are complex and diverse.

In recent years, artificial intelligence has gradually emerged and has been applied widely, such as image preprocessing [3, 4], target classification [5, 6], and understanding of image scenes [7, 8]. Chen [9] proposes a method for real-time detection of scattered garbage using low-altitude remote sensing videos captured by drones. First, the data augmentation method is used to expand Training samples, and then use the Yolov4 detection network to classify the scattered garbage areas, and finally locate them. Youme [10] proposed a method using unmanned A solution to detect hidden landfills in the St. Louis region of Senegal, West Africa. The scheme selects Single Shot Detector (SSD) as the detection algorithm, adopts the method VGG16 [11] in Convolutional Neural Networks (CNN) as the basic model, and uses convolution to detect different feature maps.

The study of deep learning in illegal garbage detection is still in its infancy. The main problem is that most of them are limited to small area close-up to distinguish and identify the types of solid waste. There is lacking of large-scale applications. Many researches often only use one image to study garbage extraction capabilities. There are doubts whether the trained networks can maintain comparable extraction capability in different regions. In order to solve the above problems, we use multiple UAV images (about 100 square kilometers) for the extraction of large-scale garbage dumps.

In this paper, a novel solid waste extraction (SWE) method for high-resolution remote sensing images is proposed to monitor the urban solid waste accumulation phenomenon. SWE uses an atrous convolution asymmetric decomposition module, which is suitable for polymorphic feature extraction, to enhance the extraction ability of spatial context information and uses a residual mechanism to maintain model gradients and prevent model degradation. For the information fusion problem, the attention mechanism and the semantic flow alignment module are used to optimize the long-range features and improve the edge extraction accuracy, respectively.

2 Method

2.1 Atrous Convolutional Group Based on Asymmetric Decomposition

The texture of the target is diverse, the scale is different, and the edge does not have a regular shape. Therefore, a atrous convolution group based on asymmetric decomposition is used in the deepest layer of the model to enhance the feature extraction ability of the model.

As shown in Fig. 1, the module uses four parallel atrous convolutions [12] based on asymmetric [13] decomposition, and the normlization and relu function are uniformly used after the two-layer convolution group. The atrous convolution uses the jumping receptive field to replace the conventional receptive field. The parameter dilation rate defines the interval of each value when the convolution kernel obtains data. This module is composed of four parameter dilation rates [3, 5, 7, 9]. The combination of different dilation rates enables each pixel of the image to extract semantic features from different scales. Each pixel gets denser information within the neighborhood spacing, which ensures the feature proportion and spatial correlation.

The 3×3 convolution in the convolution group is replaced by a 3×1 convolution and a 1×3 convolution combination after asymmetric decomposition. They have the same receptive field. The replaced combination can reduce the amount of parameters by 33% and speed up Model learning rate. In addition, the asymmetric decomposition also enables the model to learn the extension features of the target in the horizontal and vertical directions, which is helpful to locate the edge of the target.

The original input features are added with the outputs of convolutions with different dilation rates makes the module tends to learn the difference between the input features and the ground truth, which can prevent the model from degenerating.

2.2 Information Fusion Module Based on Semantic Flow

Fully convolutional network usually extracts target features of different scales in the encoding stage, and restores the original image resolution by upsampling or deconvolution in the decoding stage. Feature fusion is achieved by means of channel stacking (such as UNet [14]), point summing (such as LinkNet [15]) or attention fusion (such as Raunet [16]). When the high- level information of the decoding process is fused with the low-level information of the encoding process, there is a problem of semantic misalignment. This paper adopts a novel information fusion module to optimize the low- level features and adjust semantic flow of the high- level features. Finally, the two optimized features are added with the original low- level features as decoding features.

The specific information fusion module is shown in Fig. 2, including the attention module and the semantic flow alignment module. In the attention mechanism, the Squeeze-Excitation method is used. First, convert the feature into $1 \times 1 \times C$. The transformed feature has a global receptive field that can represent the main information of each channel. The weight of each channel is obtained by excitation and multiplied with the input feature to realize the weighted adjustment of the feature channel. Finally, it is added with the input features to complete the multi-channel integration and optimization of the original features at the channel level. In the semantic flow alignment module, the

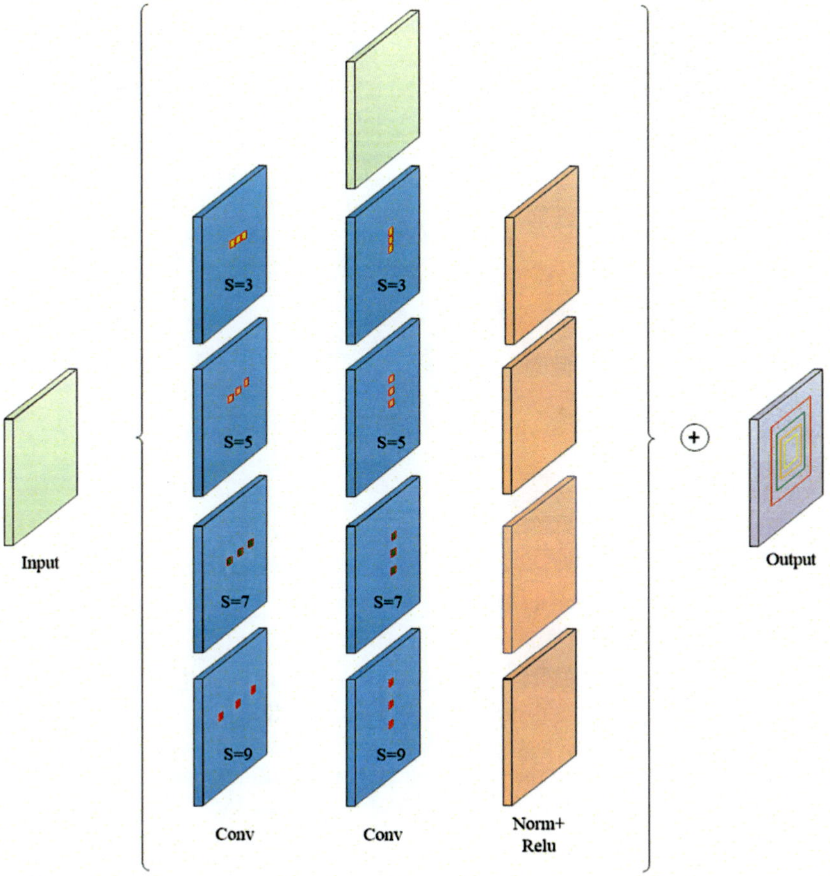

Fig. 1. Atrous convolutional group based on asymmetric decomposition

low- level features in the encoding stage and the high-level features in the decoding stage are convolved with 1×1 respectively, and the high- level features are up-sampled to obtain the up-sampling feature map, the convolution operation is used to obtain the semantic offset field between the up-sampled feature map and the low- level feature. Finally, the adjustment of high- level feature maps is guided by the semantic offset field to obtain accurate semantic feature maps. The sum of the optimized features of the two modules is the final output.

2.3 Network Architecture

As show in Fig. 3, SWE has an encoder–decoder structure and uses information fusion module to combines low-level detail information and high-level semantic information. In the figure, we simply represent this module as IFM.

In the encoder part, ResNet34 [17] network is used for feature extraction, and the pre-trained model on ImageNet [18] dataset is loaded as initial parameters. The feature

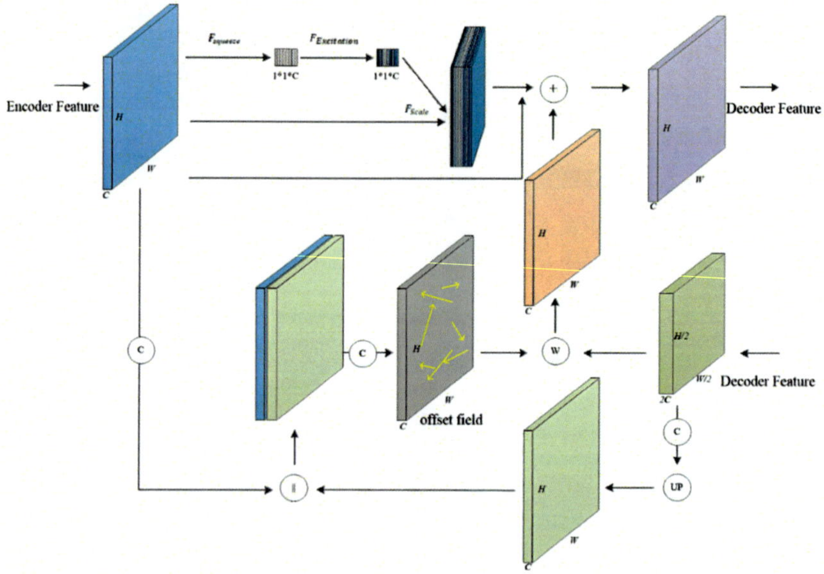

Fig. 2. Information fusion module, ⓒ conv ⓤ Upsample ⓘ concat ⊕ add ⓦ swap

map resolution in the deepest layer of the network is 16 × 16, and the atrous convolution group based on asymmetric decomposition introduced in Sect. 2.1 is used. This module does not change the feature channel and feature map size. In the decoder part, IFM is used layer by layer to restore the resolution to 128 × 128, and then deconvolution is used for up-sampling to restore the resolution from 128 to 256. Finally, the pixel-by-pixel prediction result is output, where the probability value is greater than 0.5 as the target extraction result.

3 Experiment

3.1 Datasets

The experimental data is UAV image data in some areas of Jiaxing City, Zhejiang Province, China, collected in June 2021, with a spatial resolution of 0.038 m. The original image contains two scenes of 194374 × 301992 pixels and 146917 × 160405 pixels, with a total area of about 100 square kilometers. We manually label the garbage objects larger than 0.5 square meters and crop out 256 × 256 pixel image patches centered on the objects, other images without objects are not used as datasets. No data augmentation was performed and only 637 images were extracted from 100 square kilometers. After distribution, the training set, certificate set and test set are 410, 90 and 137 respectively.

Some sample images from the datasets are displayed in Fig. 4, the garbage background is complex, with different scales, types, shapes, and sparseness. In this case, applying deep learning methods with a small amount of training data is still a huge challenge.

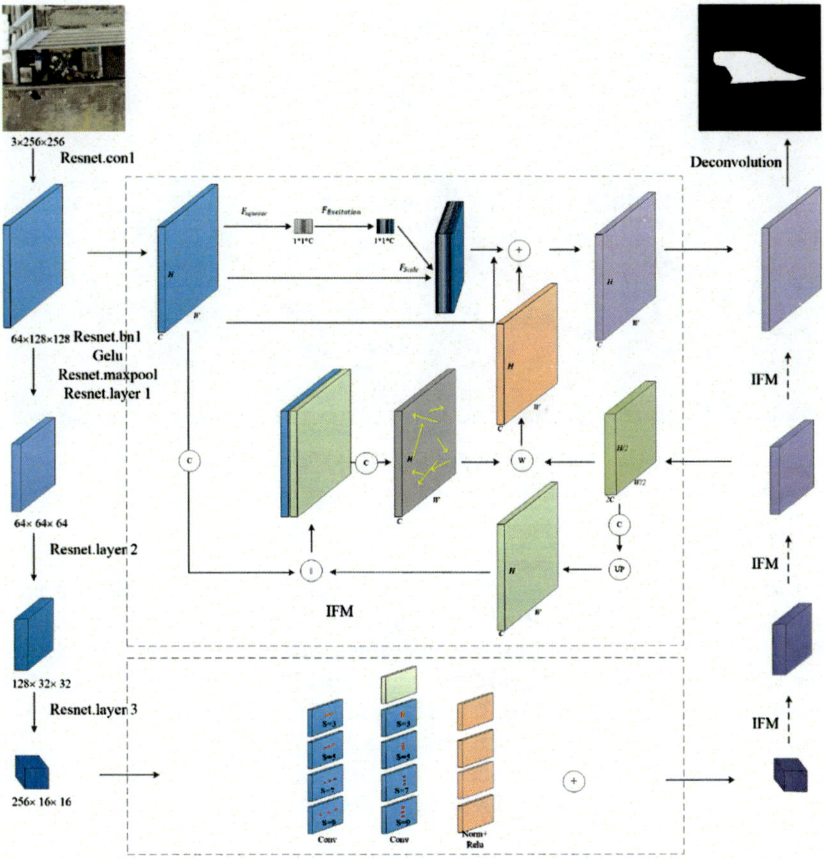

Fig. 3. Architecture of proposed SWE

3.2 Implementation Details

We implement our network on the open source deep learning framework Pytorch 1.2.0, all experiments are trained in 100 batches, the initial learning rate is set to 0.0001, the optimizer is Adam, and BCEloss is selected as the loss function. The network is trained on NVIDIA RTX 2080Ti, the batch size is usually set to 8, and the maximum runnable value is used for large networks.

The experimental comparison networks in this paper include ENet [19], UNet, NestedUnet [20], DeepLabv3+ [21], DaNet [22], PSPNet [23], SegFormer [24], ERFNet [25], DDRNet [26] and the model SWE proposed in this paper. The F1 value, overall precision, recall and precision were selected as numerical indicators for performance analysis and comparison.

Fig. 4. Example images of dataset

3.3 Result and Analysis

The accuracy evaluation results of ten different models are shown in Table 1. We order them according to the F1 value from low to high, and the highest item of each evaluation metrics is bolded. The accuracy evaluation results of ten different models are shown in Table 1. We order them according to the *F1* value from low to high, and the highest item of each evaluation index is bolded. It can be seen from the table that the lowest *F1* in the comparison method is SegFormer, which is only 74.587%, the highest is 86.56% for NestedUNet, and the *F1* of other models is around 85%. The *F1* of the method proposed in this paper is 89.467%, which is 2.9% higher than that of the NestedUNet method. In addition, the *OA* and *recall* are also the highest among all methods, only 0.5 percentage points lower than the ENet model in *Precision.* Overall, the model proposed in this paper performs best in the quantitative analysis.

In order to observe ability of SWE proposed in this paper, some typical images tested in dataset are shown in Fig. 5. The target of area 1 is construction waste, and the contrast between the background and the target is obvious. In the comparison model, there are a lot of missing extractions in the extraction results of SegFormer and PSPNet, other models can extract objects well with small edge errors.

Area 2 is dominated by domestic garbage. The upper left corner is connected to other garbage, so it is marked, while the area of the lower left corner is less than 0.5 square meters, so it is not marked. SegFormer, Deeplabv3+ and SWE can extract two small area targets, while other models have omissions.

Area 3 includes white industrial waste and small amounts of other debris, which can easily be confused and mis-extracted with the bare ground background and debris. Among the compared models, DDRNet and ERFNet have a lot of over-extraction. UNet and NestedUNet also have error extraction. PSPNet performs well in this area, but has some missing extractions. SWE has less error extraction at the edge, the overall extraction result is more accurate, and is less affected by the background.

Table 1. Performance on dataset.

Method	F1(%)	OA(%)	Recall(%)	Precision(%)
SegFormer	74.587	86.479	66.554	84.826
DDRNet	81.896	89.087	82.788	81.023
DANet	83.624	90.327	82.839	84.424
ENet	84.996	91.527	80.488	**90.037**
DeepLabv3 +	85.008	91.292	82.804	87.332
ERFNet	85.15	91.454	82.179	88.344
PSPNet	85.696	91.799	82.4	89.267
UNet	85.826	91.683	84.451	87.246
NestedUNet	86.56	91.945	86.998	86.127
SWE	**89.467**	**93.725**	**89.388**	89.547

Area 4 takes the green space as the background, and there are two scattered garbage. DDRNet, ENet, PSPNet and NestedUNet do not recognize the lower target, and other models also have errors on the edge. The model proposed in this paper is more consistent with the label on the edge.

4 Summary

In this paper, a method for urban solid waste detection from UAV images is presented. Aiming at the characteristics of different sizes and irregular edge of waste, the atrous convolution group based on asymmetric decomposition was used to extract multi-scale context information and improved the global and edge perception capabilities of the model. In addition, a feature fusion module based on semantic flow was designed to transfer long-range information and accurately recover image resolution. Experiments demonstrated that the proposed SWE effectively extracted multicategory garbage and performed better on overall precision and recall than other semantic segmentation methods. However, the method falsely classifies the debris stacked in front of residential houses into targets, which still needs to be solved for the urban-level solid waste monitoring.

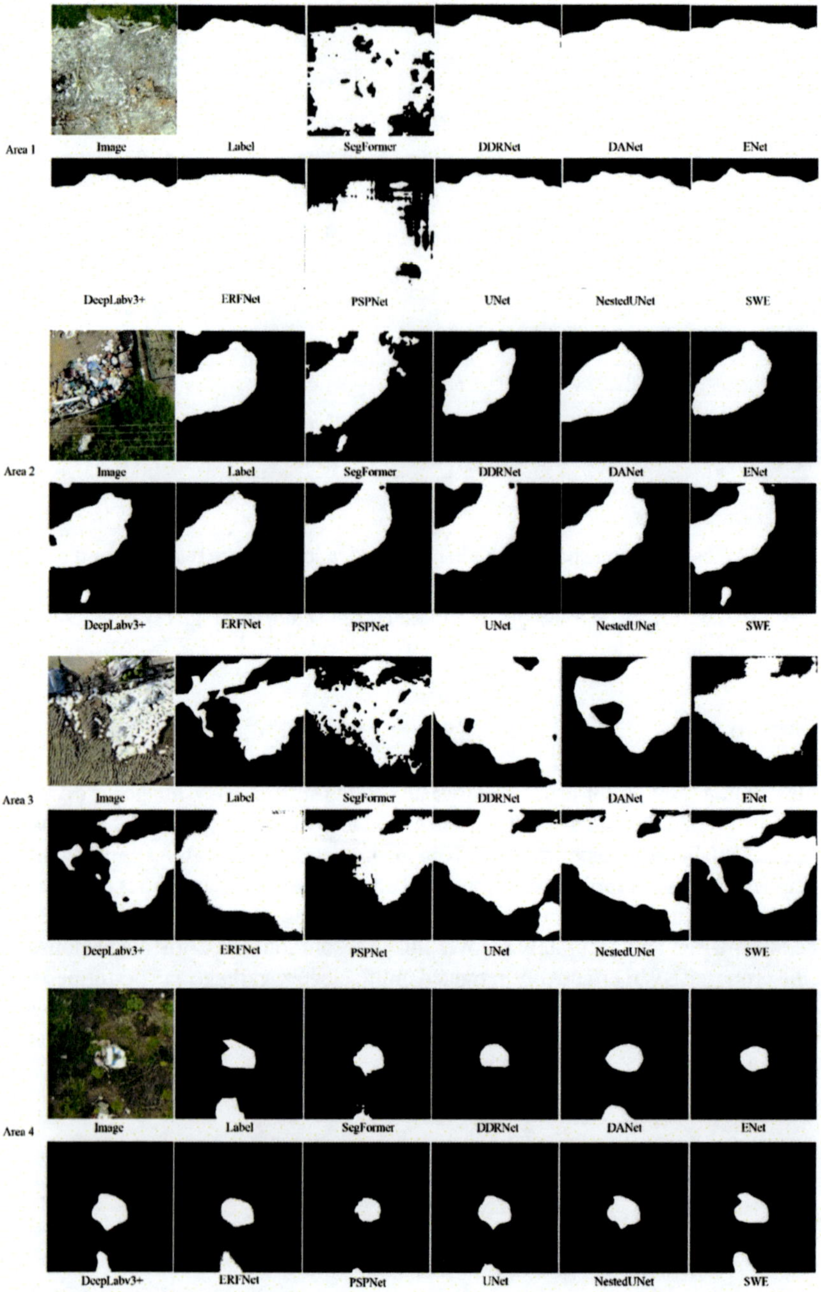

Fig. 5. Example results of dataset

Acknowledgements. This work was supported by Science and Technology Program of Zhejiang Province No. 2022C35070 and the Postgraduate Research and Practice Innovation Program of Jiangsu Province No. KYCX21_1349.

References

1. Begur, H., Dhawade, M., Gaur, N., Dureja, P., Gao, J., Mahmoud, M., et al.: An edge-based smart mobile service system for illegal dumping detection and monitoring in San Jose. In: 2017 IEEE SmartWorld, Ubiquitous Intelligence & Computing, Advanced & Trusted Computed, Scalable Computing & Communications, Cloud & Big Data Computing, Internet of People and Smart City Innovation (SmartWorld/SCALCOM/UIC/ATC/CBDCom/IOP/SCI), pp. 1–6 (2017)
2. Angelino, C.V., Focareta, M., Parrilli, S., et al.: A case study on the detection of illegal dumps with GIS and remote sensing images//Earth resources and environmental remote sensing/GIS applications IX. Int. Soc. Optics Photonics **10790**, 107900M (2018)
3. Chao, D., Chen, C.L., Tang, X.: Accelerating the Super-Resolution Convolutional Neural Network. Springer, Cham (2016)
4. He, Z., Patel, V.M.: Densely connected pyramid dehazing network. In: 2018 IEEE/CVF Conference on Computer Vision and Pattern Recognition (CVPR), pp 3194–3203 (2018)
5. Gao, H., Yuan, Y, Zheng, X.: Remote sensing road extraction by refining road topology. In: China High Resolution Earth Observation Conference. Springer, Singapore, pp. 187–197 (2019)
6. Zhang, L., Wu, J., Fan, Y., et al.: An efficient building extraction method from high spatial resolution remote sensing images based on improved mask R-CNN. Sensors **20**(5), 1465 (2020)
7. Dvornik, N., Shmelkov, K., Mairal, J., et al.: Blitznet: a real-time deep network for scene understanding. In: Proceedings of the IEEE International Conference on Computer Cision, pp 4154–4162 (2017)
8. Wang, T.C., Liu, M.Y., Zhu, J.Y., et al.: High-resolution image synthesis and semantic manipulation with conditional gans. In: Proceedings of the IEEE Conference on Computer Vision and Pattern Recognition, pp. 8798–8807 (2018)
9. Chen, W., Wang, H., Li, H., et al.: Real-time garbage object detection with data augmentation and feature fusion using SUAV low-altitude remote sensing images. IEEE Geosci. Remote Sens. Lett. **19**, 1–5 (2021)
10. Youme, O., Bayet, T., Dembele, J.M., et al.: Deep learning and remote sensing: detection of dumping waste using UAV. Procedia Comput. Sci. **185**, 361–369 (2021)
11. Simonyan, K., Zisserman, A.: Very deep convolutional networks for large-scale image recognition (2015). arXiv preprint, arXiv:1409.1556
12. Yu, F., Koltun, V.: Multi-scale context aggregation by dilated convolutions. In : Proceedings of International Conference on Learning Representations
13. Jaderberg, M., Vedaldi, A., Zisserman, A. Speeding up convolutional neural networks with low rank expansions (2014). arXiv preprint, arXiv:1405.3866
14. Ronneberger, O., Fischer, P., Brox, T.: U-net: convolutional networks for biomedical image segmentation. In: International Conference on Medical Image Computing and Computer-Assisted Intervention. Springer, Cham, pp. 234–241 (2015)
15. Woo, S., Kim, D., Cho, D., et al.: Linknet: relational embedding for scene graph. Adv. Neural Inf. Process. Syst. **31** (2018)

16. Ni, Z.L., Bian, G.B., Zhou, X.H., et al.: Raunet: residual attention u-net for semantic segmentation of cataract surgical instruments. In: International Conference on Neural Information Processing. Springer, Cham, pp. 139–149 (2019)
17. He, K., Zhang, X., Ren, S., Sun, J.: Deep residual learning for image recognition. In: Proceedings of IEEE Conference on Computer Vision and Pattern Recognition, pp. 770–778 (2016)
18. Deng, J., Dong, W., Socher, R., Li, L., Li, K., Li, F.: Imagenet: a large-scale hierarchical image database. In: Proceedings of IEEE Conference on Computer Vision and Pattern Recognition, pp. 248–255 (2009)
19. Paszke, A., Chaurasia, A., Kim, S., et al.: Enet: a deep neural network architecture for real-time semantic segmentation (2016). arXiv preprint, arXiv:1606.02147
20. Zhou, Z., Siddiquee, M.M.R., Tajbakhsh, N., et al.: A nested U-Net architecture for medical image segmentation (2018). arXiv preprint, arXiv:1807.10165
21. Chen, L.C., Zhu, Y., Papandreou, G., et al.: Encoder-decoder with atrous separable convolution for semantic image segmentation. In: Proceedings of the European conference on computer vision (ECCV), pp. 801–818 (2018)
22. Fu, J., Liu, J., Tian, H., et al.: Dual attention network for scene segmentation. In: Proceedings of the IEEE/CVF Conference on Computer Vision and Pattern Recognition, pp 3146–3154 (2019)
23. Zhao, H., Shi, J., Qi, X., et al.: Pyramid scene parsing network. In: Proceedings of the IEEE Conference on Computer Vision and Pattern Recognition, pp. 2881–2890 (2017)
24. Xie, E., Wang, W., Yu, Z., et al.: SegFormer: simple and efficient design for semantic segmentation with transformers. Adv. Neural Inf. Process. Syst. **34** (2021)
25. Romera, E., Alvarez, J.M., Bergasa, L.M., et al.: Erfnet: efficient residual factorized convnet for real-time semantic segmentation. IEEE Trans. Intell. Transp. Syst. **19**(1), 263–272 (2017)
26. Hong, Y., Pan, H., Sun, W., et al.: Deep dual-resolution networks for real-time and accurate semantic segmentation of road scenes (2021). arXiv preprint, arXiv:2101.06085

Image-Based Sliding Mode Attitude Control Research on Small Video Satellites

Mengmeng Wang[1], Chao Song[1], Caizhi Fan[1](✉), and Yonghe Zhang[2]

[1] National University of Defense Technology, Changsha, China
wmmgfkd@163.com

[2] Beijing Institute of Tracking and Telecommunications Technology, Beijing, China

Abstract. Small video satellites are a new category of tracking and observation satellites that have been developing rapidly in recent years, which can carry optical imaging payloads to continuously image specific targets or areas in real time and obtain their dynamic video information. Traditional small video satellite imaging tracking method requires the priori position information of the target and cannot observe non-cooperative ground targets. To remedy the deficiency, this paper designs a video image-based method which performs attitude adjustment based on the projection of the target in the image plane. Firstly, this paper establishes a camera model to reveal the principle of small video satellite perspective imaging; The spacecraft attitude kinematic model expressed in quaternion form and the spacecraft attitude dynamics model under the rigid body assumption are introduced; Then, the satellite attitude error and angular velocity error are calculated based on the deviation of the target real-time pixel coordinates from the desired projected position, and the error quaternion and error angular velocity are used as control feedback. In order to counteract the effects of torque disturbance and rotational inertia uncertainty on the satellite visual tracking control, a fast response overshoot-free sliding mode tracking controller without relying on the accurate model of the satellite is designed. The LaSalle invariance principle are used to demonstrate the asymptotic stability of the control system. Numerical simulation proves that the designed controller can effectively accomplish the visual tracking control of ground targets and has strong robustness against torque disturbances.

Keywords: Visual tracking control · Small video satellite · Ground target · Sliding mode control

1 Introduction

The current major natural disasters and emergencies in hotspots are frequent, which put forward higher requirements on our national earth observation capability. At present, all countries are actively researching tracking and surveillance equipment, and small video satellites are a typical category developed in recent years. Due to the advantages of fast response time and high resolution real-time continuous imaging function, small video satellites are widely used in disaster monitoring, resource census, motion target surveillance and other fields, which have high military and civilian values. Influential

L. Wang et al. (Eds.): CHREOC 2022, LNEE 969, pp. 135–149, 2023.
https://doi.org/10.1007/978-981-19-8202-6_13

optical small video satellites in the world such as LAPAN-TUBSAT satellite [1], Sumbandia satellite [2], and Tiantuo-2 satellite [3] are all operated by adjusting the attitude of the satellite in real time so that the optical axis of the on-board camera is aligned with the target or the relevant area, achieving continuous tracking video imaging of ground targets.

Usually, small video satellites use the position information-based method [4, 5] for target tracking, which solves the attitude error and angular velocity error through the target coordinates. And then the controller outputs the control torque to complete the satellite attitude maneuver and staring tracking. This method is only applicable to cooperative target tracking because of the need to obtain the priori position information of targets. However, most of the ground targets that we need to track and observe are non-cooperative and the location-based approach is no longer available. The method based on image information can make full use of the visual images acquired by the on-board camera and use the real-time deviation of the target projection in the field of view as the control feedback, which is free from the dependence on the priori position information of the target, so it can be applied to the monitoring and observation of non-cooperative ground targets.

Due to the complex space environment and the changes in its rotational inertia caused by the fuel consumption, the spacecraft is inevitably affected by the disturbing torque of the space environment and the changes in the unknown rotational inertia parameters inside the system. In order to ensure the efficient completion of the precise observation mission to the Earth target, it is necessary to address the attitude control problems caused by internal parameter variations and external disturbances. The sliding mode control method is an effective problem solver [6, 7]. The method does not rely on the accurate spacecraft model and is insensitive to parameter variations and external disturbances when the system enters the sliding mode [8], so it is highly robust to uncertainties and external disturbances. Sliding mode control methods have been widely used in spacecraft control [9–11], but no research has been conducted by scholars in small video satellite image feedback visual tracking control using sliding mode control methods for continuous monitoring attitude control of targets.

In this article, we propose a sliding mode control method based on image information feedback. The method is capable of overshoot-free fast and accurate autonomous tracking for non-cooperative ground targets, enabling the target imaging position in the image plane controlled to the desired location without decoupling the model, with strong robustness. The validity of the method is proven by comparing the results of simulations in the ideal case and in the presence of internal rotational inertia deviations and external disturbing torque.

2 Mathematical Modeling

2.1 Coordinate Frames

The geocentric inertial frame is described as $O_i - X_i Y_i Z_i$, in which the frame origin O_i is at the center of the earth; the $O_i X_i$ axis is aligned with the spring equinox; the $O_i Z_i$ axis points to the earth's north pole; and the $O_i Y_i$ axis constitutes a right-handed coordinate with the $O_i X_i$ and $O_i Z_i$ axes.

The satellite ontology frame is described as $O_b - X_bY_bZ_b$, in which the frame origin O_b is at the center of the small video satellite's mass, the three axes point to the three directions of the satellite's main inertial.

The pixel frame is described as $O' - uv$, in which the frame origin O' is in the upper left corner point of the imaging plane, with pixel as the frame unit; the $O'u$ and $O'v$ axes indicate the amount of columns and rows of the pixel point.

The camera coordinate frame is described as $O_c - X_cY_cZ_c$, in which the origin O_c is at the camera's optical center; the O_cZ_c axis is the optical axis aligned and perpendicular to the Image plane; the O_cX_c axis and O_cY_c axis are parallel to the $O'u$ axis and $O'v$ axis and have the same direction.

2.2 Perspective Projection Model

A perspective projection model is developed, using a series of equation expressions to show how the ground target described under the geocentric inertial frame is projected onto the image plane through the geometry relationship. The internal projection model of the camera is shown in Fig. 1.

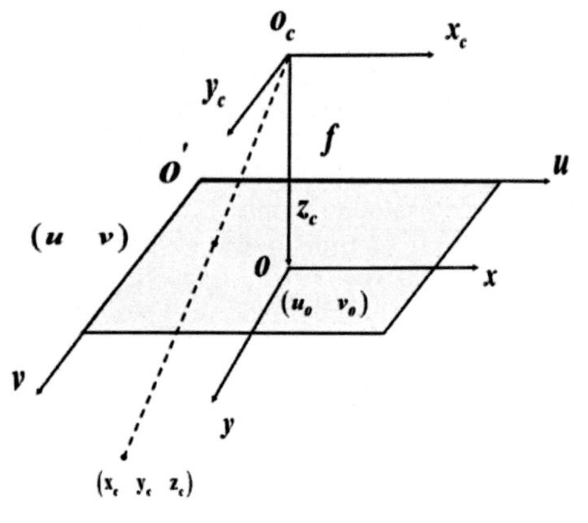

Fig. 1. The internal projection model

The focal length of the camera is f; the location of the ground target in the camera frame is ${}^{c}\mathbf{R}_{cT} = \left(x_c\ y_c\ z_c\right)^T$ and the projection location in the pixel coordinate frame is $\left(u\ v\right)^T$; the pixel size is $du \times dv$; the coordinates of the pixel point where the camera's optical axis (i.e., the O_cZ_c axis direction) intersects the image plane is $\left(u_0\ v_0\right)^T$. According to the geometric relationship, the following equation is given.

$$z_c\begin{pmatrix} u \\ v \\ 1 \end{pmatrix} = \begin{pmatrix} \frac{f}{dx} & 0 & u_0 & 0 \\ 0 & \frac{f}{dy} & v_0 & 0 \\ 0 & 0 & 1 & 0 \end{pmatrix}\begin{pmatrix} x_c \\ y_c \\ z_c \\ 1 \end{pmatrix} = \mathbf{\Pi} \cdot \begin{pmatrix} {}^{c}\mathbf{R}_{cT} \\ 1 \end{pmatrix} \tag{1}$$

As shown in Fig. 2, the external camera imaging model describes the transformation relationship between the coordinate of the target in the inertial frame and the coordinate in the camera frame.

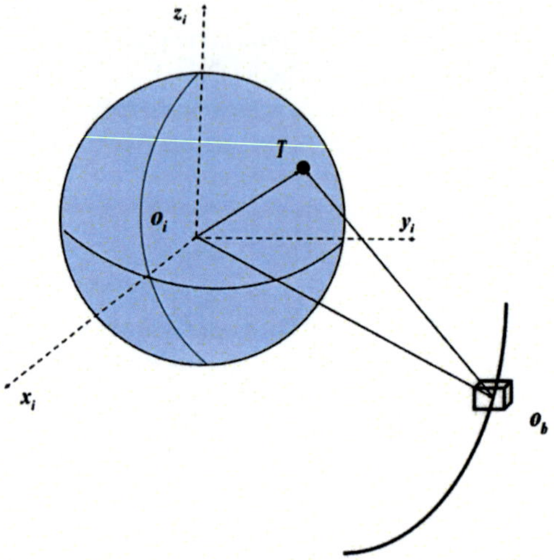

Fig. 2. The external camera imaging model

The vector of the small video satellite pointing to the ground target is ${}^{\mathbf{b}}\mathbf{R}_{\mathbf{bT}}$, which is represented in the satellite body coordinate frame; in the geocentric inertial frame, the location of the ground target is ${}^{\mathbf{i}}\mathbf{R}_{\mathbf{ib}}$. The conversion relationship is given by the following equation.

$$\begin{pmatrix} {}^{\mathbf{b}}\mathbf{R}_{\mathbf{bT}} \\ 1 \end{pmatrix} = \begin{pmatrix} {}^{\mathbf{b}}_{\mathbf{i}}\mathbf{M} & -{}^{\mathbf{b}}_{\mathbf{i}}\mathbf{M}\,{}^{\mathbf{i}}\mathbf{R}_{\mathbf{ib}} \\ 0_{1\times 3} & 1 \end{pmatrix} \begin{pmatrix} {}^{\mathbf{i}}\mathbf{R}_{\mathbf{iT}} \\ 1 \end{pmatrix} = \mathbf{T}\cdot \begin{pmatrix} {}^{\mathbf{i}}\mathbf{R}_{\mathbf{iT}} \\ 1 \end{pmatrix} \tag{2}$$

where ${}^{\mathbf{b}}_{\mathbf{i}}\mathbf{M}$ indicates the conversion matrix from the inertial frame to the satellite ontology frame.

Without losing commonality, assuming that the satellite ontology frame and the camera frame overlap, i.e. ${}^{\mathbf{c}}\mathbf{R}_{\mathbf{cT}} = {}^{\mathbf{b}}\mathbf{R}_{\mathbf{bT}}$.

Combining the Eqs. 1 and 2, we have:

$$\begin{aligned} \begin{pmatrix} u \\ v \\ 1 \end{pmatrix} &= \frac{1}{z_c} \begin{pmatrix} \frac{f}{dx} & 0 & u_0 & 0 \\ 0 & \frac{f}{dy} & v_0 & 0 \\ 0 & 0 & 1 & 0 \end{pmatrix} \begin{pmatrix} {}^{\mathbf{b}}_{\mathbf{i}}\mathbf{M} & -{}^{\mathbf{b}}_{\mathbf{i}}\mathbf{M}\,{}^{\mathbf{i}}\mathbf{R}_{\mathbf{ib}} \\ 0_{1\times 3} & 1 \end{pmatrix} \begin{pmatrix} x_i \\ y_i \\ z_i \\ 1 \end{pmatrix} \\ &= \frac{1}{z_c}\cdot \mathbf{\Pi}\cdot \mathbf{T}\cdot \begin{pmatrix} {}^{\mathbf{i}}\mathbf{R}_{\mathbf{iT}} \\ 1 \end{pmatrix} \\ &= \frac{1}{z_c}\cdot \mathbf{M}\cdot \begin{pmatrix} {}^{\mathbf{i}}\mathbf{R}_{\mathbf{iT}} \\ 1 \end{pmatrix} \end{aligned} \tag{3}$$

Let m_i^T represents the ith row vector of the array **M**, the above Eq. 3 can be rewritten as:

$$z_c = m_3^T \begin{pmatrix} {}^{\mathbf{i}}\mathbf{R_{iT}} \\ 1 \end{pmatrix} \tag{4}$$

$$\begin{pmatrix} u \\ v \end{pmatrix} = \frac{1}{z_c} \begin{pmatrix} m_1^T \\ m_2^T \end{pmatrix} \begin{pmatrix} {}^{\mathbf{i}}\mathbf{R_{iT}} \\ 1 \end{pmatrix} \tag{5}$$

2.3 Attitude Dynamics Equation

The attitude dynamics equation describes the relationship between the action moment and the attitude motion. The small video satellite has low deflection and can be used with the rigid body assumption. Then the attitude kinematic equation [12] is:

$$\mathbf{I}\dot{\omega} = -\omega \times \mathbf{I}\omega + \mathbf{T_c} + \mathbf{T_d} \tag{6}$$

where **I** is the rotational inertia matrix of the small video satellite; $\boldsymbol{\omega}$ denotes the angular velocity of the satellite ontology coordinate frame with respect to the geocentric inertial frame; $\mathbf{T_c}$ is the attitude control torque, and $\mathbf{T_d}$ represents the disturbance torque.

2.4 Attitude Kinematics Equation

The attitude kinematic equation describes the relationship between the motion parameters of the satellite. We select the geocentric inertial frame as the reference coordinate frame, and use the quaternion $\mathbf{q} = \left(q_0\ \mathbf{q_v}\right)^T$ to describe the satellite's attitude, which is related to the Eulerian axis vector **a** and the Eulerian angle θ as follows.

$$\begin{cases} q_0 = \cos\left(\frac{\theta}{2}\right) \\ \mathbf{q_v} = \mathbf{a} \cdot \sin\left(\frac{\theta}{2}\right) \end{cases} \tag{7}$$

where $\mathbf{q_v}$ is the vector part of the quaternion **q** and has;

$$\mathbf{q_v} = \left(q_1\ q_2\ q_3\right)^T \tag{8}$$

The attitude kinematics depicted by the quaternion is as follows:

$$\dot{\mathbf{q}} = \frac{1}{2}\mathbf{q} \otimes \boldsymbol{\omega} \tag{9}$$

where the symbol $\otimes$ is the quaternion multiplication symbol.

3 Error Solving

The pixel point coordinate of the target center point can be extracted by image processing and other related techniques. Based on the deviation of the pixel coordinates from the desired projection position, the error quaternion can be calculated, and the error angular velocity can be acquired by combining the relevant theory of attitude kinematics. The error quaternion and the error angular velocity are used as the control feedback quantity at the same time.

3.1 Error Quaternion

The error quaternion solution is schematically shown in Fig. 3. The target position direction vector in the camera coordinate frame is denoted as $\mathbf{r_t}$, which is co-linear with $^{\mathbf{c}}\mathbf{R_{cT}}$ and has:

$$\mathbf{r_t} = [x_{pt}\ y_{pt}\ f\ ]^T \tag{10}$$

where x_{pt} and y_{pt} are derived from the following equation:

$$\begin{cases} x_{pt} = (u - u_0) \cdot dx \\ y_{pt} = (v - v_0) \cdot dy \end{cases} \tag{11}$$

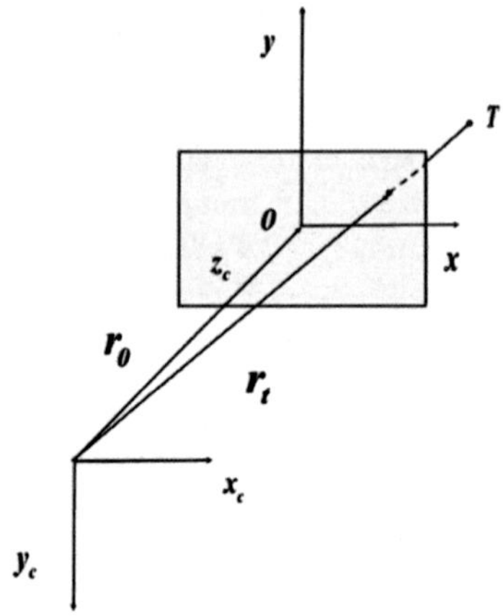

Fig. 3. Schematic diagram of the error quaternion solution

Let $\mathbf{r_o}$ be the orientation vector of the optical axis in the camera coordinate frame:

$$\mathbf{r_o} = [0\ 0\ f\ ]^T \tag{12}$$

The purpose of attitude control for small video satellite visual tracking is to keep the projection of the ground target at the center of the image plane. The attitude error can be described as the rotational attitude from the target position direction $\mathbf{r_t}$ in the camera coordinate frame to the optical axis direction $\mathbf{r_o}$. Therefore, the error quaternion $\mathbf{q_e}$ of the small video satellite is:

$$\mathbf{q_e} = \cos(\frac{\gamma}{2}) + \mathbf{e}\sin(\frac{\gamma}{2}) \tag{13}$$

where the Eulerian axis $\mathbf{e}$ and the Eulerian angle γ can both be represented by $\mathbf{r_t}$ and $\mathbf{r_o}$:

$$\mathbf{e} = \frac{\mathbf{r_t} \times \mathbf{r_o}}{\|\mathbf{r_t} \times \mathbf{r_o}\|} \tag{14}$$

$$\gamma = \arccos\left(\frac{\mathbf{r_t}\mathbf{r_o}}{\|\mathbf{r_t}\|\|\mathbf{r_o}\|}\right) \tag{15}$$

3.2 Error Angular Velocity

The error angular velocity can't be gained straightly from the deviation of the ground target's projected position in the image plane at a certain moment, but can be solved by the continuous change of the pixel coordinates over a period of time.

The attitude of the small video satellite could be obtained from the on-board sensor, and the quaternion is assumed to be $\mathbf{q_b}$. The desired satellite attitude quaternion $\mathbf{q_d}$ can be found by incorporating the error quaternion $\mathbf{q_e}$.

$$\mathbf{q_d} = \mathbf{q_b} \otimes \mathbf{q_e^{-1}} \tag{16}$$

where the inverse of the quaternion $\mathbf{q}$ is given as:

$$\mathbf{q}^{-1} = \begin{pmatrix} q_0 & -\mathbf{q_v} \end{pmatrix}^T \tag{17}$$

The Intended angular velocity $\boldsymbol{\omega}_\mathbf{d}$ can be derived according to the following equation:

$$\boldsymbol{\omega}_\mathbf{d} = 2\Xi^\mathbf{T}(\mathbf{q_d})\dot{\mathbf{q}}_\mathbf{d} \tag{18}$$

where $\dot{\mathbf{q}}_\mathbf{d}$ can be derived by differentiating $\mathbf{q_d}$, and

$$\Xi(\mathbf{q}) = \begin{bmatrix} -q_1 & -q_2 & -q_3 \\ q_0 & -q_3 & q_2 \\ q_3 & q_0 & -q_1 \\ -q_2 & q_1 & q_0 \end{bmatrix} \tag{19}$$

Hence, the error angular velocity $\boldsymbol{\omega}_\mathbf{e}$ is:

$$\boldsymbol{\omega}_\mathbf{e} = \boldsymbol{\omega}_\mathbf{b} - \mathbf{A}(\mathbf{q_e})\boldsymbol{\omega}_\mathbf{d} \tag{20}$$

where $\mathbf{A}(\mathbf{q_e})$ is determined by $\mathbf{q_e} = \begin{pmatrix} q_{e0} & \mathbf{q_{ev}} \end{pmatrix}^T$, which satisfies:

$$\mathbf{A}(\mathbf{q_e}) = \left(q_{e0}^2 - \|\mathbf{q_{ev}}\|^2\right)\mathbf{I}_3 + 2\mathbf{q_{ev}^T}\mathbf{q_{ev}} - 2q_{e0}\mathbf{q_{ev}^\times} \tag{21}$$

$$\mathbf{q_{ev}^\times} = \begin{bmatrix} 0 & -q_3 & q_2 \\ q_3 & 0 & -q_1 \\ -q_2 & q_1 & 0 \end{bmatrix} \tag{22}$$

4 Control Law Design and Stability Analysis

The benefit of sliding mode variable structure control over conventional control method is that when the system is affected by some uncertainties (such as errors in mathematical modeling, external environmental disturbances, parameter uptake, etc.), its "structure" can be changed in an instantaneous manner, according to the current state of the system in a purposeful jump change, so that the system moves according to the state trajectory of the "sliding mode" set in advance.

Both theory and practice have proven that the sliding mode design is independent of the parameters of the controlled system and torque interference, making the sliding mode variable structure control fast, insensitive to parameter uptake and environmental disturbances, simple and easy to implement, and many other essential advantages.

4.1 Sliding Mode Controller Design

The design of traditional sliding mode variable structure controller can be divided into two parts: one is to select a suitable sliding mode surface, the other one is to design the control law so that the system state can move along the sliding mode surface and eventually converge to zero.

The following linear sliding mode switching surface is selected:

$$\mathbf{s} = \lambda \mathbf{q_{ev}} + \boldsymbol{\omega}_\mathbf{e} \tag{23}$$

where $\mathbf{s}$ indicates the selected sliding mode surface and λ represents the normal number.

The input such that the sliding mode surface is equal to 0 is the equivalent input, which is a more widely used system input, i.e. we have:

$$\dot{\mathbf{s}} = \lambda \dot{\mathbf{q}}_\mathbf{ev} + \dot{\boldsymbol{\omega}}_\mathbf{e} = \lambda \dot{\mathbf{q}}_\mathbf{ev} + \mathbf{I}^{-1}\mathbf{T}_c \tag{24}$$

$$\mathbf{T}_c = -\lambda \mathbf{I} \dot{\mathbf{q}}_\mathbf{ev} \tag{25}$$

The control law ensures that if the initial state of the system is on the slip surface, it will always operate on the slip surface (The proof process will be reflected in Sect. 4.2), but when the initial state is not on the slip surface or there is interference, the system will deviate from the slip surface or not be near the slip surface at all. Therefore, we need to add a control term to make the system converge to and eventually stay on the slipform surface when the system state is not on the slipform surface, which is generally called the convergence law. Typical convergence laws are the isokinetic convergence law, exponential convergence law, power convergence law, general convergence law, etc. The isokinetic convergence law is the most commonly used, i.e. there are:

$$\dot{\mathbf{s}} = -\eta \operatorname{sgn}(\mathbf{s}) \tag{26}$$

where $\eta \geq \|\mathbf{T_d}\|$, the sliding mode variable structure attitude control law can be designed as:

$$\mathbf{T}_c = -\lambda \mathbf{I} \dot{\mathbf{q}}_\mathbf{ev} - \eta \operatorname{sgn}(\mathbf{s}) \tag{27}$$

$\operatorname{sgn}(\mathbf{s}) = \left[\operatorname{sgn}(s_x)\ \operatorname{sgn}(s_x)\ \operatorname{sgn}(s_x) \right]$, sgn denotes the sign function.

Figure 4 gives a graph of the variation of the sign function and hyperbolic tangent function, from which we can obtain that the continuous hyperbolic tangent function can approximate the discontinuous sign function. Therefore, to effectively eliminate the chattering, the hyperbolic tangent function is used instead of the sign function, i.e., we have:

$$\mathbf{T}_c = -\lambda \mathbf{I} \dot{\mathbf{q}}_\mathbf{ev} - \eta \tanh(a\mathbf{s}) \tag{28}$$

where $\tanh(a\mathbf{s}) = \left[\tanh(as_x)\ \tanh\left(as_y\right)\ \tanh(as_z) \right]$, $a > 0$. We can see that there is no decoupling term in the controller, which does not need to rely on an accurate satellite model and therefore has a strong robustness.

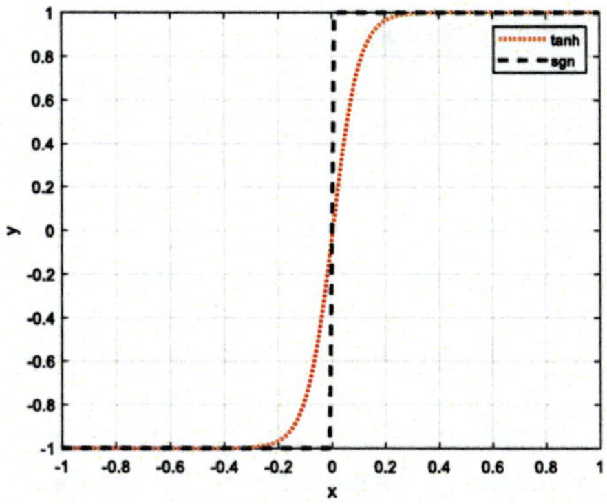

Fig. 4. The variation of the sign function and hyperbolic tangent function

4.2 Stability Analysis

Firstly, it is proved that the system can reach the desired attitude i.e. the system state converges to the equilibrium point $\left(\boldsymbol{\omega}_\mathbf{e}\ \mathbf{q}_\mathbf{ev}\right) = \left(0\ 0\right)$ when the system state moves to the sliding mode switching surface, at this time:

$$\mathbf{s} = \lambda \mathbf{q}_\mathbf{ev} + \boldsymbol{\omega}_\mathbf{e} = 0 \tag{29}$$

Bringing Eq. 29 into Eq. 9, we have:

$$\begin{aligned}
\dot{q}_{e0} &= -\frac{1}{2}\mathbf{q}_\mathbf{ev}^\mathbf{T}\boldsymbol{\omega}_\mathbf{e} \\
&= -\frac{1}{2}\mathbf{q}_\mathbf{ev}^\mathbf{T}(-\lambda \mathbf{q}_\mathbf{ev}) \\
&= \frac{1}{2}\lambda \mathbf{q}_\mathbf{ev}^\mathbf{T}\mathbf{q}_\mathbf{ev} \\
&= \frac{\lambda}{2}\left(1 - q_{e0}^2\right) \geq 0
\end{aligned} \tag{30}$$

From Eq. 30 above, q_{e0} is monotonically increasing, and according to the relationship between q_{e0} and $\mathbf{q}_\mathbf{ev}$, q_{e0} will increment to 1, i.e. $\mathbf{q}_\mathbf{ev}$ will converge to 0, and then $\boldsymbol{\omega}_\mathbf{e}$ will also converge to 0.

It is proved below that the designed control law can make $\lim_{t\to\infty}\left(\boldsymbol{\omega}_\mathbf{e}\ \mathbf{q}_\mathbf{ev}\right) = \left(0\ 0\right)$, i.e. the designed control law is stable, under the premise of choosing Eq. 23.

Derivation of Eq. 23 and then left multiplication of each side by $\mathbf{I}$, and substitution of Eq. 28 gives:

$$\begin{aligned}
\mathbf{I}\dot{\mathbf{s}} &= \lambda \mathbf{I}\dot{\mathbf{q}}_\mathbf{ev} + \mathbf{I}\dot{\boldsymbol{\omega}}_\mathbf{e} \\
&= \lambda \mathbf{I}\dot{\mathbf{q}}_\mathbf{ev} + \mathbf{T}_\mathbf{d} + \mathbf{T}_c \\
&= \lambda \mathbf{I}\dot{\mathbf{q}}_\mathbf{ev} + \mathbf{T}_\mathbf{d} - \lambda \mathbf{I}\dot{\mathbf{q}}_\mathbf{ev} - \eta \tanh(as) \\
&= -(\eta \tanh(as) - \mathbf{T}_\mathbf{d})
\end{aligned} \tag{31}$$

Take the Lyapunov function function as:

$$V = \frac{1}{2}\mathbf{s}^{\mathbf{T}}\mathbf{I}\mathbf{s} \tag{32}$$

Derivative for V yields:

$$\begin{aligned} \dot{V} &= \mathbf{s}^{\mathbf{T}}\mathbf{I}\dot{\mathbf{s}} \\ &= -\mathbf{s}^{\mathbf{T}}(\eta \tanh(a\mathbf{s}) - \mathbf{T_d}) \le 0 \end{aligned} \tag{33}$$

Because when and only when $\mathbf{s} = 0$, $\dot{V} = 0$, i.e. when $\dot{V} \equiv 0$, $\mathbf{s} \equiv 0$. According to the LaSalle invariance principle, the asymptotic stability of the closed-loop system can be proved.

5 Simulation Verification

In the simulation, the two-body orbit model of the small video satellite is created and the coordinate position of the ground target are given. The effect of the controller action is examined by the change of the target projection pixel coordinates.

The following simulations are performed in the ideal case and in the existence of internal rotational inertia deviation and disturbance torque, respectively.

The simulation parameters are shown in Table 1.

The position of the small video satellite during the simulation is obtained from the orbit forecast. Under the set simulation parameters, the optical axis is oriented to the pixel coordinate point (1750,1450), the target can appear in the image plane and satisfy observability conditions such as illumination and no occlusion.

The results of the simulation are shown as follows, (a) shows the simulation results in the ideal case, and (b) shows the simulation results in the existence of rotational inertia deviation and disturbance torque.

Figures 5 and 6 respectively display the plots of error attitude quaternion and error angular velocity with time. Figure 7 illustrates the sliding mode variation. Figure 8 shows the variation of the control torque provided by the flywheel. From these plots, we can see that all of the curves converge within 4.5 s, which means that the system can reach stability within 4.5 s for both simulation conditions. When under ideal conditions, the error quaternion, error angular velocity, and control torque can reach 2×10^{-5}, 4×10^{-5}, and 6×10^{-4}, respectively. When there are internal rotational inertia deviation, external disturbance torque conditions, the error quaternion, error angular velocity, and control torque can reach 5×10^{-5}, 2×10^{-4}, and 5×10^{-3}, respectively. The simulation results in both cases achieve sufficiently high control accuracy. It is proved that the designed sliding mode control method is non-sensitive to parameter uncertainties and external environmental disturbances.

Figure 9 shows the change curve of the target projection pixel coordinates in the simulation results under the action of the designed sliding mode controller, from which it can be seen that the projection coordinate of the target in the image plane at the initial moment is (3500,2043). Along with the control process and the continuous maneuvering of the satellite attitude, the projection position can gradually converge to the desired

Table 1. Simulation parameters

Parameter	Value	Unit
Satellite semimajor axis	6878.14	km
Satellite orbital eccentricity	4.41×10^{-8}	xx
Satellite orbital inclination	45	deg
Satellite right ascension of ascending node	358.49	deg
Satellite argument of perigee	49.62	deg
Satellite true anomaly (initial moment)	9.14	deg
Satellite initial attitude quaternion	(−0.5055, −0.1278, 0.7804, 0.3452)	
Satellite initial angular velocity	(−0.02, 0.01, 0.08)	rad/s
Satellite rotational inertia	diag (5,5,5)	$kg.m^2$
Location coordinate of the ground target	(1228.158, 5185.468, 3502.848)	km
Camera focal length	0.8	m
Size of frame	3500 × 2900	pixels
Pixel size	7 × 7	μm
Normal number gain λ	3	
Normal number gain η	1.2	
Normal number gain a	10	
The maximum control torque of the flywheel	0.4	N.m
Rotational inertia deviation	diag (1,1,1)	$kg.m^2$
External disturbance torque	$T_d = \begin{bmatrix} 0.003 \times \sin(10t/pi) \\ -0.003 \times \sin(10t/pi) \\ 0.003 \times \sin(10t/pi) \end{bmatrix}$	N.m

image center point (1750,1450) under the interference-free condition. When the condition of disturbance exists, the converged pixel coordinates will have a small fluctuation in (1750,1450), but the effect on the visual tracking control is slight, which indicates that the designed sliding mode control method could be robust to external disturbance.

Figure 10 shows the variation curves of the target projection trajectory in the simulation results with the sliding mode control method under the conditions of no disturbance and presence of disturbance. It can be seen that the initial target projection position represented by the blue solid dot is located at the edge of the image plane, and the desired coordinate position represented by the red dot is at the center of the image plane. From these two sets of plots we can see more intuitively that the control method used can make the target projection at the center of the image field of view and the control system

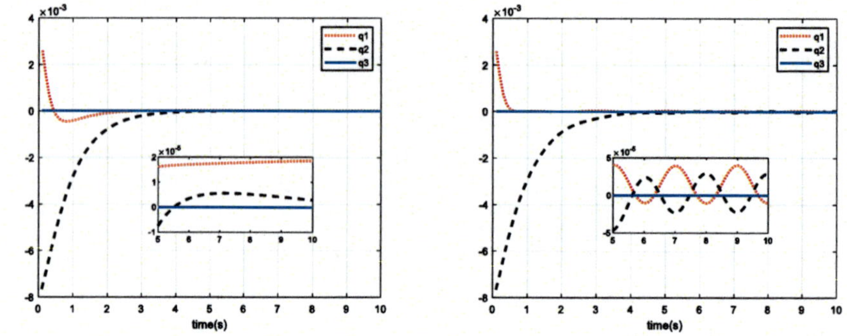

(a) (b)

Fig. 5. Variation of error quaternions

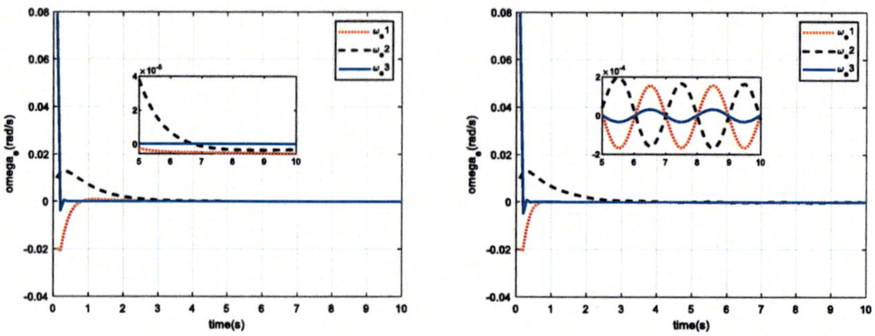

(a) (b)

Fig. 6. Variation of error angular velocity

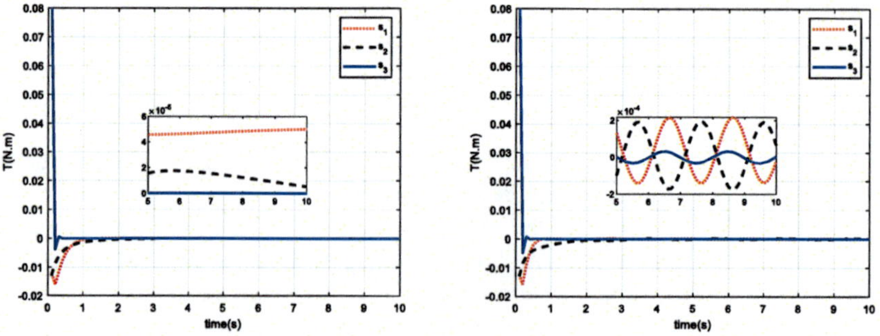

(a) (b)

Fig. 7. Sliding modal variation

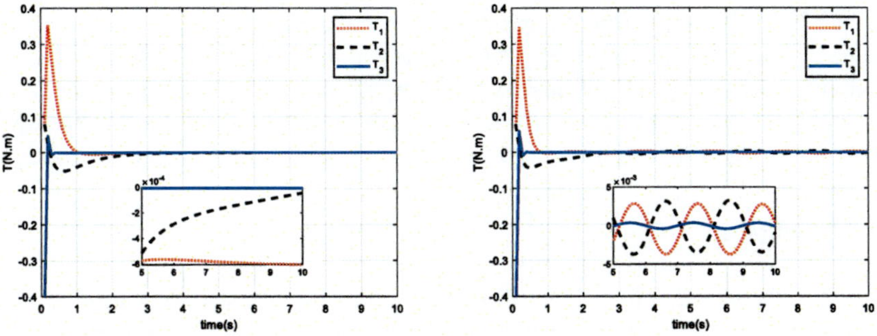

(a) (b)

Fig. 8. Variation of control torque

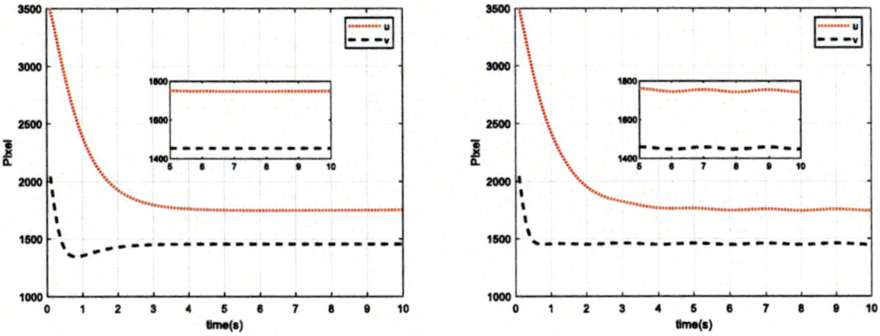

(a) (b)

Fig. 9. Variation of pixel point

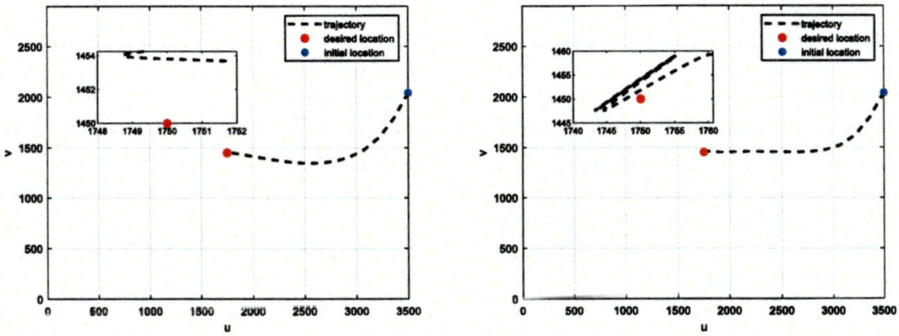

(a) (b)

Fig. 10. Target projection point trajectory

can reach a stable state. Under ideal conditions, the control error in the u-direction is within 2 pixel, and the error in the v-direction is within 4 pixels; under the conditions of interference and rotational inertia deviation, the control error in the u-direction is within 10 pixels, and the error in the v-direction is within 10 pixels. All of them achieve high control accuracy, which again confirms the effectiveness of the designed controller.

6 Conclusion

Aiming at the situation that the small video satellite cannot track non-cooperative targets independently based on position information in the ground target tracking tasks, this article designs a sliding mode tracking control method based on visual image information that does not depend on the exact model, and the proposed controller has the ability to guarantee asymptotic stability of the system under controller action. Simulation results indicate that the visual tracking sliding mode controller could effectively eliminate the attitude error to make the target projection converge to the desired position. Even under the condition of given external disturbance and rotational inertia deviation, the designed controller is able to make the system stable within 4.5 s and track attitude and angular velocity with high accuracy, which proves that the control approach has strong robustness. The attitude control method proposed in this paper provides a new idea for robust control of small video satellite flight against interference, and lays a theoretical foundation and provides technical support for expanding its engineering application capability.

References

1. Pranoto, F.S., Pramutadi, A.M.: Studies on micro satellite aerial launch system. In: Aerospace Conference, pp. 1–6. IEEE (2017)
2. Steyn, W.H., Zyl, R.V., Inggs, M., et al.: Current and future small satellite projects in South Africa. In: Geoscience Remote Sensing Symposium. IEEE (2014)
3. Xiang, J., Zeng, G., Han, D., et al.: Design and on-orbit test for video satellite: Tiantuo-2. In: Proceedings of the 67th International Astronautical Congress, IAC 2016, September 26–September 30, 2016, Guadalajara, Mexico, F. International Astronautical Federation, IAF (2016)
4. Chen, X., Steyn, W., Hashida, Y.: Ground-target tracking control of earth-pointing satellites. In: AIAA Guidance, Navigation, and Control Conference and Exhibit (2000)
5. Chen, X.Q., Ma, Y.H., Geng, Y.H., et al.: Staring imaging attitude tracking control of agile small satellite. Opt. Precis. Eng. (2012)
6. Zhang, L.Y., et al.: A New Nonsingular Terminal Sliding Mode Control for Rigid Spacecraft Attitude Tracking
7. Young, K.K.: Design of variable structure model following control system. IEEE Trans. Autom. Contr. **23**(6), 1079–1085 (1978)
8. Hung, J.Y., Gao, W.: Variable structure control: a survey. IEEE Trans. Ind. Electron. **40**(1), 2–22 (1993)
9. Bustan, D., Pariz, N., Hosseini Sani, S.K.: Robust fault-tolerant tracking control design for spacecraft under control input saturation. Isa Trans. **53**(4), 1073–1080 (2014)
10. Lo, S.C., Chen, Y.P.: Smooth sliding-mode control for spacecraft attitude tracking maneuvers. J. Guid. Control Dyn. **18**(6), 1345–1349 (1995)

11. Zhu, Z., Xia, Y., Fu, M.: Adaptive sliding mode control for attitude stabilization with actuator saturation. IEEE Trans. Ind. Electron. **58**(10), 4898–4907 (2011)
12. Wie, B.: Space vehicle dynamics and control. In: AIAA Education Series, AIAA, p. 438. Reston, VA (1988)

Optimization and Application of Lucy-Richardson Filter on Optical Remote Sensing Image Restoration

Yunsen Wang(✉), Chengzhi Ma, Panfeng Wu, Weiqiang Li, and Qi Song

Shandong Institute of Space Electronic Technology, Yantai 264670, Shandong, China
yswang513@163.com

Abstract. In complex space environment, remote sensing imaging process might be affected by different kinds factors. Lucy-Richardson (L-R) filter method is an extensive used method in image reconstruction field. In order to decrease the calculation time and promote the quality of image, a optimal semi-blind remote sensing data reconstruction algorithm based on improved L-R filter is presented. Firstly, self-correlation function is innovated to calculate the point spread function (PSF). To suppress the superposition of noise in L-R iterative process, both full-reference and no-reference evaluation parameters are applied in iteration termination conditions. Finally, Gaussian filter is adopted to filtrate low-pass signal when the noise is constantly amplified. Experiments and simulation results explanation that the improved Lucy-Richardson filter algorithm could gain relatively satisfactory reconstruction effects. And the proposed processing method could be used for the restoring and reconstruction of those remote sensing data degraded by rectilinear motion and random noise interference.

Keywords: Lucy-Richardson filter · Point spread function · Remote sensing image reconstruction · Image evaluation parameter

1 Introduction

As many of you known, remote sensing data is frequently degenerated by various factors under complex imaging conditions in space [1–3]. So the restoration of remote sensing data plays a significant role if it is difficult to control the load platform itself [4].

Classical reconstruction algorithms routinely employ filter models according to the degradation principle, such as Wiener filter, inverse filtering and Laplacian constraint [5–8]. These algorithms are always based on MTF (means modulation transfer function) or PSF (means point spread function) [9]. However, the PSF parameter is frequently unable to obtained exactly. In addition, another widely used method is Richardson–Lucy filter [10, 11].

Recently, machine learning is widely used in image processing field [12], CNN (Convolution Neural Network) in particular. Refs. [13, 14] adopted CNN to revert degenerative images. Hopfield neural network is utilized in Refs. [15–17] to realize image restoration based on optimization reason. Moreover, many other prevalent algorithms

L. Wang et al. (Eds.): CHREOC 2022, LNEE 969, pp. 150–160, 2023.
https://doi.org/10.1007/978-981-19-8202-6_14

such as U-net, GAN-net and residual net [18–20] are applied in image restoration field. However, neural network or deep learning algorithms need a great number of samples.

Based on the literature above, a optimal semi-blind remote sensing data reconstruction method based on improved L-R filter is presented to reduce the calculation time and improve the image's quality. Self-correlation function is cited to calculate the point spread function. To suppress the superposition of noise in L-R iterative process, SNR (signal to noise ratio) and GMG (grayscale mean gradient) are applied in iteration termination conditions. And Gaussian filter is adopted to filtrate low-pass signal when the noise is larger than the preset threshold. Experiments and simulation results declare that the proposed Lucy-Richardson filter model could acquire relatively satisfactory reconstruction effects and time consumption.

2 Data Degradation Model and Lucy-Richardson Filter

2.1 Principle of Image Restoration and Degradation

Universal image degeneration and reduction model is presented firstly, as shown in the Fig. 1. Where $f(x, y)$ is the source remote sensing image, $g(x, y)$ is the degeneration data, $h(x, y)$ is the modulation transfer function, and $n(x, y)$ is the random noise.

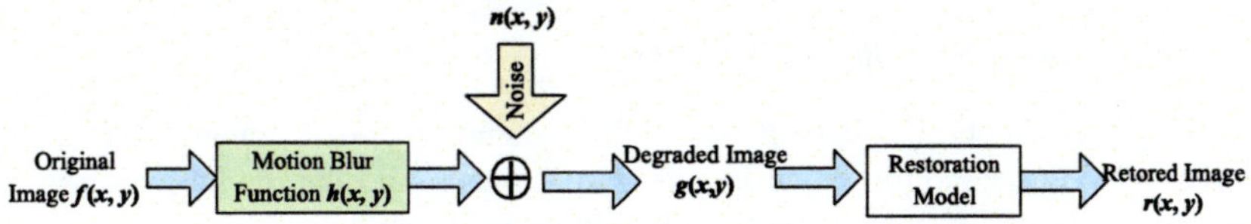

Fig. 1. Universal image restoration and degradation models

To improve the operation speed, the modulation transfer function $h(x, y)$ is often considered to be a space-unchanging and linear equation. And the degeneration process of remote sensing data could be represented by

$$g(x, y) = f(x, y) * h(x, y) + n(x, y) \tag{1}$$

In Fig. 1, $r(x, y)$ represents the restored remote sensing data. Then the goal of image reconstruction is to qualify the following formula.

$$r(x, y) = f(x, y) \tag{2}$$

2.2 Lucy-Richardson Filter Algorithm

Lucy-Richardson filter is an iterative algorithm for image restoration in the background of Poisson distribution, Bayesian theory and maximum likelihood estimation. After the blurred data is convoluted with the given modulation transfer function, and the image without degradation might be obtained, even if the image noise is unknown. Lucy-Richardson filter's principle could be given by the following equation.

$$r_{k+1}(x, y) = r_k(x, y)\left[h(-x, -y) * \frac{g(x, y)}{h(x, y) * r_k(x, y)}\right] \tag{3}$$

where $r_k(x, y)$ represents the restored image after k-th iteration, $h(x, y)$ is point spread function, $g(x, y)$ is the degraded image. And the symbol * represents convolution operation. To start the iteration process, the initial value of $r_0(x, y)$ is set as follow.

$$r_0(x, y) = g(x, y) \tag{4}$$

And the remaining questions are the calculation of point spread function and the determination of iteration stop condition.

3 Design of Remote Sensing Image Restoration Method

3.1 Estimation of Point Spread Function

According to the motion characteristics of the satellite, degraded remote sensing data influenced by linear motion mode is take into account. And the modulation transfer function of degraded system meets with the following formula.

$$h(x, y) = \frac{1}{d}, \quad 0 \le x \le d\cos\theta, \ y = x\sin\theta \tag{5}$$

where d is fuzzy scale, θ is the fuzzy direction. To simplify operation steps, the fuzzy scales of horizontal direction and vertical direction are calculated respectively. In horizontal direction, $h(x)$ is given as

$$h(x) = \frac{1}{d}, \quad 0 \le x \le d - 1 \tag{6}$$

In this paper, the self-correlation function is applied to calculate the fuzzy scale d. The detailed calculation procedures are as below.

(1) Implement first-order difference on the vertical orientation of blurred image $g(x, y)$.

$$g_y(i, j) = g(i, j) - g(i, j - 1) \tag{7}$$

(2) Perform first-order difference operation on the horizontal orientation.

$$g_{yx}(i, j) = g_y(i, j) - g_y(i - 1, j) \tag{8}$$

(3) Compute the degenerative image's self-correlation function on every row. In case there are l pixels on the j-th row, the self-correlation function would be given by

$$r(k) = \sum_{i=l}^{h-k-l} \sum_{m=-l}^{l} g(i+k+m,j)g(i+m,j) \quad k \in [-1, 1] \tag{9}$$

where h is the height scale of the source image.

Estimate the average value of self-correlation equation and delineate the self-correlation equation profile. The minimal value of the profile location the fuzzy scale.

Similarly, the fuzzy scale in vertical direction could be computed soon. Then the point spread function *h(x, y)* is estimated.

3.2 Noise Suppression and Iteration Termination Condition

During the iterative process of image restoration, noise information might be amplified. So the value of *SNR* (Signal to Noise Ratio) is quoted here and evaluate the random noise level of restored remote sensing images. The expression of *SNR* could be given as

$$SNR = 10\ \lg\left[\frac{\sum_{x=1}^{w}\sum_{y=1}^{h} f(x,y)^2}{\sum_{x=1}^{w}\sum_{y=1}^{h} [f(x,y) - f_p(x,y)]^2}\right] \tag{10}$$

where *f(x,y)* is the image, $f_p(x, y)$ is the denoised data, h and w are the height and width of the data, respectively. In the process of Lucy-Richardson filter, there is no reference image. In this paper, the degraded image $g(x, y)$ is firstly preprocessed with Gaussian filter. And the result is $g_G(x, y)$. Then the *SNR* could be calculated by

$$SNR = 10\ \lg\left[\frac{\sum_{x=1}^{w}\sum_{y=1}^{h} g_G(x,y)^2}{\sum_{x=1}^{w}\sum_{y=1}^{h} [g_G(x,y) - r(x,y)]^2}\right] \tag{11}$$

The larger the SNR_G, the smaller the noise, and vice versa.

Another index of image restoration evaluation is *GMG* (Grayscale Mean Gradient). And in one sense, the value *GMG* parameter delivers the image's texture contrast and features.

$$GMG = \frac{1}{(w-1)(h-1)} = \sum_{i=1}^{w}\sum_{j=1}^{h}\sqrt{\frac{[f(i,j+1) - f(i,j)]^2 + [f(i+1,j) - f(i,j)]^2}{2}} \tag{12}$$

Ideally, both *GMG* and *SNR* become larger as the iteration progresses. When this condition cannot be satisfied, the minimum threshold of *SNR* is fixed in the proposed method. On the other hand, the specified threshold of GMG is also assigned to reduce iteration time.

3.3 Remote Sensing Data Restoration Algorithm

TO reduce the superposition of noise, Gaussian filter is adopted to filtrate low-pass signal. The remote sensing image t restoration process is shown in the Fig. 2.

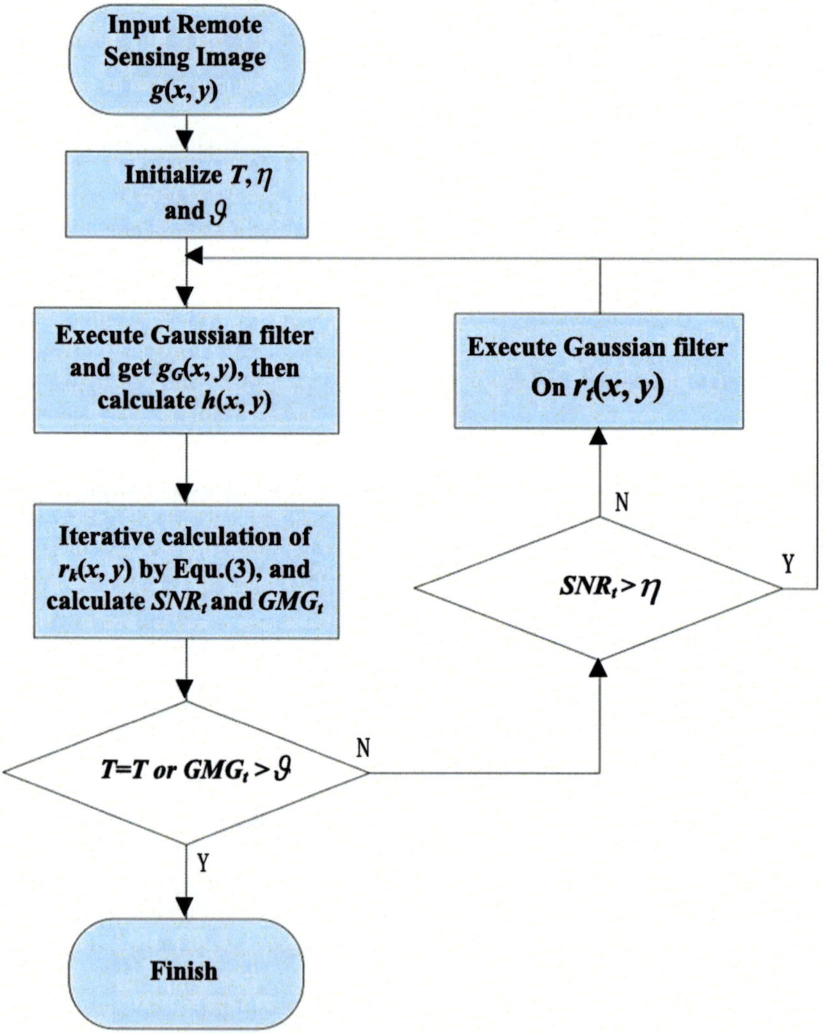

Fig. 2. Remote sensing data restoration process

The implementation steps are as below.

Step 1: Perform initialization operation of these values of T, η and ϑ, here T is the maximal iteration number, η is the minimum threshold of *SNR*, and ϑ is the specified threshold of *GMG*. And $r_0(x, y) = g(x, y)$. The number of iterations $t = 0$.

Step 2: Carry out Gaussian filter on $g(x, y)$, assume the result is $g_G(x, y)$.

Step 3: Determine the point spread function $h(x, y)$ through autocorrelation function.

Step 4: Execute the iterative Eq. (3). The number of iterations $t = t + 1$. Calculate SNR_t by Eq. (11). Calculate GMG_t by Eq. (12).

Step 5: If $t = T$ or $GMG_t = \vartheta$, finish processing. Otherwise, compare the size of SNR_t and η. Go to Step 6.

Step 6: If $SNR_t > \eta$, return to Step 4. Else, implement Gaussian filter on $r_t(x, y)$. Return to Step 4.

4 Experiments and Simulations

4.1 Simulations

To testify the availability and significance of the proposed algorithm, the checkboard figure with the size 512 × 512 is picked as the test data, which is shown in Fig. 3a. And Fig. 3b shows the degenerate image. And the image is influenced by linear motion modulation transfer function and Gaussian noise. And the value of fuzzy scale is 15.

Figure 3c–f are the restoration results when the number of iterations equals to 5, 10, 20, and 50. It can be concluded from these figures that the higher the number of iterations, the better the image restoration effect. However, after iterating to a certain extent, the improvement effect is not obvious at all. Furthermore, L-R algorithm has good noise suppression effect at the beginning of filtering. With the increase of filtering times, the sides of the data are getting clearer and clearer, and the noise is becoming more and more obvious.

Conventional classic Wiener filter method, constrained least squares method and blind deconvolution method are chosen to compare to the proposed L-R filter method, the test results are shown in Fig. 4.

Through human vision, it is clearly that Wiener filter's process effects have lower noise and more fuzzy details, as shown in Fig. 4a. And the restoration result of constrained least squares method has higher noise and more clear details. Figure 4c shows the treatment results of blind deconvolution. It seems that new noises and uncertain factors are brought into the data.

Table 1 displays two assess arguments of the initial data, degenerate data and four methods' processing results. As analyzed above, Wiener filter method can obtain larger SNR, but the GMG value is relatively low. It seems that the L-R filter method has more balanced performance between SNR and GMG when the number of iterations reaches to 20.

4.2 Experimental Tests

To further test the advantages of the algorithm, this paper chooses five more degenerative remote sensing data to implement reconstruction testing, as displayed in the five

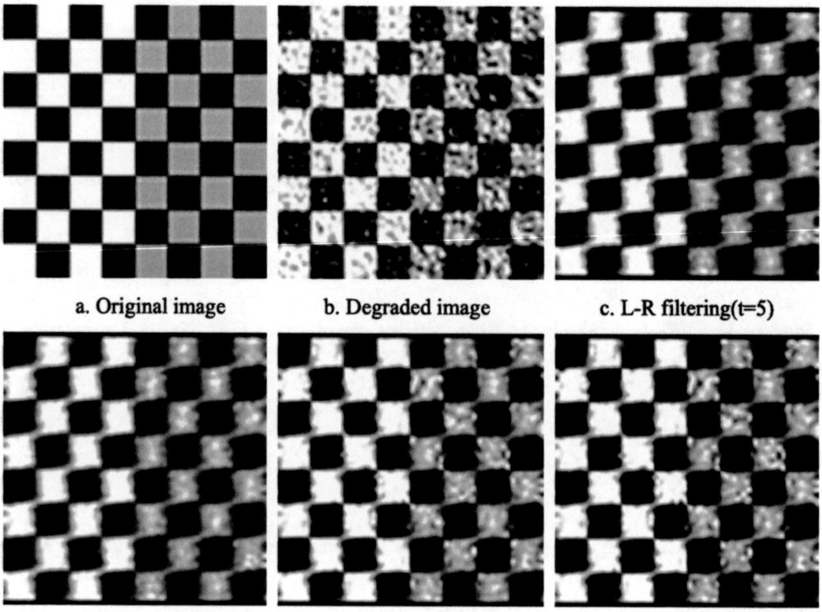

a. Original image b. Degraded image c. L-R filtering(t=5)

d. L-R filtering(t=10) e. L-R filtering(t=20) f. L-R filtering(t=50)

Fig. 3. Restoration results of L-R filter on checkboard

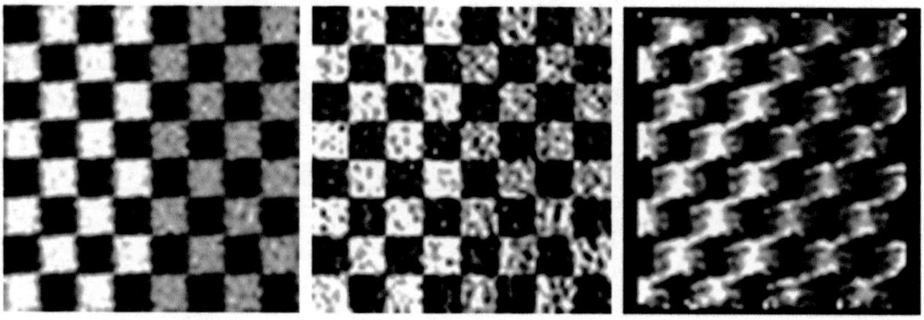

a. Wiener filtering b. Constrained least squares c. blind deconvolution

Fig. 4. Processing effects of three methods

subgraphs on the left side of Fig. 5. These remote images' size is 4096 × 4096. And the initial calculation parameters are as follows, $T = 15$, $\eta = 16$ and $\vartheta = 5$.

The right five images of Fig. 5 are recovered data. By comparing five groups of data, it is apparent that the details and contours of the recovered data is clearer, and the features of surface seem more notable. By implementing the presented L-R filter, these remote sensing data's GMG features are promoted to some extent, which is shown in Table 2. The SNR index could be limited within the specified scope. And the number of iterations is no larger than T.

Table 1. Checkboard's processing results

	SNR	GMG
Initial image	/	32.46
Degraded image	/	26.92
L-R filtering(t = 5)	9.92	34.02
L-R filtering(t = 10)	8.95	35.36
L-R filtering(t = 20)	**7.21**	**36.84**
L-R filtering(t = 50)	6.96	**37.57**
Wiener filtering	**9.96**	33.56
Constrained least squares	3.71	35.03
Blind deconvolution	−68.9	34.86

4.3 Analysis of Experimental Results

Lucy-Richardson filter is based on Poisson distribution, Bayesian theory and maximum likelihood estimation. Theoretically, as the number of iterations increases, the image quality will be better and better without considering the noise. After integrating SNR and GMG factors, the presented improved L-R model could contain more useful details. Therefore, the algorithm would achieve more approving remote sensing restored results.

5 Conclusions

To decrease the calculation time and improve the level of image, this paper puts forward a novel semi-blind remote sensing data reconstruction algorithm on the basis of an improved L-R filter. Compared with other methods, our presented method has two following advantages: (I) Autocorrelation functions are quoted to calculate the point spread function, which simplify the calculation steps. (II) Gaussian filter and SNR/GMG parameters are applied in iteration termination conditions to suppress the superposition of noise in L-R iterative process. Experiments and simulation results explanation that this modified Lucy-Richardson filter algorithm can gain relatively satisfactory reconstruction results.

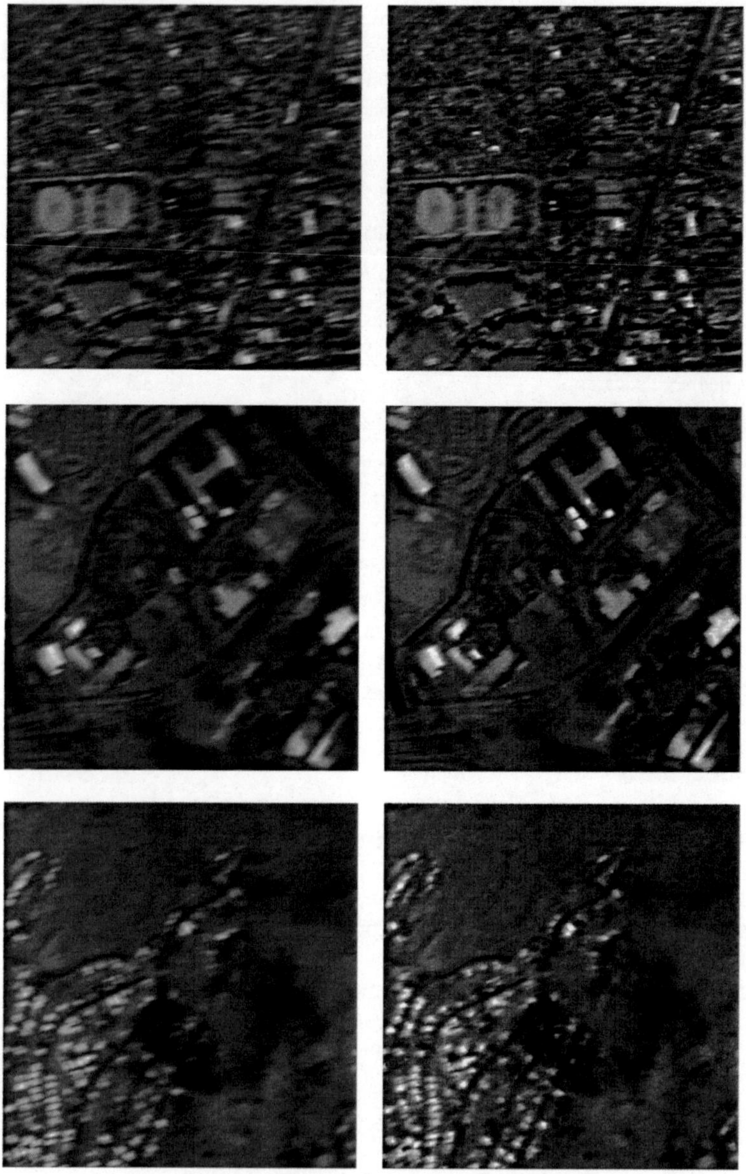

Fig. 5. Restoration effects of proposed L-R algorithm

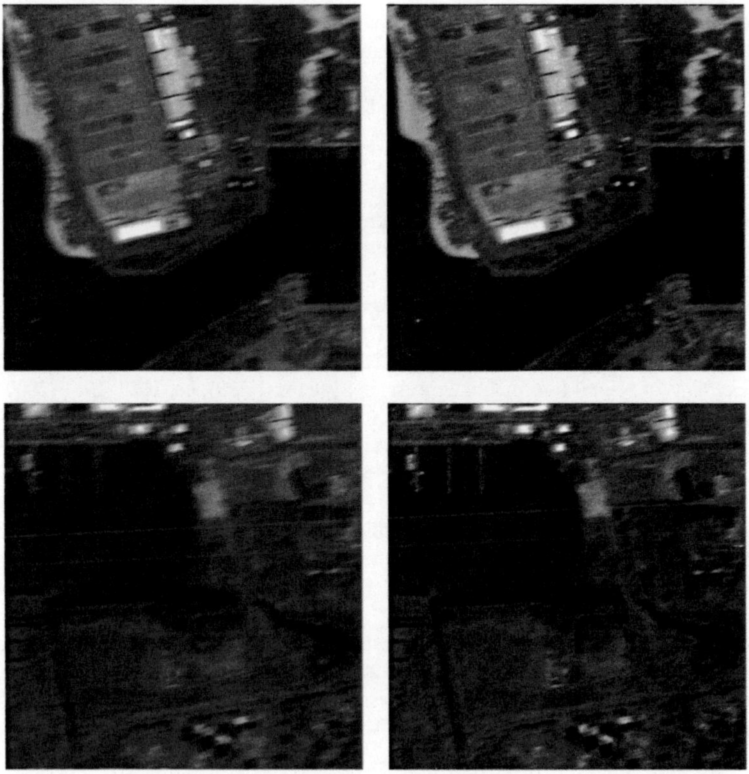

Fig. 5. (*continued*)

Table 2. Restoration parameters of proposed L-R method

Image No.	SNR	GMG		Number of iterations
		Before	After	
1	16.92	4.56	5.23	12
2	17.98	4.12	5.81	13
3	17.67	4.16	6.13	12
4	20.39	4.12	5.47	11
5	18.62	4.07	5.66	14

References

1. Jia, T., Shi, Y., Zhu, Y., Wang, L.: An image restoration model combining mixed L1/L2 fidelity terms. J. Vis. Commun. Image R. **38**, 461–473 (2016)
2. Cai, X., Chan, R., Zeng, T.: A two-stage images segmentation method using a convex variant of the Mumford-Shah model and thresholding. SIAM J. Imag. Sci. **6**, 368–390 (2013)

3. He, L., Cui, G., Feng, H., Xu, Z., Li, Q., Chen, Y.: The optimal code searching method with an improved criterion of coded exposure for remote sensing image restoration. Opt. Commun. **338**, 540–550 (2015)
4. Ye, Z.: Optical Remote Sensing Motion Degration Simulation and Image Restoration Technology. Zhejiang University (2017)
5. Wu, X., Wu, J., Zhang, H.: Research on image restoration techniques based on inverse filtering algorithm. Inf. Technol. **35**, 183–185 (2011)
6. Khan, M., Nizami, I.F., Majid, M.: No-reference image quality assessment using gradient magnitude and wiener filtered wavelet features. Multim. Tools Appl. **78**(11), 14485–14509 (2018). https://doi.org/10.1007/s11042-018-6797-4
7. Levin, A., Fergus, R., Durand, F., et al.: Image and depth from a conventional camera with a coded aperture. ACM Trans. Graph. **26,** 70 (2007)
8. Krishnan, D., Fergus, R.: Fast image deconvolution using hyper-Laplacian priors. In: Proceedings of Neural Information Processing Systems (NIPS), pp. 1033–1041 (2009)
9. Wang, Y., Mu, W., Du, X., Ma, C., Shen, X.: Remote sensing image on-board restoration based on adaptive wiener filter. Lecture Notes Electr. Eng. **657**, 271–282 (2020)
10. Richardson, W.H.: Bayesian-based iterative method of image restoration. J. Opt. Soc. Am. **62**, 55–59 (1972)
11. Lucy, L.: An iterative technique for the rectification of observed distributions. Astron. J. **79**, 745–753 (1974)
12. Liu, L., Li, S., Lai, S.: Advance of neural network in degraded image restoration. J. Graph. **40**, 213–224 (2019)
13. Sun, J., Cao, W., Xu, Z., et al.: Learning a convolutional neural network for non-uniform motion blur removal. In: Proceedings of the 2015 IEEE Conference on Computer Vision and Pattern Recognition, pp. 769–777. IEEE Press, New York (2015)
14. Nah, S., Kim, T.H., Lee, K.M.: Deep multi-scale convolutional neural network for dynamic scene deblurring. In: 2017 IEEE Conference on Computer Vision and Pattern Recognition (CVPR), p. 3. IEEE Press (2017)
15. Zhou, Y.T., Chellappa, R.: Image restoration using a network. IEEE Trans. Acoust, Speech, Signal Process. **36**(1988), 1141–1151 (1988)
16. Han, Y., Wu, L.: Image restoration using a modified Hopfield neural network of continuous state change. Signal Process. **20**, 431–435 (2002)
17. Perry, S.W., Guan, L.: Weight assignment for adaptive image restoration by neural networks. IEEE Trans. Neural Netw. **11**, 156–170 (2000)
18. Kupyn, O., Budzan, V., Mykhailych, M., et al.: DeblurGAN: Blind motion deblurring using conditional adversarial networks. In: 2018 IEEE/CVF Conference on Computer Vision and Pattern Recognition, pp. 8183–8192 (2018)
19. Tang, J., Wang, K., Zhang, W., et al.: A deep learning-based image restoration method in optical synthetic aperture imaging system. Acta Optica Sinica **40**(2020), 2111001
20. Li, Y., Xu, Q., Li, K.: New method of residual dense generative adversarial networks for image restoration. J. Chin. Comput. Syst. **41**, 830–836 (2020)

Hyperspectral Image Data Construction and Expansion Method of Ground Object

Zhao Jiale[1], Zhou Bing[1(✉)], Wang Guanglong[2], Ying Jiaju[1], Deng Lei[1], and Wang Qianghui[1]

[1] Department of Electronic and Optical Engineering, Army Engineering University, Shijiazhuang 050003, China
zhbgxgc@163.com

[2] Department of Missile Engineering, Army Engineering University, Shijiazhuang 050003, China

Abstract. Hyperspectral image is an image that contains both geometric space information and spectral information, however, the cost of acquiring a large number of hyperspectral image data samples is very high due to the limitations of imaging technology and detector level.With the development of artificial intelligence technology and the advent of the era of big data, data sample build and extend the method arises at the historic moment. In order to obtain sufficient sample data and establish the hyperspectral data information database of specific ground object required for research, the obtained hyperspectral image data needs to be expanded which can provide basic data support for the study of spatial spectral characteristics of ground objects and subsequent hyperspectral image classification and target detection. In this article, Firstly, the data structure and characteristics of hyperspectral image are analyzed, and then the common methods and significance of hyperspectral image data sample expansion under the condition of land-based imaging are systematically described from the two aspects of spatial dimension data and spectral dimension data.Finally, Aiming at the main problems existing in the process of hyperspectral image data expansion, the development direction in the future is prospected.

Keywords: Hyperspectral image · Data expansion · Space · Spectrum

Classifications: **CLC number**: TP751 · **Document code**: A

1 Introduction

Imaging spectral technology appeared in the early 1980s. It is a new technology that integrates imaging technology and spectral technology.With the advent of imaging spectroscopy technology, the spectral information of ground objects determined by material composition and surface materials is completely combined with the spatial information reflecting the spatial morphological relationship between ground objects, that is, the concept of "Atlas integration" of the obtained image is realized.While acquiring the spatial information of ground objects, hyperspectral images can give each pixel in space

L. Wang et al. (Eds.): CHREOC 2022, LNEE 969, pp. 161–169, 2023.
https://doi.org/10.1007/978-981-19-8202-6_15

fine spectral information with its own characteristics. Due to the complete and rich spatial and spectral information of targets in hyperspectral images, hyperspectral imaging technology is widely used in agriculture [1], environmental monitoring [2], military [3] and other fields.

Traditional hyperspectral imaging obtains the spatial spectral information of the image by scanning. This imaging method has poor real-time performance, which leads to the relative scarcity of hyperspectral image data suitable for multi task research. In recent years, with the development of artificial intelligence technology and the advent of the era of big data, data-driven hyperspectral image processing methods began to develop rapidly [4]. The accuracy of hyperspectral image feature classification and target detection no longer only depends on the optimized model structure and algorithm flow, but also depends on the number and type of data samples. Therefore, the construction and expansion of hyperspectral image data has become the basis and premise of hyperspectral image processing, which is of great significance.

2 Data Structure of Hyperspectral Image

Hyperspectral imaging can simultaneously obtain the spatial and spectral information of the scene area to form a three-dimensional data cube. From the perspective of spatial dimension, hyperspectral image is the superposition of two-dimensional images formed under different narrow band light channels, which can reflect the morphological structure relationship between ground objects. From the perspective of spectral dimension, hyperspectral image is a spatial arrangement of spectral features of continuous adjacent structural pixels, which can reflect the fine spectral information of a single pixel. The data structure of hyperspectral image is shown in Fig. 1.

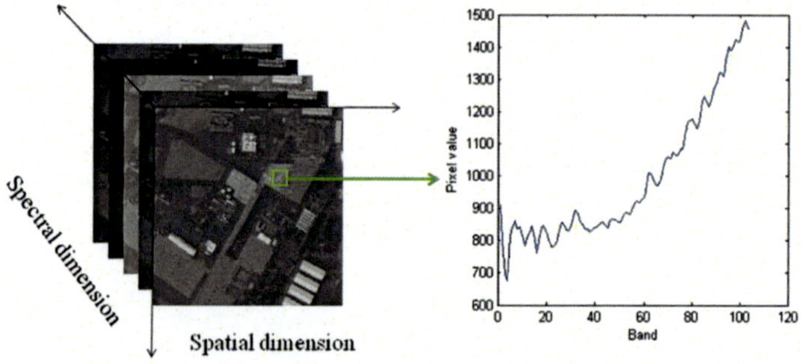

Fig. 1. Hyperspectral image data structure

Due to the special data structure form of hyperspectral image and the unique imaging mode when obtaining the image, hyperspectral image has the following characteristics: (a) it has the characteristics of map unity. Hyperspectral image contains rich radiation, spatial and spectral information, which is a comprehensive carrier of a variety of information. (b) High spectral resolution. Hyperspectral image can obtain the continuous narrow band spectral characteristic curve of ground objects, and can select or extract specific bands to highlight the target features according to needs. (c) High correlation between bands. This leads to a great increase in the number of training samples required for classification and detection tasks.

3 Construction and Expansion of Hyperspectral Image Data of Ground Object

Hyperspectral data has rich spatial spectral information and has extraordinary advantages in ground object target positioning and classification. However, it is difficult to effectively improve the accuracy of the results simply relying on the model structure and algorithm optimization. With the advent of the era of big data, scholars have realized the importance of data samples and achieved good results in many aspects of hyperspectral image processing [5]. The construction and expansion of hyperspectral image data mainly start from two aspects: spatial dimension and spectral dimension.

3.1 Spatial Dimension Data Expansion

With the continuous development of imaging technology and spectroscopic detection technology, and the wide application of imaging spectrometer on the ground and near ground platform, the spatial resolution of hyperspectral image has been greatly improved, and hyperspectral image processing no longer depends only on fine spectral information. In recent years, due to the increasingly obvious role of spatial morphological information in the application of hyperspectral image processing, the demand for hyperspectral image spatial data samples is increasing [6].

3.1.1 Traditional Hyperspectral Image Spatial Data Expansion

Hyperspectral image spatial data is very limited. The number of training samples can be expanded by data enhancement and transformation of the original data to generate new data. Using the traditional image expansion method to expand hyperspectral image spatial data is a simple and common way. The number and diversity of samples can be increased by translating, cutting, flipping, rotating and other data enhancement strategies for the single band gray image of hyperspectral image. In addition, the amount of sample data can be increased by adding noise to the existing image and transforming the contrast and principal component analysis. The traditional spatial data expansion of hyperspectral image is to process the two-dimensional spatial image and change its geometry, color characteristics, noise level and so on. However, experiments show that the improvement effect of traditional hyperspectral image spatial data expansion on classification and detection performance is still not obvious [7].

Another way of hyperspectral image expansion is to fuse multiple optical images of targets. Due to the serious shortage of the amount of original data, when completing the task of image classification and target detection, a kind of fusion image in Google Earth or other types of UAV and remote sensing image can be constructed to expand the sample data. Figure 2 shows the construction of hyperspectral image composite data set. This method can increase the number of training samples and obtain better classification and detection effect [8].

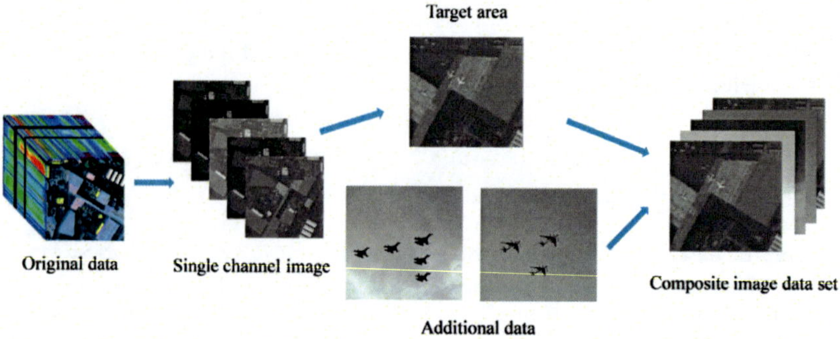

Fig. 2. Construction of composite data set of hyperspectral images

3.1.2 Hyperspectral Image Spatial Data Expansion Based on Deep Learning

The hyperspectral image spatial data expansion method based on deep learning is superior to the traditional data expansion method in both processing speed and generation effect. At present, it is mainly divided into two ways: learning the existing hyperspectral images to complete the generation of new samples; The hyperspectral image is obtained by inversion and restoration of RGB image. The first method mainly uses to generate countermeasure network. The generation of countermeasure network can generate more new data according to the characteristics of this small number of data sets in the case of small and insufficient data sets, so as to achieve the purpose of data set expansion. Compared with the real data, the data generated by this method not only have semantic similarity, but also show the diversity of spatial types. After adding the synthetic data, the neural network can be trained more stably, and the classification accuracy has been further improved, which verifies the feasibility and effectiveness of this technology [9]. Another way is to obtain hyperspectral images directly from RGB images through depth learning method. RGB image has the advantages of fast imaging speed, high spatial resolution, low acquisition cost and easy integration. The RGB image for training is established from hyperspectral data through the camera spectral sensitivity function, but as long as the camera spectral response function is fixed, the method of deep learning is also suitable. For example, the NTIRE 2018 and NTIRE 2020 challenges have put forward many excellent in-depth learning models. In 2018, HSCNN + proposed a method based on residual network and another method based on dense network to extract missing spectral channel spatial information from RGB images [10]. The NTIRE 2020 challenge provides the most extensive evaluation of RGB image restoration methods so far, provides participants with a larger natural hyperspectral image data set than ever before, and proposes a novel depth adaptive weighted attention network (AWAN) for reconstruction from RGB images [11]. Generally speaking, the effect of hyperspectral image spatial data expansion based on deep learning can meet the needs of application, but the recovery effect is relatively dependent on data, which is contradictory to a small amount of original data that needs to be expanded.

3.2 Spectral Dimension Data Expansion

The core of hyperspectral imaging technology is that the image has high spectral resolution, the wavelength interval of the formed multi-channel image reaches the nanometer level, and the purpose of identifying ground objects is achieved through fine spectral information. Each kind of ground object has specific spectral reflection characteristics, that is, "the same spectrum of the same object and the different spectrum of the foreign object". However, with the change of the imaging platform environment and imaging conditions, the spectral reflection characteristics of the object will also produce some fluctuations. By analyzing the factors affecting the spectral reflectance characteristics of objects, the spectral reflectance database of typical targets under different conditions is established, which is of great significance to the subsequent hyperspectral image processing based on spectral dimension.

3.2.1 Factors Affecting the Spectral Reflection Characteristics of Objects

Object reflectance is the inherent physical property of an object. The acquisition of target reflectance information from hyperspectral images obtained by spaceborne or airborne means requires a complex preprocessing process. This mainly includes atmospheric correction, geometric correction and radiometric calibration. Under the condition of ground or near ground imaging, the detection distance is relatively short, and the process from the interaction between light and ground objects to the acquisition of reflection information by the detector is relatively simple. Therefore, the spectral reflection characteristics of ground objects are mainly analyzed from the following three aspects: the characteristics of ground objects, directional characteristics and other reasons.

(a) Features of ground features: mainly including the composition, shape and structure, roughness, etc. The nature of the feature itself is the fundamental factor to distinguish different features. Different types of features have distinct distinguishable spectral reflectance characteristics of such features, and each pixel can obtain a continuous spectral curve containing diagnostic spectral characteristics of the feature [12].
(b) Directional characteristic factors; The detection distance of remote sensing imaging is much larger than the target scale, and the detection of ground objects is similar to vertical detection. In land-based application, the relative position of light source, imaging spectrometer and target ground objects needs to be considered. The research on the directional characteristics of spectral reflectance of ground objects refers to the relationship between the reflectance of ground objects and the detection zenith angle, the zenith angle of incident light and their relative azimuth. Most ground objects in nature are non Lambertian bodies, so the reflection of light and ground objects mostly belongs to anisotropic reflection. BRDF model is often used to describe the basic characteristics of surface anisotropic reflection. It specifically refers to how the incident light is distributed in each exit direction after being reflected by a surface. By establishing a suitable BRDF model, the directional characteristics of features can be studied. The model diagram is shown in Fig. 3.

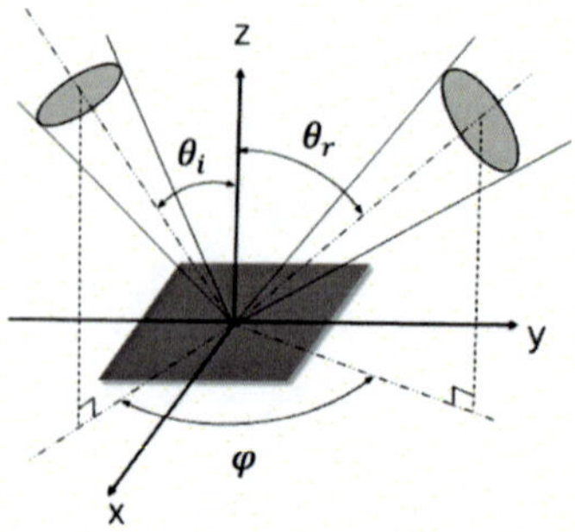

Fig. 3. BRDF model diagram

(c) Other reasons; It refers to any other factors that affect the light receiving direction and light receiving intensity in the whole process of light emission of the light source, light transmission in the medium and the interaction between light and ground objects. For example, the influence of the surrounding environment and the weather [13].

3.2.2 Spectral Data Expansion Based on Parametric Model

Establishing the spectral reflection characteristic model of typical ground objects is an important method to expand the target spectral data. In 1981, Li Xiaowen and others studied the spectral characteristics of vegetation by establishing a reversible conifer canopy reflectance model, and then further completed the inversion of plant spectral reflectance by establishing a geometric optical bidirectional reflectance model [14]. The spectral reflection characteristics of the research target under certain conditions need to go through data analysis, model establishment, parameter fitting, inversion verification and other steps. The specific steps are shown in Fig. 4.

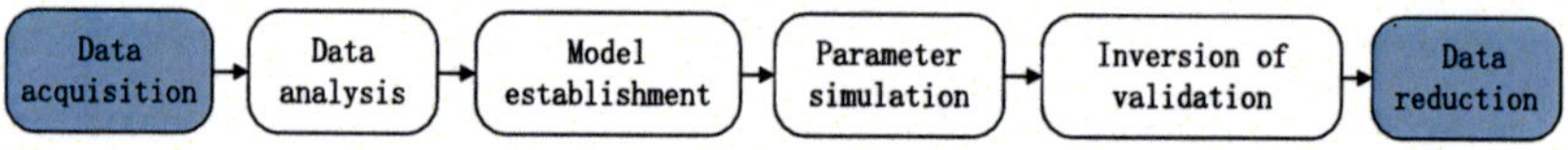

Fig. 4. Data expansion method based on parameter mode

BRDF model can be used to expand spectral dimension data. Many scholars have studied the establishment and optimization of BRDF model of ground features [15]. Among them, the semi empirical kernel driven model uses the linear combination of empirical coefficients and has simple expression and higher application range. It can not only obtain high accuracy within the range of shooting data, but also extrapolate to obtain the bidirectional reflection of arbitrary incident light and observation angle. The model has simple structure and fast inversion speed [16]. Rossthick-Lisparser model is a typical semi empirical kernel driven model. The BRDF described is relatively accurate and the result of inversion reflectivity is relatively accurate. The expression of Rossthick-Lisparser is:

$$R = k_1 R_S + k_2 R_D \tag{1}$$

$$R_S = f_{iso}(\lambda) + f_{vol}(\lambda) \times K_{vol} + f_{geo}(\lambda) \times K_{geo} \tag{2}$$

Considering that the light radiated to the object includes the directional incident light of the sun and the scattered light in the air, the target reflectivity R can be divided into two parts: the target reflectivity factor caused by direct light R_S and the target reflectivity factor caused by scattered light R_D. The spectral reflectance of ground objects in different weather environments can be realized by changing the ratio of k_1 to k_2. K_{vol} and K_{geo} are volume scattering nuclei and geometrical optics nuclei, respectively, These two quantities are only related to the zenith angle of detection, the zenith angle of incident light, and the relative azimuth between the light source and the detector. The three constant coefficients $f_{iso}(\lambda)$,$f_{vol}(\lambda)$ and $f_{geo}(\lambda)$ are only related to the wavelength. Therefore, the $f_{iso}(\lambda)$,$f_{vol}(\lambda)$ and $f_{geo}(\lambda)$ of typical features can be measured through known data, and the spectral reflectance of features can be retrieved in combination with the environmental angle conditions of the measured target, so as to expand the spectral data.

3.2.3 Spectral Data Expansion Based on Sample Learning

Compared with the method of establishing parameter model, the spectral data expansion method based on sample learning does not need to manually specify the learning feature mode, so it needs to input more a priori spectral data. Common learning based methods include decision tree method, Bayesian method, support vector machine, artificial neural network, deep learning and so on. Deep learning is a learning method based on deep neural network, which was officially started in 2006 [17, 18]. Since then, deep learning model has attracted extensive attention and developed rapidly. The commonly used deep neural network models include deep belief network (DBN), restricted boltzmann machine (RBM), convolutional neural network (CNN), etc. Deep learning is an end-to-end learning method. Relying on different models, it can extract deep-seated abstract features and complete various tasks through the combination of multi-layer features. Zhang Jingcheng et al. Proposed a near earth hyperspectral data expansion method applied to deep learning. Specifically, two methods of illumination disturbance and adding noise are used to simulate the spectrum of the original near earth hyperspectral data, which realizes the expansion of near earth hyperspectral data and effectively increases the amount of spectral data [19]. Gan Haiming and others analyzed the correlation between the texture features of the characteristic band image and the chlorophyll content, and combined the spectral features and texture features into the sparse self coding model in deep learning to predict the chlorophyll content of longan leaves. Experiments show that this method has good accuracy and robustness [20]. However, the spectral data expansion based on sample learning still needs a large number of a priori samples, and does not make full use of the physical and chemical characteristics of the object itself. It still needs to be developed in small sample learning and fusion of target physical attributes.

4 Conclusion

Hyperspectral image data of ground object includes one-dimensional spectral information and two-dimensional spatial information. Limited by imaging mode and detector level, it is very difficult to obtain a large number of hyperspectral image data samples of the same target in reality. However, in the tasks of hyperspectral image classification and target detection, sufficient samples are required as a priori knowledge to effectively improve the processing effect. Therefore, it is necessary to obtain more samples through data reconstruction and expansion. This paper summarizes the common methods of hyperspectral image spatial and spectral data expansion, but there are still many directions worthy of further research and improvement. Firstly, the data expansion method based on small samples needs to be improved. The current data expansion method still needs a large number of a priori data, but the purpose of data expansion is to use a small amount of measured data to obtain more samples as the basis for subsequent processing. The key to solve these two contradictions is to improve the data expansion method of small samples in addition to collecting more original data. The second is to improve the combination of a priori physical characteristics and later learning characteristics. The method of data inversion based on the model has a good effect on a certain kind of objects, but its universality is poor. The learning based method does not make full use of a priori physical information. Therefore, the integration of the two methods can effectively improve the integrity of features and the expanded data is more effective and stable.

References

1. Yu, R., Luo, Y.Q., Li, H.N., et al.: Three-dimensional convolutional neural network model for early detection of pine wilt disease using UAV-based hyperspectral images. Remote Sens **13**(20) (2021)
2. Zdravcheva, N.D.: Hyperspectral environmental monitoring. In: IOP Conference Series: Materials Science and Engineering, pp. 614 (2019)
3. Wang, J.C., Zhu, M.: Development of hyperspectral reconnaissance technology. Aerosp. Electron. Warfare **35**(03), 37–45 (2019)
4. Yang, Q.Q., Jin, C.Y., Li, T.W., et al.: Research progress and challenges of datadriven quantitative remote sensing. Natl. Remote Sens. Bull. **26**(2), 268–285 (2022)
5. Liu, X.F., Liu, J.M., Fu, M.: Generating countermeasure network extended samples for hyperspectral image classification. Electron. Meas. Technol. **45**(03), 146–152 (2022)
6. Zhang, Y., Hua, W.S., Huang, F.Y., et al.: Hyperspectral anomaly target detection based on spatial spectrum joint anomaly degree. Spectroscopy Spectral Anal. **40**(06), 1902–1908 (2020)
7. Li, W., Chen, C., Zhang, M., et al.: Data augmentation for hyperspectral image classification with deep CNN. IEEE Geosci. Remote Sens. Lett. 1–5 (2018)
8. Liu, X., Wang, C., Wang, H., et al.: Target detection of hyperspectral image based on faster R-CNN with data set adjustment and parameter turning. OCEANS 2019-Marseille (2019)
9. Shorten, C., Khoshgoftaar, T.M.: A survey on image data augmentation for deep learning. J. Big Data **6**(1) (2019)
10. Arad, B., Ben-Shahar, O., Timofte, R., et al.: NTIRE 2018 Challenge on Spectral Reconstruction from RGB Images. In: 2018 IEEE/CVF Conference on Computer Vision and Pattern Recognition Workshops (CVPRW). IEEE (2018)

11. Arad, B., Timofte, R., Ben-Shahar, O., et al.: NTIRE 2020 Challenge on Spectral Reconstruction from an RGB Image (2020)
12. Wang, Q.H., Hua, W.S., Huang, F.Y., et al.: Based on the spectral Angle background purification of hyperspectral anomaly detection algorithm. Laser Technol. **44**(5), 623–627 (2020). (in chinese)
13. Yang, L.H., Xu, J., Jiang, S.P.: Vacuum cryogenic environment reflector spectrum reflectance in situ measurement techniques. J. Appl. Opt. **36**(04), 559–565 (2015)
14. Li, X., Strahler, A.H.: Geometric-optical modeling of a conifer forest canopy. IEEE Trans. Geosci. Remote Sens. **GE-23**(5), 705–721 (1985)
15. Ding, A.X., Qiao, Z.D., Dong, Y.D., et al.: BRDF model integration and case analysis basedon linear kernel driven model. Remote Sens. Technol. Appl. **33**(03), 545–554 (2018)
16. Yan, G.J., Wu, J., Wang, J.D., et al.: Spectral prior knowledge in the application of remote sensing inversion of vegetation structure. J. Remote Sens. (01), 1–6 (2002)
17. Hinton, G.E., Osindero, S., The, Y.: A fast learning algorithm for deep belief nets. Neural Comput. **18**(7), 1527–1554 (2006)
18. Hinton, G.E., Salakhutdinov, R.R.: Reducing the dimensionality of data with neural networks. Science **313**, 504–507 (2006)
19. Zhang, J.C., Liu, P., Che, H.S., et al.: A near earth hyperspectral data expansion method for deep learning. CN 110070004A[P] (2019)
20. Gan, H.M., Yue, X.J., Hong, T.S., et al.: Hyperspectral inversion model for predicting chlorophyll content in Longan leaves based on deep learning. J. South China Agri. Univ. **39**(03), 102–110 (2018)

Quad-Polarimetric Reconstruction with Adaptive Volume Scattering Model from Compact PolSAR Data

Xiaopeng Zhao[1], Rui Guo[1(✉)], Hui Wang[2], and Liang Guo[3]

[1] School of Automation, Northwestern Polytechnical University, Xi'an 710072, China
gr2003@nwpu.edu.cn
[2] Shanghai Institute of Satellite Engineering, Shanghai 200240, China
[3] School of Physics and Optoelectronic Engineering, Xidian University, Xi'an 710071, China

Abstract. Compact polarimetric (CP) SAR has already been widely studied and instrumented in some spaceborne systems because it can achieve approximated quad-polarimetric information by reconstruction. In this paper, a new method based on three-component decomposition is proposed for the reconstruction which avoids the iteration operation and keeps better performance comparing to the traditional reconstruction models. In addition, the volume scattering model is improved to suit different terrain types. Experiment on the Gaofen-3 data validates the effectiveness of the proposed method with the improved model.

Keywords: Compact Polarimetric Synthetic Aperture Radar (CP SAR) · Reconstruction · Three-component decomposition

1 Introduction

To break through the limitation of system complexity and swath width of quad-pol SAR, CP SAR has developed rapidly in recent years both in theoretical research and system. To date, some satellites fixed CP mode have launched, such as RISAT-1 launched by India in 2012, ALOS-2 PALSAR by Japan in 2014, and RADARSAT Constellation Mission (RCM) by Canada in 2019 [1].

Although in practice most CP systems adopt the HP (hybrid polarization, transmits circular and receives linear) mode as abovementioned satellite systems, most theoretical research has been carried out on liner $\pi/4$ mode. The primary reconstruction model was proposed by Souyris et al. by using iteration calculation [2]. And then Nord et al. introduced a variable parameter to improve Souyris' model [3]. Collins got an empirically fitted parameter expression for this model from ocean data [4]. Yin et al. considered the spiral scattering component in the reconstruction for non-reflection symmetry case [5]. These methods are all solved by iterating cross-polarization and co-polarization coherence coefficient. To simplify the reconstruction iteration and improve the reconstruction accuracy, a new method based on three-component decomposition is proposed, meanwhile an adaptive volume scattering model is also introduced to fit different terrain types. Gaofen-3 data is employed for experiment to validate the effectiveness of the method and the model.

L. Wang et al. (Eds.): CHREOC 2022, LNEE 969, pp. 170–177, 2023.
https://doi.org/10.1007/978-981-19-8202-6_16

2 Relative Principles

Based on the assumption of reflection symmetry $\langle S_{HH} \cdot S_{HV}^* \rangle = \langle S_{VV} \cdot S_{HV}^* \rangle = 0$ and the following relationship as Eq. (1), Souyris et al. proposed the quad-pol data reconstruction from $\pi/4$ CP mode [2].

$$\frac{\langle |S_{HV}|^2 \rangle}{\langle |S_{HH}|^2 \rangle + \langle |S_{VV}|^2 \rangle} = \frac{1 - |\rho|}{N} \ and \ \rho = \frac{\langle S_{HH} \cdot S_{VV}^* \rangle}{\sqrt{\langle |S_{HH}|^2 \rangle \rho \langle |S_{VV}|^2 \rangle}} \tag{1}$$

where ρ reflects the coherence between the co-polarization channels, and $N = 4$ in Souyris' reconstruction model. However, $N = 4$ only corresponds to two special cases which are fully polarized state and completely nonpolarized state. Hence, Eq. (2) was proposed by Nord et al. to estimate the value of N, in order to optimize the quad-pol data reconstruction model [3].

$$N = \frac{\langle |S_{HH} - S_{VV}|^2 \rangle}{\langle |S_{HV}|^2 \rangle}. \tag{2}$$

3 The Adaptive Reconstruction Model for CP Data

In Souyris' reconstruction model, the core is the estimation of $|S_{HV}|^2$ and ρ in iteration way. To simply and accelerate the estimation of ρ while keeping the estimation accuracy, different polarization decomposition models has been employed in the reconstruction process [5]. As the classical model, the three-component decomposition by Freeman-Durden [6] is introduced here, where the polarimetric covariance matrix is decomposed into three terms corresponding to surface, double-bounce and volume scattering mechanism. And the covariance matrices of the three scattering mechanisms are as follows:

$$[C_s] = \begin{vmatrix} |\beta|^2 & 0 & \beta \\ 0 & 0 & 0 \\ \beta^* & 0 & 1 \end{vmatrix}, [C_d] = \begin{bmatrix} |\alpha|^2 & 0 & \alpha \\ 0 & 0 & 0 \\ \alpha^* & 0 & 1 \end{bmatrix}, [C_v] = \begin{bmatrix} 1 & 0 & 1/3 \\ 0 & 2/3 & 0 \\ 1/3 & 0 & 1 \end{bmatrix} \tag{3}$$

where C_s, C_v and C_d are for surface, double-bounce and volume respectively, α and β denote the value of $S_{HH} \cdot S_{VV}^*$ belonging to double-bounce and surface terms. From Eqs. (1) and (3), it achieves that the value of ρ for the three scattering terms are $\frac{\beta}{|\beta|} \cdot \frac{\alpha}{|\alpha|}$, and $\frac{1}{3}$. Then ρ in Eq. (1) can be represented as:

$$\rho = \frac{P_s}{span} \cdot \frac{\beta}{|\beta|} + \frac{P_d}{span} \cdot \frac{\alpha}{|\alpha|} + \frac{P_v}{span} \cdot \frac{1}{3}. \tag{4}$$

where *span* is the total scattering power, P_s, P_d, P_v are power for surface, double-bounce and volume respectively. All the parameters can be obtained by CP decomposition [7].

Then, substituting the estimated ρ from Eq. (4) into Eq. (1), $|S_{HV}|^2$ can be calculated. As ρ is a complex, it may loss the sign information by using $|\rho|$ in Eq. (1). Equation (1) can be refined as follows:

$$\frac{\langle |S_{HV}|^2 \rangle}{\langle |S_{HV}|^2 \rangle + \langle |S_{HV}|^2 \rangle} = \frac{1 - \mathrm{Re}(\rho)}{N}. \tag{5}$$

where $\mathrm{Re}(\cdot)$ means the real part of a complex.

3.1 Proposed Method

From above, our proposed reconstruction method can be summarized as shown in Fig. 1.

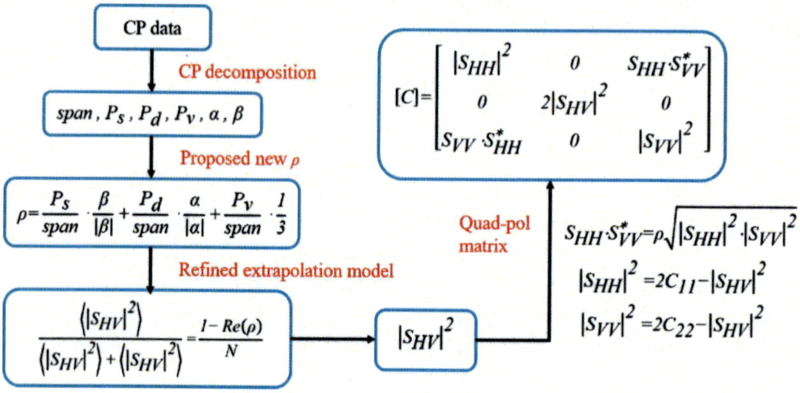

Fig. 1. Reconstruction flowchart of our proposed method.

3.2 Adaptive Volume Scattering Model

Since volume scattering term is a quite complicated mechanism, variable models have been considered to deal with it [8]. In the paper the following volume scattering model in Eq. (6) is adopted to improve the method in Sect. 3.1.

$$[C_V]_{quad} = \begin{bmatrix} 1 & 0 & b \\ 0 & 1-b & 0 \\ b & 0 & 1 \end{bmatrix}. \tag{6}$$

In this volume scattering model, an adaptive parameter b is utilized to suit different terrain types. b is defined as the degree of freedom parameter, and $b \in [0, 1]$. In Freeman-Durden decomposition, b equals to 1/3. And b is set as 0 for applying to the four-component decomposition [5]. So ρ in Eq. (4) can be redefined as

$$\rho = \frac{P_s}{span} \cdot \frac{\beta}{|\beta|} + \frac{P_d}{span} \cdot \frac{\alpha}{|\alpha|} + \frac{P_v}{span} \cdot b \tag{7}$$

Specially, in this paper the value of b is set depending on the polarization degree defined by Stokes vector [1], because $|S_{HV}|^2$ is validated correlated to the polarization degree in the researches [9]. To achieve adaptive parameter to suit different terrain, b is adjusted according to the polarization degree to increase or decrease the proportion of $|S_{HV}|^2$ components of volume scattering model. In the proposed model, b is set as 0 if Dop < 0.5 and 2/3 if Dop > 0.9. Otherwise, $\mathrm{b} = 1/3$ in accordance to Freeman's model.

4 Validation Experiment

The proposed method and the adaptive volume model are experimented on the Gaofen-3 data. The data is collected from the San Francisco with the size of 3000×2500 pixels as shown in Fig.2. The spatial resolution is 5.36m in azimuth and 2.25m in range. Fig.2 is the pseudo-color image based on Pauli basis. The CP data is simulated from the quad-pol data according to its specific scattering matrix [3].

Fig. 2. Pseudo-color image of GF3 data over San Francisco. Sited areas: city with red rectangle, ocean with green rectangle, forest with blue rectangle.

4.1 Model Validation

The proposed method in Eq. (4) is verified on the data. Figure 3 shows the comparison between the original value of ρ in Eq. (1) and the proposed ρ in Eq. (4). The scatter points should lie along the diagonal line to support the validity of the polarization state model. And it can be observed in Fig. 3 that most scatter points fall near the diagonal.

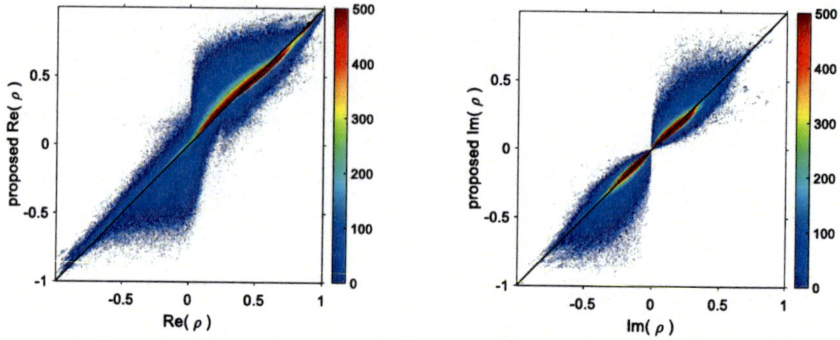

(a) The real part of ρ. (b) The imaginary part of ρ.

Fig. 3. Verification of the proposed method.

4.2 Reconstruction Result

The quad-pol reconstruction results with different models are illustrated in Fig. 4 in form of pseudo-color image. Compared with Fig. 2, the reconstruction results show a little poor performance in city area. In other area such as ocean and forest, all the results are more consistent with Fig. 2.

(a) Souyris' model. (b) Proposed method. (c) Adaptive model.

Fig. 4. Pseudo-color image of the reconstruction results.

4.3 Result Discussion

Since the performance of $|S_{HV}|^2$ reconstruction result is of great importance, the evaluation is given in Fig. 5 in form of the comparison between the real value and the reconstructed value. As more scatter points lying along the diagonal line implies better reconstruction performance of $|S_{HV}|^2$. It is obvious that our proposed method and model perform all better than Souyris' typical model, and the adaptive volume scattering model shows best in Fig. 5c.

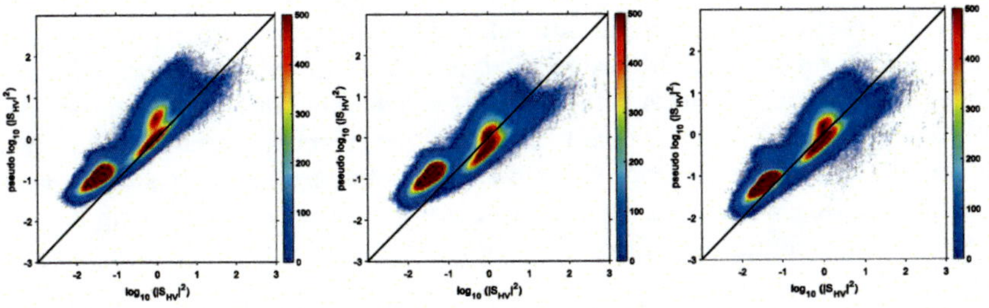

(a) Souyris' model. (b) Proposed method. (c) Adaptive model.

Fig. 5. Reconstruction result of $|S_{HV}|^2$.

For better evaluation, the logarithm relative error defined as Eq. (8) is calculated and given in Fig. 6. From the vision, more error value closes to zero in Fig.6c. Comparing to Fig.6a, the error value in the urban area is reduced in both Fig.6b, c.

$$\text{error}_{\log} = \left| \frac{\left(\log_{10}\left(|S_{HV}|^2_{\text{quad}}\right)\right) - \left(\log_{10}\left(|S_{HV}|^2_{\text{CP}}\right)\right)}{\log_{10}\left(|S_{HV}|^2_{\text{quad}}\right)} \right|. \tag{8}$$

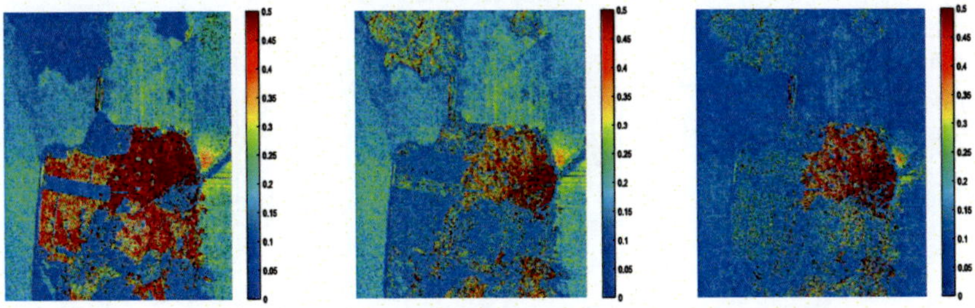

(a) Souyris' model. (b) Proposed method. (c) Adaptive model.

Fig. 6. Relative error image of $|S_{HV}|^2$.

Then three terrain types are selected in Fig. 2 for quantitative evaluation, where city is framed with red rectangle, ocean with green rectangle, forest with blue rectangle. Table 1 shows the mean value and standard variance value of the reconstruction error between the real and the reconstructed quad-pol parameters, and four parameters are involved, including $|S_{HH}|^2$, $|S_{VV}|^2$, $|S_{HV}|^2$ and $|\rho|$. The relative error is given by $\left|\left(|S_{HV}|^2_{\text{quad}} - |S_{HV}|^2_{\text{CP}}\right)/|S_{HV}|^2_{\text{quad}}\right|$, the same as other parameters. In Table 1, it is easy to conclude that the proposed methods could improve the reconstruction results. The error in proposed method is smaller and more stable. Herein, the adaptive model reduced the mean value of $|S_{HV}|^2$ reconstruction error from 2.2377 to 0.9455, meanwhile std. of this error is also quite low.

Table 2 illustrates the mean and std. Value of reconstruction error by using adaptive volume model on the three different terrain types. Compared with the averaged value in Table 1, the ocean area achieves the best reconstruction results due to the single scattering mechanism of ocean scene. Notably, the forest reconstruction result achieves better evaluation value than the average value in Table 1, implying that the model of adaptive volume scattering is more consistent with the real scene. However, more improvements need to be made for the city scene.

Table 1. Mean and Std. by using the three methods.

Model	Quantitive measurement	$\lvert S_{HH}\rvert^2$	$\lvert S_{VV}\rvert^2$	$\lvert S_{HV}\rvert^2$	$\lvert\rho\rvert$
Souyris' model	Mean	0.0729	0.0969	2.2377	0.0853
	Std	0.1387	0.1610	2.6897	0.1272
Proposed method	Mean	0.0686	0.0768	1.6277	0.0649
	Std	0.1512	0.1512	1.9006	0.1121
Adaptive model	Mean	0.0657	0.0745	0.9455	0.0641
	Std	0.1466	0.1461	1.3940	0.1120

Table 2. Mean and Std. on three terrain types

Terrain type	Quantitive measurement	$\lvert S_{HH}\rvert^2$	$\lvert S_{VV}\rvert^2$	$\lvert S_{HV}\rvert^2$	$\lvert\rho\rvert$
City	Mean	0.0667	0.1013	2.8150	0.1549
	Std	0.0694	0.0908	2.4301	0.1508
Ocean	Mean	0.0129	0.0116	0.8387	0.0041
	Std	0.0151	0.0146	0.5356	0.0028
Forest	Mean	0.1644	0.1751	0.3970	0.1217
	Std	0.1862	0.1898	0.1697	0.1044

5 Summary

A new pseudo-quad-pol reconstruction method with adaptive volume scattering model is proposed in this paper. According to experiment on Gaofen-3 data, the proposed method can improve the reconstruction performance while simplifying the reconstruction iteration and enhancing the reconstruction accuracy.

Acknowledgement. This work was supported by the Shanghai Aerospace Science and Technology Fund under grant SAST2020–028.

References

1. Xu, L., Hong, Z., Chao, W., et al.: Progress in the processing and application of compact polarimetric SAR. J. Radars **9**(1), 55–72 (2020)
2. Souyris, J.C., Imbo, P., Fjortoft, R., et al.: Compact polarimetry based on symmetry properties of geophysical media: the π/4 model. IEEE Trans. Geosci. Remote Sens. **43**(3), 634–646 (2005)
3. Nord, M.E., Ainsworth, T.L., Lee, J.S., et al.: Comparison of compact polarimetric synthetic aperture radar modes. IEEE Trans. Geosci. Remote Sens. **47**(1), 174–188 (2009)
4. Yin, J., Yang, J., Zhang, X.: On the ship detection performance with compact Polarity. In: 2011 IEEE Radar Con (RADAR), pp. 675–680. IEEE (2011)
5. Collins, M.J., Denbina, M., Atteia, G.: On the reconstruction of quad-pol SAR data from compact polarimetry data for ocean target detection. IEEE Trans. Geosci. Remote Sens. **51**(1), 591–600 (2013)
6. Freeman, A., Durden, S.L.: A three-component scattering model for polarimetric SAR data. IEEE Trans. Geosci. Remote Sens. **36**(3), 963–973 (1998)
7. Rui, G., Wei, H., Shuangxi, Z., et al.: Analysis of three-component decomposition to compact polarimetric synthetic aperture radar. IET Radar Sonar Navig. **8**(6), 685–691 (2014)
8. Freeman, A.: Fitting a two-component scattering model to polarimetric SAR data from forests. IEEE Trans. Geosci. Remote Sens. **45**(8), 2583–2592 (2007)
9. Espeseth, M.M., Brekke, C., Anfinsen, S.N.: Hybrid-polarity and reconstruction methods for sea ice with L- and C-Band SAR. IEEE Geosci. Remote Sens. Lett. **13**(3), 467–471 (2016)

A High-Precision Monolithic 3D Modeling Method Based on UAV Tilt Photography

Changliang Ha[1], Guangrui Yu[1](✉), Lili Yu[2], and Shuqi Wang[3]

[1] 32023, Troops, Dalian 116023, People's Republic of China
823836547@qq.com

[2] China Nuclear Industry Fifth Construction Co Ltd., Shanghai 201512, People's Republic of China

[3] The Yellow River Delta Chambroad Institute Co Ltd., Binzhou 256600, People's Republic of China

Abstract. In the construction and application of UAV tilt photography model, managing and operating a single building model is a new requirement in the engineering field. However, when traditional tilt photogrammetry obtains orthophoto and tilt image data at the same time, the collection of low altitude facade texture of buildings is often not fine enough, and most buildings or feature models are adjacent, which is not conducive to separate management and use. In order to solve the above problems, a high-precision 3D monomer modeling method is proposed in this paper. Firstly, combined with the geometric characteristics and resolution requirements of buildings, the design scheme of three-dimensional surround route is studied, and the global high-resolution image data is collected. Then, in the monomer modeling process, the point cloud of the tilt photography model is divided into virtual grids, and the point cloud in each virtual grid is regarded as a whole. The point cloud of the building facade is extracted by setting the height difference, and the contour information is further extracted after the point cloud classification processing. Finally, the 3D construction of a single building model is realized by extracting the intersection triangle of contour lines. The experimental results show that this method can automatically and efficiently realize the monomer construction of tilt photography building model, and has a certain application value for engineering construction and development.

Keywords: UAV · Tilt photogrammetry · Monomer model · Three dimensional surround · Virtual grid

1 Introduction

With the continuous breakthrough and development of aerospace and sensor technology, the difficulty of remote sensing data acquisition is gradually reduced, and geographic information products are also developing in multiple directions. As a product of the new era, 3D building model has played an extremely important role in the fields of digital city, disaster assessment, urban planning and so on [1–3]. Because the three-dimensional model needs to accurately describe the external geometry and texture features of the

L. Wang et al. (Eds.): CHREOC 2022, LNEE 969, pp. 178–192, 2023.
https://doi.org/10.1007/978-981-19-8202-6_17

building, the difficulty of monomer fine modeling increases. In particular, the rapid acquisition of the complete image data of the building is also the key to determine the modeling quality [4]. Generally, according to the different types of remote sensing data, there are three three-dimensional model construction methods in the market: one is to use two-dimensional vector data to construct the three-dimensional model, which can realize the three-dimensional stretching of the two-dimensional plane base map to a certain height. At the same time, it can also add side texture information to make the model realistic. Representative software include 3ds Max and Auto CAD. This method can not express accurate building details, And the texture adding process is too much human participation and time-consuming [5]; Second, the three-dimensional model is constructed by using lidar technology. After the high-precision building white model is produced by using lidar data, the three-dimensional model production is completed by using texture mapping. The main workload of this method is to obtain and match the texture information of the top and side of the building, which is very difficult [6]; Third, the tilt photogrammetry technology is used to construct the three-dimensional model. The UAV platform is equipped with a camera to obtain the multi view aerial image data of the building with overlapping range, and the three-dimensional model is constructed through photogrammetry and machine vision theory. Compared with this method, this method has low cost and high efficiency, but the tilt photogrammetry of UAV mostly relies on the flight mode of strip route to collect data, This will undoubtedly miss the detailed features of the side part of the building, resulting in local distortion of the model, which needs to be repaired in the later stage to meet the modeling requirements [7, 8].

From the above three-dimensional model construction methods, it can be seen that the application of UAV tilt photogrammetry technology is a more convenient and efficient means of image acquisition and model construction, and has been applied to many fields, such as positioning services, disaster assessment and archaeological research [9]. In recent years, there have been many successful cases of oblique photography to build a three-dimensional city abroad, and the reasonable use of ground street scenes and manual image acquisition have corrected the problems of unclear texture and thin near ground structure of the three-dimensional model. However, this undoubtedly increases the means of data acquisition and leads to a great increase in the workload. Moreover, the three-dimensional model is a continuum containing buildings and other ground objects, It is impossible to edit and manage a single target, which is limited in the application of relevant GIS fields [10]. Some domestic scholars put forward the concepts of physical monomer and logical monomer, and use relevant cutting methods to complete the stripping of monomer targets. However, to cut the building model, it is necessary to obtain the building contour through human-computer interaction. However, the existing contour extraction methods are greatly disturbed by ground object factors such as shadow, occlusion and vegetation [11].

In view of the lack of acquisition of near ground high-resolution image data in the tilt photography process of UAV, and the inability of automatic construction of a single three-dimensional building model, this paper proposes a three-dimensional surround route planning and design combined with the structural characteristics of buildings, so as to obtain the multi-directional image data of highland space at one time. In the process of 3D model construction, a method of automatically extracting a single entity from 3D model is studied, which avoids the complex process of manual drawing range, and the accuracy of extracting contour is higher. Finally, the purpose of managing 3D building model alone is achieved.

2 Three Dimensional Surround Route Planning and Design

In order to collect all the texture information of the building surface to the greatest extent, this paper proposes a three-dimensional surround route planning and design scheme. Taking the single lens rotor UAV as an example, tilt photography is implemented on the building, and the surround flight mode is adopted to obtain uniform high-resolution image data.

2.1 Project Design

In order to reasonably plan the flight route of UAV, first conduct field survey on the building to determine the minimum cuboid size that can accommodate the building and its ancillary facilities, and set the length, width and height of the rectangle as a, b, c respectively, and then determine the minimum external semi sphere according to the geometric parameters of the cuboids. As shown in Fig. 1, the radius R_0 can be calculated by the following formula:

$$R_0 = \frac{\sqrt{a^2 + b^2 + (2c)^2}}{2} \tag{1}$$

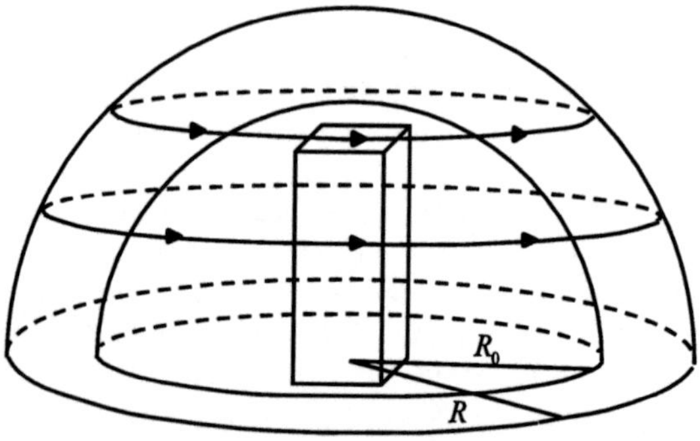

Fig. 1. Building external hemisphere and route planning

Taking the outer surface of the above minimum external hemisphere as the projection benchmark, the photographing distance is determined by imitating the ground sampling standard. Assuming that the sampling resolution of the building is GSD, the focal length of the camera is f and the pixel size is μ, the photographing distance D_0 can be expressed as:

$$D_0 = \frac{f \times GSD}{\mu} \tag{2}$$

Thus, it can be seen from Fig. 1 that the layout surface of the UAV inclined photography route is a new hemispherical surface formed by the outward moving distance D_0, and the radius is R, which can be calculated by the following formula:

$$R = R_0 + D_0 \tag{3}$$

2.2 Route Planning

After the hemispherical geometric size, camera parameters, overlap and flight speed are known, the inclined route planning parameters such as camera tilt angle, aerial photography altitude, flight radius, photography baseline and exposure interval can be calculated [12].

The first is the calculation of the initial route parameters. In order to obtain the complete near ground image of the building, the critical state is that the photographic light can just reach the intersection of the minimum external sphere and the ground. The profile of the imaging conditions is shown in Fig. 2. If the camera lens angle is α, the angle value corresponding to the main optical axis can be obtained from the sine theorem, and then the sum of the internal angles of the triangle is 180 degrees, and then the camera tilt angle α_1 can be calculated, as shown in the following formula:

$$\alpha_1 = 180° - \arcsin\frac{R \times \sin\frac{\alpha}{2}}{R_0} - \frac{\alpha}{2} \tag{4}$$

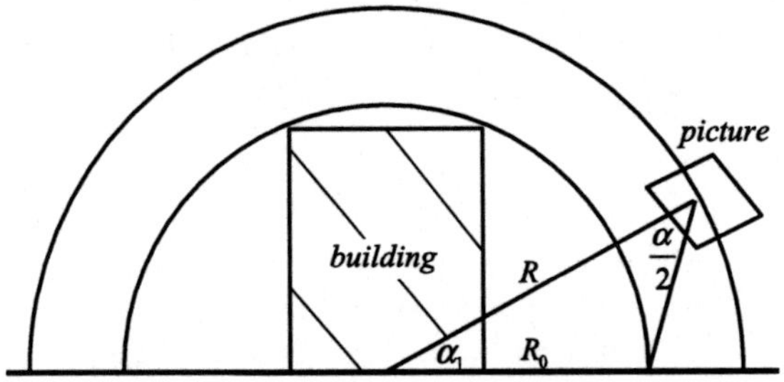

Fig. 2. Oblique photography imaging condition section

Thus, the flight altitude H_1 and surrounding radius R_1 of the UAV can be calculated, as shown in the following formula:

$$H_1 = R \times \sin\alpha_1 \tag{5}$$

$$R_1 = R \times \cos\alpha_1 \tag{6}$$

Assuming that the number of heading pixels of UAV image is P_x and the overlap is O_x, the photography baseline length B_1 can be calculated by the following formula:

$$B_1 = \frac{(1 - O_x)\mu P_x D_0}{f} \tag{7}$$

On the initial route, the UAV flies along the circle with radius R_1, and the top view is shown in Fig. 3. When the UAV flies from one exposure point to another, the photography interval angle β_1 can be calculated from the photography baseline and route radius:

$$\beta_1 = 2\arcsin(\frac{B_1}{2R_1}) \tag{8}$$

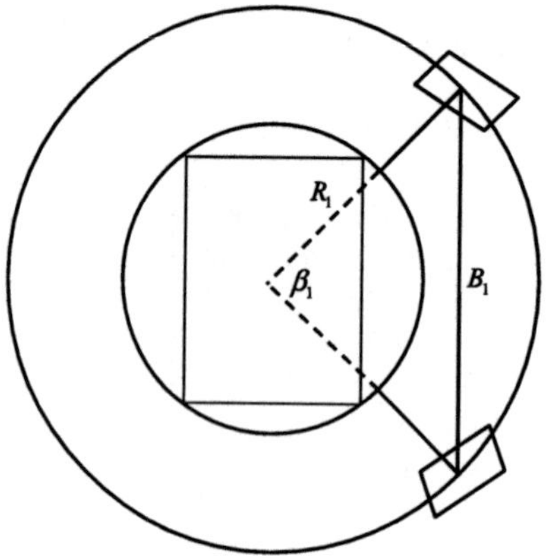

Fig. 3. Top view of inclined photography route

When the UAV takes aerial photos at approximately constant speed, in order to ensure that the overlap meets the requirements, set the maximum flight speed as V_{max}, and the photography interval t can be calculated by the following formula:

$$t = \frac{\pi \times R_1 \times \beta_1}{V_{max} \times 180^{\circ}} \tag{9}$$

Then it is the calculation of other route parameters, which mainly determines the relevant route information by setting the lateral overlap degree on the basis of the previous circular route. The camera tilt angle α_n can be calculated by the following formula:

$$\alpha_n = \alpha_{n-1} + \alpha_1(1 - O_y) \tag{10}$$

where, α_n, α_{n-1} represent the inclination angle of the current route and the previous route respectively, and O_y is the lateral overlap. After calculating the inclination angle of any other route, other route parameters such as flight height, surrounding radius, baseline length and photography interval can be calculated according to Eqs. (7)–(11). Until the image acquisition at the top of the building is completed, the flight ends. The mathematical judgment criteria for the end of flight are as follows:

$$H_n > c \tag{11}$$

$$R_n < \sqrt{R^2 - c^2} \tag{12}$$

3 Realization Method of Building Model Monomer

4 Monomer Modeling Process

The monomer modeling method in this paper is divided into two parts, that is, extracting the building contour and realizing a single entity by cutting triangular mesh. Figure 4 shows the whole process of modeling. In order to extract the building contour, the point cloud data extracted from the tilt photography model is divided into fixed size virtual grids. Then, the height difference of point cloud in each grid is calculated, and the ground and vegetation are removed by setting the height difference threshold. After clustering the point cloud data, the building boundary point cloud can be drawn according to the clustered point cloud data. In order to obtain a single entity from the tilt photography model, the cutting polygon is determined by the constructed building contour. The boundary box of the cutting polygon intersects with the boundary box of the first level of detail model (LOD) of the tilt photography model, and the intersection model is extracted. The triangle of the intersecting model is divided into three parts, namely, the inner part, the outer part and the part intersecting with the cutting polygon. Among them, the triangle generated by cutting is internally merged into a single model, and the external model is modified into the original model file. The construction of a single solid model can be realized by traversing all LODs of the tilt photography model.

4.1 Building Contour Extraction Based on Tilt Photography Model

4.1.1 Virtual Grid Generation

Point clouds are dense in areas with obvious features and sparse in areas with no obvious features. Unevenly distributed point clouds will affect the extraction of building contour [13]. In order to improve the calculation efficiency and accuracy of building contour, it is necessary to construct the neighborhood relationship of point cloud. In this paper, the virtual grid is constructed according to the point cloud for subsequent analysis and processing.

Enter the boundary box of the point cloud to obtain the minimum point coordinate $(x_{\min}, y_{\min}, z_{\min})$ and the maximum point coordinate $(x_{\max}, y_{\max}, z_{\max})$ of the point cloud area. The number of rows and columns of the whole grid is calculated as follows:

$$\begin{cases} r = \dfrac{(x_{\max} - x_{\min})}{w} \\ c = \dfrac{(y_{\max} - y_{\min})}{w} \end{cases} \tag{13}$$

where w is the size of the virtual grid considering the point cloud density. The number of rows and columns of the virtual grid at any point $i(x_i, y_i)$ can be calculated as:

$$\begin{cases} r_i = \dfrac{(x_i - x_{\min})}{w} \\ c_i = \dfrac{(y_i - y_{\min})}{w} \end{cases} \tag{14}$$

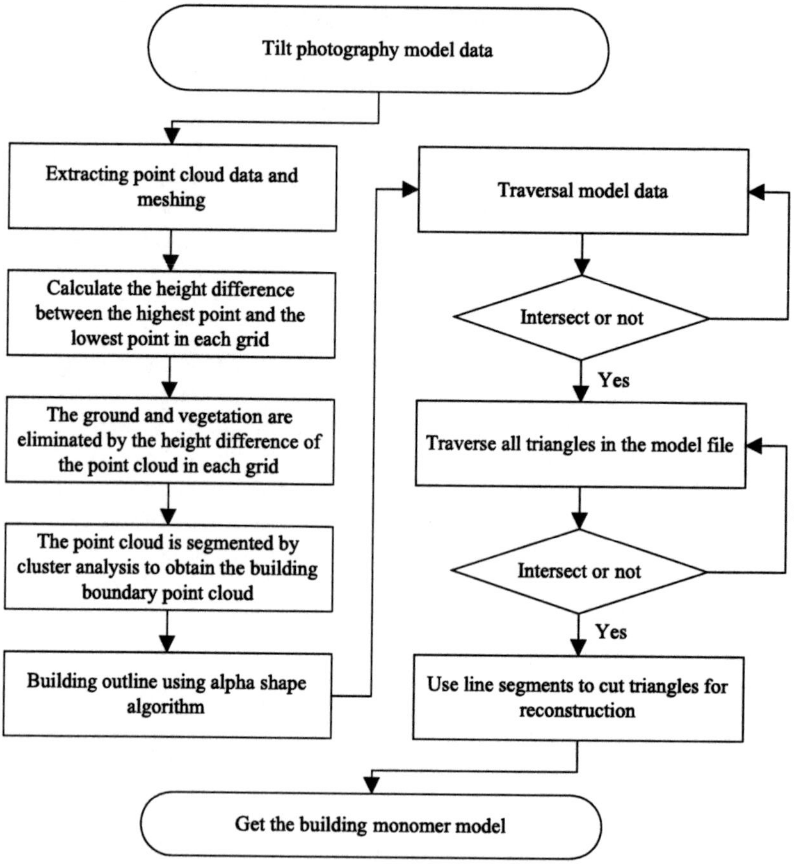

Fig. 4. Monomer modeling flow chart

It can be seen that all point cloud data are finally divided into several grids of the same size. The point cloud data contained in each grid is represented by the central point cloud of the grid, which can solve the problem of large amount and uneven distribution of point cloud data.

4.1.2 Building Elevation Point Cloud Extraction Based on Height Difference

The biggest difference between the building and the ground is reflected in the elevation. In the inclined photography model, the elevation change between adjacent points in the plane is not obvious, but the elevation of the building edge will change significantly due to the influence of the building wall [14]. Therefore, by setting the elevation difference threshold and traversing all grids, the point cloud with the elevation difference lower than the threshold can be deleted from the grid, which eliminates most plane point clouds, including ground points, building top surface and most vegetation, and then saves the facade point cloud data of the building.

4.1.3 Building Elevation Point Cloud Extraction Based on Height Difference

Existing point cloud classification algorithms include DBSCAN, K-means, etc. [15]. This paper mainly uses the density based DBSCAN clustering method to divide the region with high enough density into multiple clusters. Different from K-means, this method does not need to know the number of clusters in advance, and can identify noise points, and then use Alpha Shapes algorithm to construct the external contour of each building from the classified point cloud.

4.2 Obtain a Single Entity by Cutting Triangulation

4.2.1 Triangle Intersection Detection and Classification

The oblique photography 3D model contains a large number of triangles. If each of these triangles intersects the cutting polygon formed by the building contour, the processing efficiency will be greatly affected [16]. Therefore, this paper deals with the following processes: (1) determine the intersection relationship between polygon bounding box and model bounding box; (2) Identify the model where the bounding box intersects with the cutting polygon bounding box; (3) Detect the intersection of triangular surfaces; (4) These triangles can be divided into three types: within region, outside region and intersecting part; (5) The intersecting triangular surfaces are cut, and the results are divided into two categories: within range and out of range. The part outside the triangle area will be deleted, and the part inside the area will be retained as a single solid model.

4.2.2 Cutting Triangulation

The relationship between cutting line and triangle can be divided into three cases: triangle and cutting line are tangent, triangle and cutting line intersect, and triangle and two cutting lines intersect. Figures 5 and 6 show all cases of cutting triangles with a single cutting line and two cutting lines respectively.

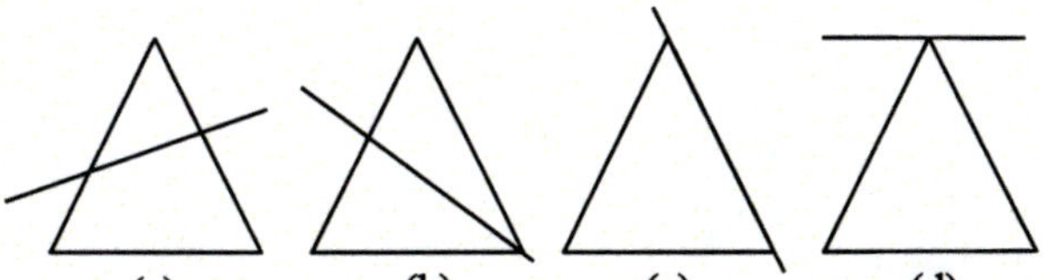

(a) (b) (c) (d)

Fig. 5. Cutting triangles with a single cutting line

The cases where a single cutting line cuts a triangle include: (1) the cutting line intersects two sides of the triangle; (2) The cutting line intersects one vertex and one side of the triangle; (3) The cutting line is tangent to one side of the triangle; (4) The cutting line is tangent to a point of the triangle. The cases of two cutting lines cutting a triangle include: (1) the vertex of the cutting polygon is located inside the triangle, and the two edges intersect with the two points of the triangle; (2) The vertex of the cutting polygon is located inside the triangle, and the two edges intersect one side of the

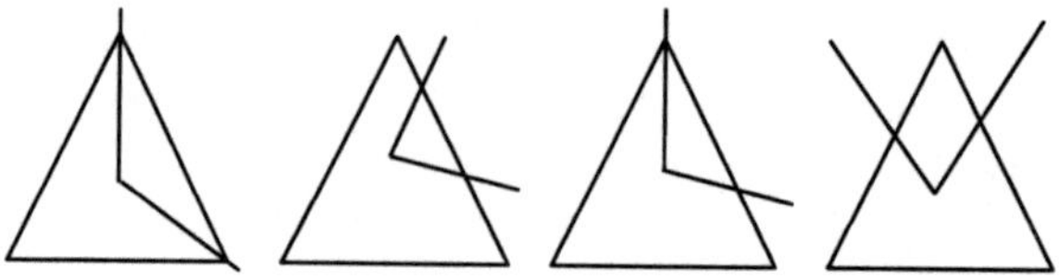

Fig. 6. Two cutting lines cut the triangle

triangle; (3) The vertex of the cutting polygon is located inside the triangle, and the two edges intersect with one point and one side of the triangle; (4) The vertices of the cut polygon are located inside the triangle, the vertices of the triangle are located inside the cut polygon, and both sides intersect with both sides of the triangle.

In the cutting process, the principle of cutting the smallest geometric element shall be followed to ensure that the subsequent geometric features are still met. Since the oblique photography model used in the text contains triangular slices, the topological relationship between triangular slices should remain unchanged after cutting the minimum unit.

4.2.3 Construction of Monomer 3D Model

Since a single entity cutting triangles is a physical entity, it is necessary to increase the number of vertices and triangles and reconstruct the data in the 3D model. Firstly, the above intersection points generated in the cutting process will become the new vertices of the 3D model, and a single triangular slice will be divided into multiple slices. In order to achieve the display effect of the 3D model, it is necessary to calculate the texture coordinates of the newly added vertices. Then, the intersecting triangle slices are deleted from the model, and the newly generated triangle slices are divided into two types, namely, outside the region and inside the region. All triangles in the cutting polygon will be fused into a single model, and the external triangles will be fused into the original scene. In the case of single entity modeling, it does not affect the overall display of the original model.

5 Experiment and Analysis

5.1 Data Acquisition

This paper uses Dajiang spirit 4 RTK rotor UAV as the aerial photography platform, the camera focal length is 8mm and the image resolution is 4864 × 3648, a real three-dimensional modeling experiment was carried out on a high-rise building. Firstly, through field investigation, understand the basic geometric information of the building, and determine the minimum external cuboid size and the geometric center position of the bottom; Then, in order to meet the requirements of imaging and modeling accuracy, set the image spatial resolution to 0.15 cm, the heading overlap to 80%, and the side overlap to 60%. Combined with the above three-dimensional surround route planning model, complete the calculation of flight parameters, and the UAV will complete the acquisition of building tilt images according to the designed route; Finally, the image and POS data are imported into the context capture center software. Through data preprocessing, aerial triangulation, dense matching and other processes, the dense point cloud and tin triangulation of buildings can be obtained, and the texture is mapped to the corresponding white model to produce the three-dimensional model in osgb data format. The osgb model file stores multi-source data information, such as model vertex coordinates, texture coordinates and texture pictures. The model vertex coordinates and triangulation data are easy to extract, which is convenient for subsequent model processing.

5.2 Monomer Model Construction

5.2.1 Building Contour Extraction Results

The vertex information is extracted from the original data as the initial point cloud, and then the ground and vegetation points are eliminated by setting the virtual mesh and height difference threshold. According to experience, there is no intersection and inclusion between the contours of independent buildings [17]. Therefore, the clustering results affected by the internal feature information of buildings can be eliminated, and other interference clusters can be eliminated according to the shape and area of the contour polygon.

Different mesh size and height difference threshold will have a great impact on the extraction effect of point cloud. Through the control variable method, the experiment studied the influence of height difference threshold and grid size on the point cloud extraction effect. The height difference threshold was set as 12 m, 15 m, 18 m, 21 m and 24 m respectively, and the grid size was set as 0.5 m, 1 m and 2 m respectively, so as to obtain different point cloud extraction effects, as shown in Table 1.

It can be seen that in this study, setting the height difference threshold to 18m and the grid size to 1m can achieve the effect of point cloud extraction with clear outline. Then, the DBSCAN algorithm is used to segment the above point cloud, and the alpha shape is used to construct the peripheral contour for each point cloud cluster. Combine the building contour information with the elevation difference information of each virtual grid, draw the building boundary box, and overlay it on the oblique photography 3D model. The effect is shown in Fig. 7.

5.2.2 Implementation of Monomer Model Based on Tilt Photography Model

Using the above method, the outline of the building can be effectively obtained, which can be used as a polygon cutting inclined photography model to separate a single building model from the original scene and realize single building modeling. Figure 8 shows the traditional tilt photogrammetry model and building monomer model. It can bc seen that the monomer model produced by this method is relatively regular, and the monomer model is separated from the original scene.

5.3 Quality Analysis

In order to verify the advantages of this method in the three-dimensional monomer modeling process, the quality evaluation is carried out from three aspects: model integrity,

Table 1. Set the grid size and elevation difference threshold to obtain different point cloud extraction effects

Height difference threshold (m)	Grid size		
	0.5 m	1 m	2 m
12	Many noise point clouds	Many noise point clouds	The point cloud is too sparse
15	Many noise point clouds	Many noise point clouds	The point cloud is too sparse
18	The point cloud density is too high, but the outline point cloud is clear	The point cloud has moderate density and clear outline	The point cloud is too sparse
21	The contour point cloud is obviously missing	The contour point cloud is obviously missing	The contour point cloud is obviously missing
24	The contour point cloud is obviously missing	The contour point cloud is obviously missing	The contour point cloud is obviously missing

(a) Global bounding box

(b) Local bounding box

Fig. 7. Draw a bounding box using outline lines

detail characteristics and relative accuracy [18]. First, browse the details of the monomer model shown in Fig. 8b from a visual point of view. The three-dimensional model covers the overall structure of the building, there is no lack of texture, and the tone is uniform, as shown in Fig. 9. It can be seen that the three-dimensional surround tilt photogrammetry proposed in this paper solves the problem of obtaining high and low altitude high-resolution images, and improves the texture integrity of buildings.

In the process of single building model management, while ensuring the integrity and accuracy of texture features, the mathematical accuracy of the model is often the focus

(a) Tilt photography model (b) Monomer model

Fig. 8. Achieve the effect of a single entity

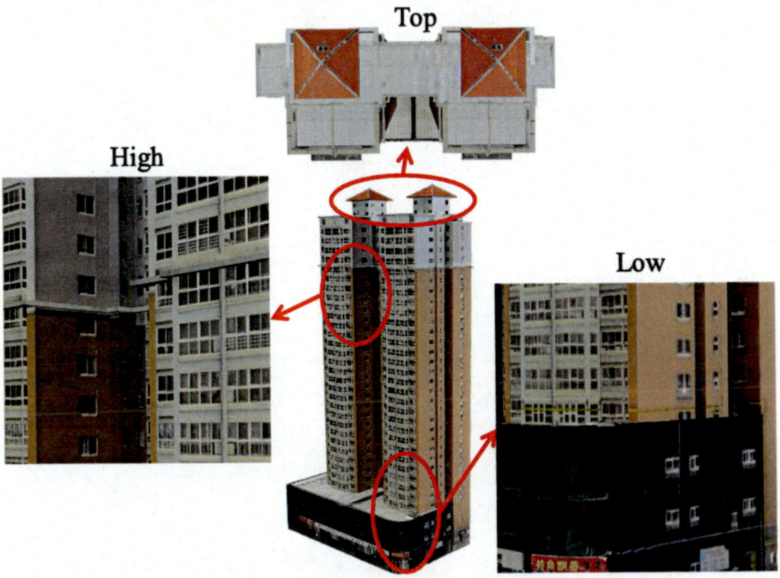

Fig. 9. Detail display of single building model

of attention [19]. In order to verify the relative accuracy of the model, this paper makes a comparative analysis combined with the comparison of indoor and field measurements. Firstly, the model is proportional to the actual building, a certain number of feature points are picked up, and the horizontal distance and height difference between 8 two points are obtained through field measurement; Then, the same points are picked up in the model through the office work, and the corresponding horizontal distance and height difference are read; Finally, calculate the difference between the two groups of data, and the statistical results are shown in Fig. 10.

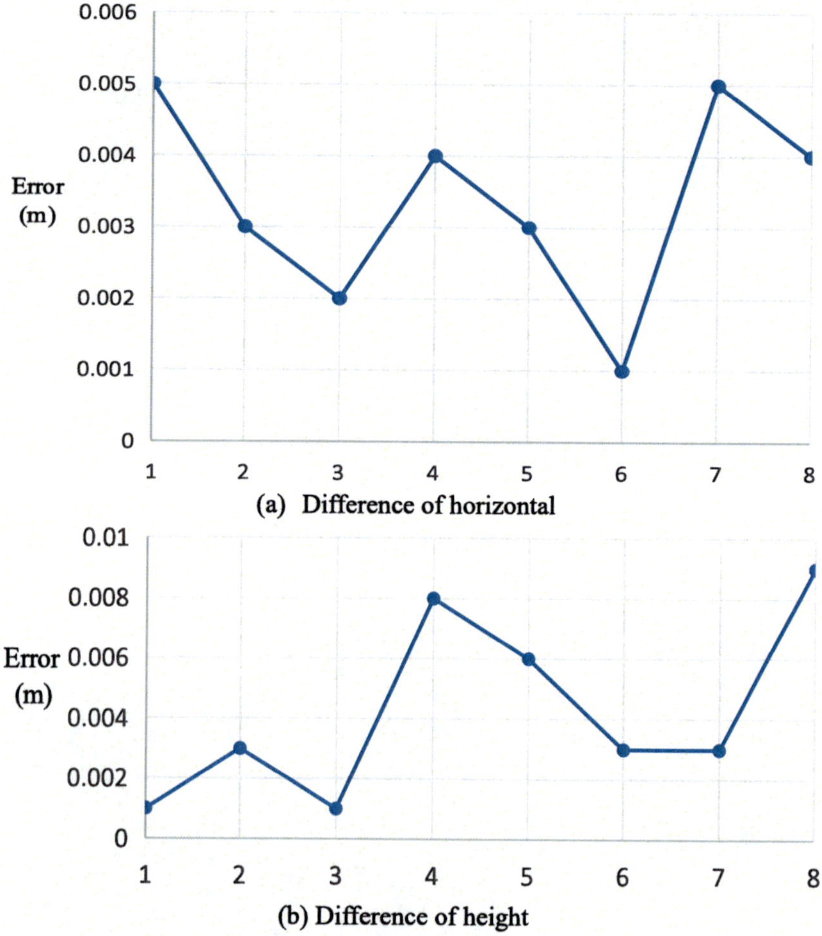

Fig. 10. Statistical results of accuracy of single building model

According to the above statistical results, the average value of the horizontal distance difference is 0.0034 m and the standard deviation is 0.0013 m; The average value of elevation difference is 0.0043 m and the standard deviation is 0.0029 m. If the measured value is taken as the true value, according to formula (15) [20], it can be calculated that the mean square error of horizontal distance measurement of single model is ±0.0036 m and the mean square error of height difference measurement is ±0.0051 m.

$$m = \pm\left(\sqrt{V^T V / n}\right) \tag{15}$$

Therefore, the three-dimensional surround route planning and design and virtual grid construction strategy adopted in this method can better realize the construction of single building model, with complete expression and detailed texture. The mathematical accuracy of the model meets the standard design requirements, which is conducive to project management and further application.

6 Conclusion

In order to solve the problems of imprecise image acquisition and the connection between building model and ground objects in the process of single building modeling of high-rise buildings, this paper puts forward solutions from two aspects: image acquisition and model construction. In the process of image acquisition, a three-dimensional surround route planning scheme is designed, which improves the resolution of the image of the building from near ground to high altitude, and promotes the accuracy and integrity of the three-dimensional model. Then, aiming at the multi-source information stored in the OSGB panoramic 3D model, the single building model is separated from the original scene by means of virtual grid division, height difference setting, contour extraction and so on. Experiments show that the monomer modeling method in this paper avoids most human intervention and has high efficiency. The constructed monomer model has good integrity and texture features, and the geometric accuracy has reached the engineering application standard.

References

1. Sanz-Ablanedo, E., Chandler, J.H., Rodriguez-Perez, et al.: Accuracy of Unmanned Aerial Vehicle (UAV) and SfM photogrammetry survey as a function of the number and location of ground control points used. Remote Sens.**10**(10), 1606 (2018)
2. Su, M.S., Hwang, W.L., Cheng, K.Y.: Analysis on multiresolution mosaic images. IEEE Trans. Image Process. **13**(7), 952–959 (2004)
3. Turner, D., Lucieer, A., Wallace, L.: Direct georeferencing of ultrahigh-resolution UAV imagery. IEEE Trans. Geosci. Remote Sens. **52**(5), 2738–2745 (2014)
4. Peng, L., Jianfang, M.: Research on 3D monomer modeling technology of fusion point cloud and tilt photography. Bull. Surv. Map. **S1**, 29–32 (2021)
5. Xiaogang, S.: Research on real estate 3D monomer modeling and application of UAV tilt photogrammetry. Anhui University of Technology (2020)
6. Ying, C., He, H., Yi, L., et al.: Elevation construction of urban buildings based on multi baseline close range photogrammetry. Beijing Surv. Map. **34**(12), 1746–1749 (2020)
7. Zheng, X., Wang, F., Li, Z.: A multi-UAV cooperative route planning methodology for 3D fine-resolution building model reconstruction. ISPRS J. Photogramm. Remote. Sens. **146**, 483–494 (2018)

8. Zan, N., Lin, W., Huangshan et al.: Study on monomer modeling of tilt photography. Geospatial Inf **17**(03), 113–114 (2019)
9. Ning, L., Jian, M., Yongchong, Y. et al.: Monomer fine 3D modeling technology of tilt photography. Remote Sens. Inf. **35**(06), 44–48 (2020)
10. Kim, J., Kim, T., Shin, D., et al.: Fast and robust geometric correction for mosaicking UAV images with narrow overlaps. Int. J. Remote Sens. 2557–2576 (2017)
11. Jhan, J.P., Rau, J., Huang, C.: Band-to-band registration and ortho-rectification of multi-lens/multispectral imagery: a case study of MiniMCA-12 acquired by a fixed-wing UAS. ISPRS J. Photogramm. Remote. Sens. **114**, 66–77 (2016)
12. Masek, J.: Automated registration and orthorectification package for Landsat and Landsat-like data processing. J. Appl. Remote Sens. **3**(1), 1–20 (2009)
13. Turner, D., Lucieer, A., Watson, C.: An automated technique for generating georectified mosaics from ultra-high resolution Unmanned Aerial Vehicle (UAV) imagery, based on Structure from Motion (SfM) point clouds. Remote Sens. **4**(5), 1392–1410 (2012)
14. Remondino, F., Barazzetti, L., Nex, F., et al.: UAV Photogrammetry for mapping and 3D modeling–current status and future perspectives. Int. Arch. Photogrammetry Remote Sens. Spatial Inf. Sci. **38–1**(C22), 25–31 (2011)
15. Zhimin, H., Yunfei, S., Lingling, Z., et al.: Improved prism model expression of building cluster objects. Surv. Map. Sci. **46**(10), 137–144 (2021)
16. Luyuan, L., Qingsong, Y., Yingjie, Q. et al.: 3D modeling of urban buildings with multi view three-dimensional grid. Surv. Map. Sci. **46**(08), 112–119 (2021)
17. Leutenegger, S., Chli, M., Siegwart, R.Y.: BRISK: Binary Robust Invariant Scalable Keypoints. ICCV, Barcelona, Spain (2011)
18. Yaoyao, Y.: Research on key technology of UAV image fast mosaic. PLA Information Engineering University (2012)
19. Xue, W.: Research on UAV image positioning optimization technology. PLA Information Engineering University (2017)
20. Fang, T., Jiahao, S., Haoshao, L.: Three dimensional geographic information modeling and analysis of UAV aerial survey tilt photogrammetry. Energy Environ. Protect. **43**(05), 79–83 (2021)

SAR Image Classification by Combining Transformer and Convolutional Neural Networks

Zi-An Yang, Nai-Rong Zheng, and Feng Wang(✉)

Key Laboratory for Information Science of Electromagnetic Waves (Ministry of Education), School of Information Science and Technology, Fudan University, Shanghai 200433, China
fengwang@fudan.edu.cn

Abstract. SAR image land cover classification can achieve the pixel-level category according to different characteristics of the target. A novel multi-scale feature extraction network combining transformer and convolutional neural networks (CNN) is proposed and verified on the high-resolution Gaofen-3 dataset labeled manually. The experimental results demonstrate that our method can improve the semantic segmentation performance compared with other algorithms. The improvement of classification accuracy reaches nearly five percent.

Keywords: SAR · Gaofen-3 · Transformer · CNN · Classification

1 Introduction

Synthetic aperture radar (SAR) can obtain rich information of earth surface under all-day and all-weather conditions. Land cover classification of SAR images plays a vital role in many areas, such as disaster monitoring, urban planning, and mapping [1, 2]. It is crucial to achieving scene understanding, which generates predictions by assigning each pixel to a target category. For the feature segmentation of SAR images, traditional machine learning methods have different implementation principles. But most of them use low-level semantic features such as color and texture, which cannot achieve satisfactory results when facing complex actual scenes. In recent years, semantic segmentation has emerged with the rapid development of the deep learning method. Since Long et al. [3] proposed fully convolutional networks (FCN) in 2015, various network models for segmentation have been proposed. Teichmann et al. [4] proposed a method of combining conditional random fields and convolutional neural networks to modify the edge information of segmentation.

However, when dealing with the feature of different scales, an appropriate receptive field should be selected according to the size of the object.Choosing different scales to capture image features can avoid the obstacles encountered in pixel-level dense prediction. The improved deep convolution neural network (DCNN) can segment a single terrain object layer by layer. To master the global semantic information with high-resolution SAR images, DCNN often requires stacking convolutional layers and brings

L. Wang et al. (Eds.): CHREOC 2022, LNEE 969, pp. 193–200, 2023.
https://doi.org/10.1007/978-981-19-8202-6_18

heavy computation. Transformer is an encoder-decoder structure based on the attention mechanism, which can enhance the long-range feature dependencies of the backbone of CNN. It is introduced into the task of land cover classification due to the superior modeling ability to handle global information. Following the idea of constructing global information dependence and fusing it with deep convolutional neural networks, we can achieve performance improvements in image segmentation.

The paper is organized as follows. Section 2 describes the related methods, Sect. 3 introduces the new algorithm combing CNN and transformer. In Sect. 4, datasets and the experimental results are demonstrated. The conclusion is given in Sect. 5.

2 Related Work

The SAR classification methods based on CNN have achieved better performance in terms of efficiency and accuracy. In 2016, Kampffmeyer et al. [5] used FCN for semantic segmentation of remote sensing images to realize pixel-level segmentation. He et al. [6] proposed a multi-scale spatial pyramid pooling method for pixel-level classification of the whole image. Guo et al. [7] fused ensemble learning to improve the segmentation accuracy of small and medium-sized targets in remote sensing images. Zhao et al. [8] proposed a superior pixel-level prediction framework PSPNet, which obtains global context information by aggregating context based on different regions.Chen et al. [9] proposed the atrous convolution operation, which enlarged the receptive field and strengthened the contextual semantic combinations. All network models are dedicated to increasing the relevance of semantic features and expanding the receptive field. Lin et al. [10] proposed a high-precision semantic segmentation model RefineNet which can use the features of each level. The mentioned papers have realized pixel-level classification of land cover and built a multi-scale feature extraction network with different receptive fields. However, the urgent problem to be solved is how to efficiently use global information to classify high-resolution SAR images.

3 Network Combining CNN and Transformer

We propose a novel encoder-decoder network, which realizes feature classification by semantic segmentation of SAR images. The proposed network is illustrated as Fig. 1.

In the encoder, the DCNN is employed to extract the low-level semantic features of the input images. The Xception [11] module replaces the traditional convolutional network with point convolution and deep convolution,ensuring feature extraction accuracy and improving efficiency. Besides, the middle layer of the Xception module can be freely changed and more advanced semantic feature information can be extracted. In this paper, it is set as 16 layers. After the DCNN operation, ordinary global pooling, adaptive pooling, atrous convolution, and global feature convolution are performed on the feature maps. The input size of the SAR images is 256 * 256 * 3, and the size of the feature map obtained after the DCNN operation is 16 * 16 * 2048. Therefore, the adaptive pooling output parameters are 3 and 5, and the atrous convolution rate is 3. We stack these feature maps and use 1 * 1 convolution to adjust the number of channels to 256. The highlight of this encoder adopts the construction idea of the transformer module. We use this feature

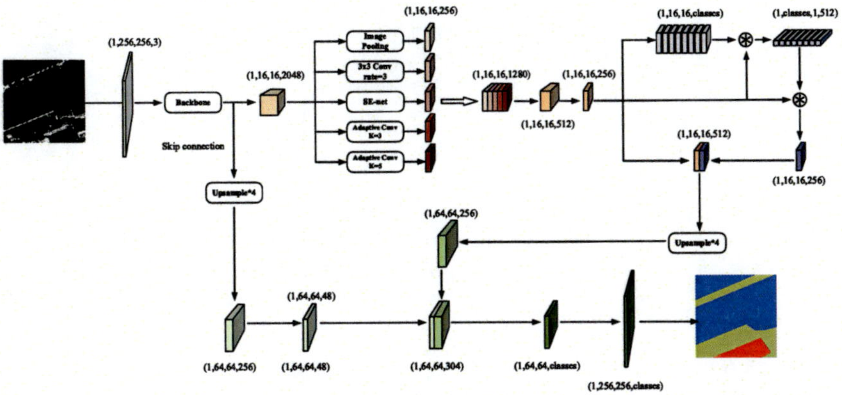

Fig. 1. The proposed network.

map as the global coarse segmentation result,which is denoted as q. Then, we extract each type of target feature and calculate the similarity with the global feature map to get the possibility of belonging to each category. This possibility is recorded as k.In the next step, the enhanced feature representation of the current pixel called v is obtained by classifying and weighting each region of the global result according to the possibility [12]. Obtaining q, k, v is an essential step for the transformer module to extract global information. At last, this result is superimposed with the rough global segmentation.In this way, the expression performance of the model can be improved in the encoder part. In order to explore the function of this module, the function of the ablation experiment exploration is also set.

In the decoder, we upsample the feature map output by four times to obtain the same spatial size as the underlying feature. Then, the connection also uses the output feature to perform four-fold upsampling by stacking the DCNN layer jump connection. Finally, the feature maps are restored to the same size as the model input after softmax activation, so as to realize the pixel classification of the image.

The loss function used is the categorical cross-entropy function integrated by TensorFlow, which can be written as,

$$\text{Loss} = -\sum_{i=1}^{n}\left[yi \log \hat{y}i + (1 - yi)\log\left(1 - \hat{y}i\right)\right] \tag{1}$$

where n represents the total number of categories contained in the dataset, y_i is the ground truth of class i, and $\hat{y}i$ is the prediction result of class i.

In this paper, we combined the advantage of CNN to extract plentiful detailed features and the attention mechanism of Transformer to obtain the weights of different categories.

Table 1. The experimental results compared with other methods.

Algorithms	IoU									FWIoU	PA
	Water	Bare soil	Road	Industry	Resident	Vegetation	Woodland	Human-built	Other		
Deeplabv3+	0.8943	0.0158	0.0451	**0.1223**	0.4047	0.2936	0.1061	0.1524	0	0.5418	0.6387
DMNet	0.9208	0.0099	0.0224	0.0151	0.2968	0.3521	0.1142	0.153	0	0.5393	0.6363
Ours	**0.9384**	0	0.0016	0.0037	0.4383	0.1882	0.07	0.1362	0	0.5338	0.6123
Ours-TRM	0.9262	0.0072	**0.046**	0.0375	**0.4307**	0.3307	0.1247	0.1621	0	0.5635	0.6517
Ours-TRM-Identity	0.9244	**0.0198**	0.0444	0.0239	0.3739	**0.4086**	**0.1604**	0.1609	0	**0.569**	**0.6637**

4 Experimental Results

The Gaofen-3 satellite images are used as experimental data. The single-polarization images covering more than 1000 km^2 are selected from ten regions such as Hampton, Omaha, and Kaohsiung, with a resolution of 1 meter. Assisted by optical images of the same region, the labels were generated manually. We set nine label categories: water, bare soil, road, industry, residential, vegetable, woodland, human-built, and other. Examples of single-polarization images, artificially labeled images, and the color code of the data set are shown in Fig. 2.

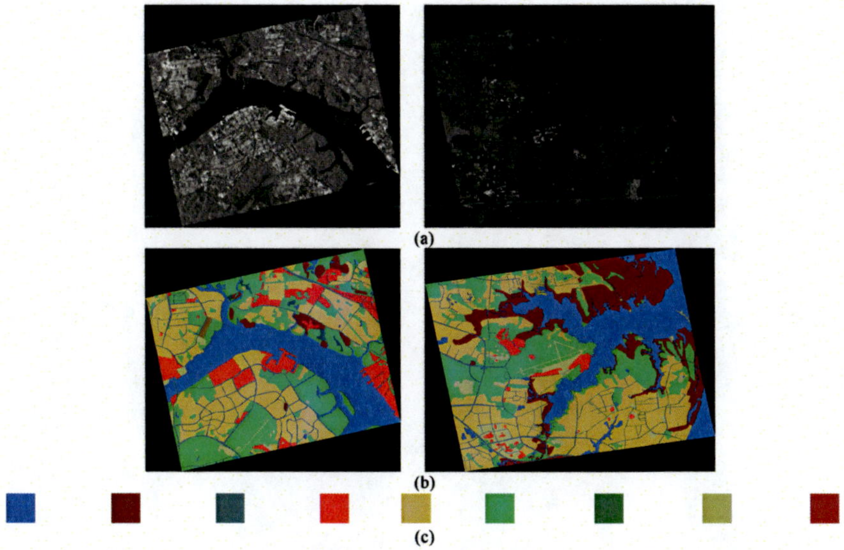

Fig. 2. Gaofen-3 dataset: **a** SAR images; **b** Ground Truth; **c** The color code: water, bare soil, road, industry, residence, vegetation, planting area, human-built, other

Since the original data cannot be directly sent to the model for training, the images must be truncated and stretched before normalization and mapped to 0–255. The input image of the network is cut into the size of 256 * 256. In the end, 17407 training samples and 5568 test samples are generated, each of which is a 256 * 256 single-polarization image and its corresponding label image.

In order to evaluate the results of semantic segmentation, pixel accuracy (PA) and frequency weighted intersection over union (FWIoU) are used as evaluation indicators. For further verification of our algorithm performance, the proposed algorithm is compared with some mainstream semantic segmentation networks, such as DMNet [13], Deeplabv3+ [14], etc. The comparison networks and test results are shown in Table 1. Among them, ours represents the proposed network without the transformer, and ours-TRM-Identity is the network concatenated with a skip connection from the output of DCNN. The experimental environment for this paper was implemented on a Linux system with four NVIDIA GeForce RTX 3090 Ti GPUs. The average training time for each network is 10 h, and our model achieves the best balance between performance and time consumption.

The predicted result shows that all networks can accurately predict the water area on this dataset, as shown in Fig. 3. The proposed network structure is better than the classic segmentation networks in classifying other types of features, and the total accuracy rate is increased by up to 5%.

We compare the structure of DCNN based on the Xception module with or without residual side connections and list the model results of adding residual side branches in the table. The experimental results show that the network with residual side connections slightly improves in the segmentation of bare soil and vegetation because the residual connection can integrate more features extracted in Xception. The adjusted network with residual connections can further improve the segmentation accuracy by 1%.

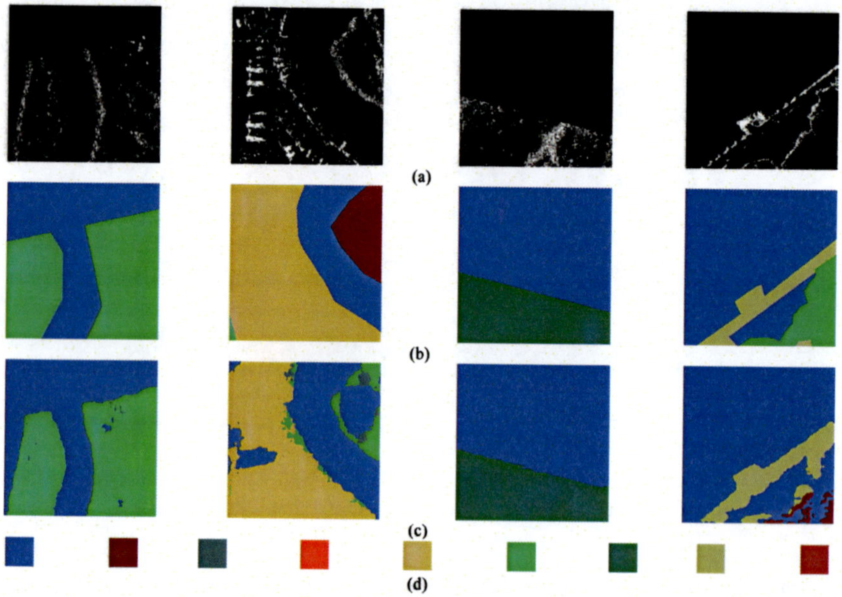

Fig. 3. The test results of our networks: **a** SAR images; **b** Ground Truth; **c** The test results; **d** The color code: water, bare soil, road, industry, residence, vegetation, planting area, human-built, other

The segmentation accuracy of all categories has been improved with Transformer. While expanding the information attention, it strengthens the connection between the current pixel and the surrounding features and accurately extracts some neglected small features. The total segmentation accuracy has been improved by 5% compared with the initial network.

5 Conclusion

In this paper, a novel semantic segmentation method is proposed for SAR image classification. The proposed algorithm can get an accuracy improvement of 5% compared with other semantic segmentation algorithms. The classification prediction results are more accurate in the visual comparison and get a performance boost in the water, bare soil, vegetation, residential, and woodland categories. In the ablation experiment, the model added the transformer structure has a significant performance improvement in all types of segmentation.

Acknowledgment. This work was supported in part by the National Natural Science Foundation of China (Grant No. 61901122), the Natural Science Foundation of Shanghai (Grant No. 20ZR1406300) and the China High-resolution Earth Observation System (CHEOS)-Aerial Observation System Project (30-H30C01-9004-19/21).

References

1. De, S., Bhattacharya, A.: Urban classification using PolSAR data and deep learning. In: 2015 IEEE International Geoscience and Remote Sensing Symposium (IGARSS), pp. 353–356 (2015)
2. Jafari, M., Maghsoudi, Y., Zoej, M.J.V.A.: New method for land cover characterization and classification of polarimetric SAR data using polarimetric signatures. IEEE J. Selected Top. Appl. Earth Observ. Remote Sens. **8**(7), 3595–3607 (2015)
3. Long, J., Shelhamer, E., Darrell, T.: Fully convolutional networks for semantic segmentation. In: 2015 IEEE Conference on Computer Vision and Pattern Recognition (CVPR), pp. 3431–3440 (2015)
4. Teichmann, M.T.T., Cipolla, R.: Convolutional CRFs for semantic segmentation (2018). arXiv:1805.04777
5. Kampffmeyer, M., Salberg, A.B., Jenssen, R.: Semantic segmentation of small objects and modeling of uncertainty in urban remote sensing images using deep convolutional neural networks. In: 2016 IEEE Conference on Computer Vision and Pattern Recognition Workshops (CVPRW), pp. 680–688 (2016)
6. He, K., Zhang, X., Ren, S., et al.: Spatial pyramid pooling in deep convolutional networks for visual recognition. IEEE Trans. Pattern Anal. Mach. Intell. **37**(9), 1904–1916 (2015)
7. Guo, R., Liu, J., Li, N., et al.: Pixel-wise classification method for high resolution remote sensing imagery using deep neural networks. ISPRS Int. J. Geo-Inf. **7**(3) (2018)
8. Zhao, H., Shi, J., Qi, X., et al.: Pyramid scene parsing network. In: 2017 IEEE Conference on Computer Vision and Pattern Recognition (CVPR), pp. 6230–6239 (2017)
9. Chen, L., Papandreou, G., Kokkinos, I., et al.: DeepLab: semantic image segmentation with deep convolutional nets, atrous convolution, and fully connected CRFs. IEEE Trans. Pattern Anal. Mach. Intell. **40**(4), 834–848 (2018)
10. Lin, G., Liu, F., Milan, A., et al.: RefineNet: multi-path refinement networks for dense prediction. IEEE Trans. Pattern Anal. Mach. Intell. **42**(5), 1228–1242 (2020)
11. Chollet, F.: Xception: deep learning with depthwise separable convolutions. In: 2017 IEEE Conference on Computer Vision and Pattern Recognition (CVPR), pp. 1800–1807 (2017)
12. Yuan, Y., Chen, X., Wang, J.: Object-contextual representations for semantic segmentation. In: Vedaldi, A., Bischof, H., Brox, T., Frahm, J.-M. (eds.) ECCV 2020. LNCS, vol. 12351, pp. 173–190. Springer, Cham (2020). https://doi.org/10.1007/978-3-030-58539-6_11

13. He, J., Deng, Z., Qiao, Y.: Dynamic multi-scale filters for semantic segmentation. In: 2019 IEEE/CVF International Conference on Computer Vision (ICCV), pp. 3561–3571 (2019)
14. Chen, L.-C., Zhu, Y., Papandreou, G., Schroff, F., Adam, H.: Encoder-decoder with atrous separable convolution for semantic image segmentation. In: Ferrari, V., Hebert, M., Sminchisescu, C., Weiss, Y. (eds.) ECCV 2018. LNCS, vol. 11211, pp. 833–851. Springer, Cham (2018). https://doi.org/10.1007/978-3-030-01234-2_49

Simulation of a Typical Meiyu Case Over the Yangtze-Huai River in a Record-Breaking Meiyu Period of 2020

J. I. N. Yang, Wang Ying(✉), and J. I. N. Meibing

School of Marine Sciences, Nanjing University of Information Science and Technology, Nanjing 210044, China
wangying_thw@nuist.edu.cn

Abstract. In the summer of 2020, Jiangsu Province has experienced a long Meiyu period with heavy rainfall, long time and wide range. In order to explore the characteristics of the Meiyu and its related atmospheric circulation background and water vapor transport background, this paper numerically simulates the Meiyu process in Jianghuai region from July 1 to 15, 2020 based on WRF (the Weather Research and Forecasting) model, and analyzes its precipitation characteristics and related causes by selecting the six hour precipitation process from 1800 UTC on July 2. The results show that the heavy rainfall belts in Jiangsu Province are triggered by influence systems in this analyzed period: (1) under the control of cyclonic circulation, the cold air flows to the front of the trough along with the northerly airflow behind the trough, so Jiangsu Province, which is located in front of the trough of low pressure, is attacked by the cold air from the south, and the warm air is forced to rise. During the process, the water vapor is supersaturated and condenses to rain, resulting in the precipitation of the Yangtze River Heavy rainfall in Northern Jiangsu. (2) The West Pacific subtropical high is generally stronger, with the west ridge reaching westwardly and northwardly than usual. The East Asian summer monsoon is weaker, which is conducive to the continuous transportation of water vapor to Jiangsu and Zhejiang. In addition, the warm air moves northward and meets with the cold air southward in southern Jiangsu, Northern Zhejiang and central and southern Anhui, forming the Meiyu front, which maintains near Jiangsu and Zhejiang for a long time, causing the southern Jiangsu A heavy rain in the rainy season. (3) The low-level jet, which has been kept in Jiangsu Province for a long time, transports enough water vapor for the occurrence of precipitation in Meiyu period. With the strong upward movement, the water vapor can reach the upper level, which affects the occurrence of convection in Meiyu rainstorm and the coupling of high-level jet and low-level jet, and provides favorable environmental conditions for the precipitation weather in Meiyu period.

Keywords: Meiyu · The west pacific subtropical high · Meiyu front · Cyclonic circulation

L. Wang et al. (Eds.): CHREOC 2022, LNEE 969, pp. 201–215, 2023.
https://doi.org/10.1007/978-981-19-8202-6_19

1 Introduction

Meiyu is a long continuous rain weather phenomenon in Taiwan, the middle and lower reaches of the Yangtze River, southern Korea and central and southern Japan. It is called Meiyu because its occurrence time coincides with the ripening time of plums in South of the Yangtze River. The large-scale precipitation in meiyu period directly affects the drought and flood, cold and warm weather in summer in South of the Yangtze River in China. Therefore, it is of great research value to explore the characteristics and the strong signals of influencing factors of Meiyu rainstorm, to improve the level of short-term climate forecast of Meiyu. Many studies have summarized the multi-time scale variation characteristics and causes of Asian summer Meiyu [1, 8] (Pei et al. 2020; Mo et al. 2020; Ding et al. 2020; Tang et al. 2020; Liu et al. 2020). Zhang et al. [8] conducted numerical simulation of 24 h accumulated precipitation in Yangtze-Huai river basin since July 5, 2013, it is found that the precipitation of the rainstorm is distributed in east-west bands, and the key weather systems affecting the large-scale precipitation are the northeast cold vortex and the Western Pacific subtropical high. Pei et al. (2020) studied a large-scale precipitation event during Meiyu period in the Yangtze-Huaihe valley started on June 26, 2015 and pointed out that the water vapor source of the rainstorm is the South China Sea and the Bay of Bengal. Mo et al. (2020) analyzed the background characteristics of abnormal precipitation and water vapor during the Meiyu period in 2016, and found that some factors such as the subtropical high lifted to the north, the upper South Asian high to the east, the abnormal sea surface temperature anomalies provided a favorable background field for the precipitation in the Meiyu period, resulting in more precipitation in the Meiyu period and the precipitation area to the north. Ding et al. (2020) analyzed the reasons for the great spatial difference in the distribution of drought and flood in China in the summer of 2019 by using the atmospheric circulation data. They believe that the continuous low-pressure trough from the Yellow Sea to the Sea of Japan leads to an obvious positive anomaly in the intensity of the Western Pacific subtropical high compared with previous years, leading to less precipitation in the middle and lower reaches of the Yangtze River and more precipitation in the southern part of the Yangtze River. Basing on atmospheric circulation and SST data, Liu et al. [11] further analyzed that the summer precipitation anomaly in 2020 might be related to the continuous warm SST in the tropical Indian Ocean. Tang et al. (2020) selected key monsoon regions (32° ~ 34°N, 112° ~ 120°E) from 1960 to 2020 to explore the relationship between abnormal precipitation weather and monsoon during Meiyu period in Jianghuai region and its causes. They found that when the summer monsoon arrived earlier (later) than expected in key areas, the plum day was later (earlier) and the plum rainfall was more (less). It is explained that monsoon will affect the intensity and location of blocking high and western Pacific subtropical high, thus affecting the precipitation in plum rain period. Li et al. (2021) conducted scientific observation and research on the precipitation weather in the Meiyu period in 2020, and obtained the observation data of many large-scale abnormal precipitation processes (multiple information of precipitation background such as precipitation, water vapor conditions, environmental temperature and cloud amount) through satellite, ground meteorological station and aircraft cloud observation, which provides the environmental conditions for the later study of the evolution of rainstorm clouds in the Meiyu period, sufficient preparations have been made to reveal the influence mechanism of water vapor conditions on precipitation clouds of Meiyu front. Hui et al. [13] studied the abnormal characteristics and generation conditions of weather phenomena such as abnormal precipitation during Meiyu period in East Asian climate in 2020, and discussed the future development direction of marine meteorology.

In 2020, the middle and lower reaches of the Yangtze River will enter the plum early and leave the plum late. The Meiyu period will last for a long time, with a large amount of Meiyu and frequent extreme precipitation events. According to statistics, total meiyu rainfall in 2020 is 759.2 mm, 1.2 times more than annual (343.4 mm), the highest in history since 1961. Meanwhile, the Meiyu period is the longest since 1961, up to 62 days, 22 days longer than usual (40 days). In July, there were many precipitation processes in Jiangsu Province, especially in the Meiyu period in the first ten days, and there were five heavy precipitation processes, which were July 2–3, July 5–7, July 11–13, July 14–17 and July 17–20. After simulating and analyzing the precipitation from July 1 to 15, 2020 based on WRF model, a six hour typical heavy precipitation process from 18:00 UTC on July 2 to 00:00 UTC on July 3 in Jiangsu Province is selected for analysis, to explore the characteristics and causes of Meiyu rainstorm.

2 Data, Model and Experimental Design

Yangtze-Huai river basin is selected as the research region in this paper. Two types of data are utilized:

(1) the precipitation monitoring data adopts the hourly precipitation grid data set integrated by China automatic station and CPC MORPHing technique (CMORPH) precipitation products. The in-situ data is the hourly monitoring of more than 30000–40000 automatic weather stations nationwide after quality control. The satellite derived precipitation products are based on the CMORPH satellite derived precipitation products developed by the US Climate Prediction Center with a global 30 min resolution of 8 km. The probability density function matching (PDF) error correction method is used to correct the systematic precipitation product error obtained by CMORPH satellite in China. Using the optimal interpolation (OI) method, Using the optimal interpolation (OI) method, the satellite product is used as the initial field, and the analysis product is used to correct the initial field, so as to obtain the data of the two types of products, and finally obtain the fused hourly precipitation data product and the corresponding graphic product. The overall product error is kept within 10%, and the error for large-scale precipitation and areas with few stations is kept within 20%, which is better than other products in China [26, 35].

(2) The FNL (Final Operational Global Analysis) global reanalysis data provided by NCEP (National Centers for Environmental Prediction) once every 6 h with a horizontal resolution of $1° \times 1°$ provides the initial field and boundary conditions for WRF model.

WRF model is a standardized medium-sized weather prediction model jointly developed by the National Center for Atmospheric Environment Prediction (NCEP) and the

National Center for Atmospheric Research (NCAR), as well as many research institutions, universities and business departments, it has significant advantages in simulating convection and mesoscale precipitation [1]. In this paper, the WRF model is used to simulate the Meiyu process from July 1 to 15, 2020. The WRF model adopts the single-layer nesting scheme, selects the Lambert projection, and the scope of the simulated grid area is shown in Fig. 1. The grid center is (32.5°N, 119.0°E), and the grid spacing is 3 km × 3 km, the number of grid points is 450 × 550, there are 51 layers in the vertical direction, the top layer of the mode is 10 hPa, and the frequency of regional output simulation results is once an hour. The parameterization scheme adopted is shown in Table 1.

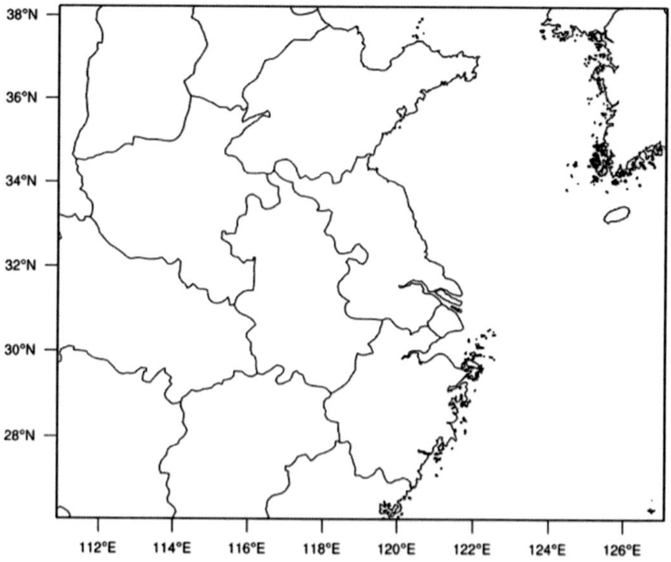

Fig. 1. Scope of simulation grid area

Table 1. Selected WRF parameterization scheme for simulation

WRF parameterization scheme		
Microphysical process	Thompson	Thompson [29]
Long-wave radiation	RRTM	Lawer et al. [32]
Short-wave radiation	Goddard	Chou and Suarez [34]
Boundary layer	YSU	Hong et al. [31]
Surface layer	Monin-Obukhov	Moinn et al. [33]
Land surface process	Noah-MP	Niu et al. [30]

3 Validation of Simulated Precipitation

In China's middle and lower reaches of the Yangtze River and other regions, the abnormal weather phenomenon of continuous rain often occurs from June to July every year. In the summer of 2020, Jiangsu Province has experienced a long Meiyu period with heavy rainfall, long time and wide range [10]. Why is the precipitation intensity of this Meiyu weather so large, and what causes the long time maintenance of Meiyu belt? These problems deserve our in-depth exploration. Many studies show that the abnormal precipitation during the Meiyu period is affected by a variety of atmospheric circulation situations. The Western Pacific subtropical high, blocking height, cold vortex and high-low jet are the key factors for the duration of the Meiyu period, the maintenance, development and decline of the rain belt in the middle and lower reaches of the Yangtze River [15]. In order to explore the characteristics and causes of atmospheric circulation and water vapor transport related to the Meiyu, the precipitation, wind field, relative humidity and other variables from July 1 to 15, 2020 are simulated and analyzed based on simulations via WRF model.

Figure 2a shows the simulation of 6 h accumulated precipitation distribution from 18:00 on July 2 based on WRF model. It can be seen that there are two rain belts in Jiangsu Province. Whether the influencing factors are the same remains to be further explored. One of the rain belts 1 (not all shown in the figure) is mainly distributed in Northern Jiangsu and extends to its offshore sea, covering a large area and affecting a wide area. There is local precipitation of more than 60 mm near Northern Jiangsu (33.8°N, 120°E). Another rain belt 2 is distributed in southern Anhui, Northern Zhejiang, Jiangxi and a few areas in southern Jiangsu, showing an east-west zonal distribution. The areas with precipitation of more than 60 mm are concentrated in the south of central Anhui, with a large range and wide impact. Figure 2b shows the actual 6 h accumulated precipitation at 18:00 on July 2. It can be seen from the figure that the distribution of the two rain belts affecting the precipitation in Jiangsu Province is consistent with the simulation diagram. The rain belt 1 is distributed in Jiangxi and Northern Jiangsu Province and extends to the sea, and the other rain belt 2 is distributed in southern Anhui, northern Jiangxi, Zhejiang and a few areas in southern Jiangsu, showing a strip distribution from east to west. There is a maximum area of precipitation in the east-west direction in the north of Jiangsu, with a center of more than 50 mm. There is also a maximum area in the middle of Anhui, but the central precipitation does not reach more than 50 mm, which is less than that in the simulated one. Through the comparison between the simulation and observation, it can be seen that the rain belt position, morphological characteristics and precipitation of the simulation map are generally consistent with the observation.

4 Weather Characteristics and Analysis of Meiyu Period in 2020

4.1 Climate Background

Under the influence of El Niño, the SST of the tropical Indian Ocean began to rise significantly in January 2020, and played a positive feedback role on the El Niño event, thus affecting China's climate change, which is one of the main reasons for the anomaly of atmospheric circulation in low latitudes in summer in China [11]. The snow anomaly

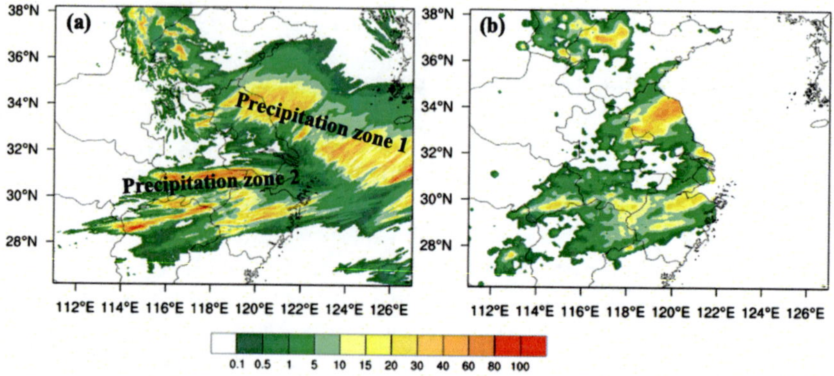

Fig. 2. Distribution of simulated and observed precipitation (unit: mm). **a** The six hour accumulated precipitation distribution at 18:00 on July 2, 2020 is obtained through WRF simulation; **b** actual distribution of six hour accumulated precipitation at 18:00 on July 2, 2020.

on the Qinghai Tibet Plateau has a significant impact on the changes of atmospheric circulation in China. In 2020, the snow cover on the Qinghai Tibet Plateau showed an obvious positive anomaly compared with the rest of the year, resulting in an obvious negative anomaly of the summer monsoon, which made the Yangtze River Basin prone to flood disasters in China [12]. The anomaly of summer monsoon is also the reason for the abnormal weather phenomenon in summer Meiyu period in China. In 2020, the South China Sea summer monsoon was produced earlier and weaker than usual, while the abnormal deepening of the East Asian Trough sent a strong cold air flow to the South and restrained the northward jump of the summer monsoon [13].

4.2 Characteristics and Causes of Meiyu

The abnormal precipitation in Meiyu period is affected by many systems. The abnormal West Pacific subtropical high affects the abnormal precipitation in Meiyu period, the influence of trough ridge system in troposphere and the interaction of high and low level jet system jointly trigger the occurrence and development of abnormal precipitation in Meiyu period [7]. The following paper analyzes the characteristics of the key influence systems such as the location and intensity anomaly of the West Pacific subtropical high, the coupling of high and low-level jet flow, cyclonic circulation and so on.

Figure 3 is an analysis of the circulation situation in Jiangsu Province and its surrounding areas during the occurrence of Meiyu based on WRF model. It can be seen from this figure of 500 hPa geopotential height field distribution at 18:00 on July 2, 2020 that Jiangsu Province is controlled by the low-pressure trough, and the trough line extends from northern Shandong to the junction of Henan and Hubei. Jiangsu Province is located in front of the low-pressure trough, and the West Air flow prevails. The cold air flows in front of the trough with the north air flow behind the trough, and flows southward into Jiangsu, with a high intensity. At the same time, the geopotential height field of Jiangsu, Zhejiang and its surrounding areas is positively deviated from that in previous years. The west ridge of the West Pacific subtropical high is more West than usual, with the 588 line extends to the inland of China through Fujian and Anhui in the West. The position of the low latitude West Pacific subtropical high is stable, the overall position is more south than usual, and the East Asian subtropical summer monsoon is abnormally weak, which is conducive to the continuous transmission of water vapor to Jiangsu and Zhejiang. In addition, the sea surface off the coast of Jiangsu Province is controlled by the strong west Pacific subtropical high, resulting in the northward movement of warm air. The cold and warm air masses converge in southern Jiangsu and Northern Zhejiang, but have no obvious shear characteristics, forming a quasi-stationary Front, which is conducive to the formation of precipitation in Jiangsu and Zhejiang [9]. As can be seen from Fig. 3c of the distribution of 700 hPa geopotential height field, the low-pressure trough is inclined to the East relative to the high-altitude position, so the rain belt brought by the low-pressure trough will move eastward with time, the temperature field will increase compared with the high-altitude temperature, the upper cold and lower warm stratification is unstable, and precipitation is easy to form. The temperature tank lags behind the height tank, and this system will continue to develop. There is a strong warm and humid air flow in the southwest of Jiangsu Province, the cold and warm air flows in southern Jiangsu, Northern Zhejiang and central Anhui intersect obviously, the Quasi-Stationary Front formed will stay here for a long time, resulting in a wide range of heavy precipitation. The transportation, intensity and maintenance time of low-level water vapor determine the distribution and magnitude of Meiyu abnormal precipitation area, and the source of low-level water vapor is closely related to the maintenance of low-level jet [16]. Combined with the observation of the geopotential height of 700 and 850 hPa, it can be seen that the low-level jet is roughly distributed in the range of 28°–32°N, with high intensity, which is consistent with the rain belt affecting the heavy precipitation in southern Jiangsu (Fig. 2). The long-term maintenance of low-level jet in Jianghuai area during precipitation provides sufficient water vapor transport for the occurrence of abnormal precipitation during Meiyu period. The low-pressure trough on the 850 hPa geopotential height map continues to tilt eastward, and the trough line tends to be parallel to the zonal circle. Combined with the 500 hPa and 700 hPa geopotential height, the low-pressure trough extending from northern Shandong to the junction of Henan and Hubei is a backward inclined trough. With the enhancement of the low-pressure trough, the advection also increases. Therefore, under the influence of the strong updraft in front of the trough, heavy precipitation is formed in Northern Jiangsu, which corresponds to the rain belt 1 in Fig. 2. Therefore, the cyclonic circulation in the northwest of Jiangsu Province is the main cause of precipitation in the rain belt 1. At this time, under the influence of the West Pacific subtropical high system, the water vapor in the lower reaches of the Yangtze River is continuously transported to Jiangsu and Zhejiang. The southeast air flow over the South China Sea and the Pacific transforms into southwest air flow near the Western ridge of the subtropical high, together with the warm and humid air flow, which transports warm and humid air to the Yangtze-Huai River area from the southwest of the Bay of Bengal, meets the cold air from the Nouth and forms a quasi-stationary front in the area of Jiangsu and Zhejiang, resulting in the distribution in the south of Anhui Another rain belt in Northern Zhejiang, Jiangxi and a few areas in southern Jiangsu [16]. Figure 3a shows the geopotential height of 300 hPa. It can be seen that the intensity of the subtropical westerly jet is very high. Jiangsu Province is in the outlet area of the upper level jet. By analyzing the zonal wind time-varying equation of the upper jet zone, it can be seen that the upper air divergence will appear in the south of the jet outlet area, and the upper air convergence will appear in the north. Furthermore, there is a sinking movement in the north and an ascending movement in the south. The high and low altitude atmosphere is coupled, and the circulation in the vertical direction is generated. According to this theory, the high-altitude jet changes the convergence, divergence and vertical coupling movement of the surrounding environmental field through its dynamic mechanism and secondary cycle, so as to provide sufficient environmental conditions for large-scale precipitation weather [10]. The vertical upward movement of high and low altitude constitutes a dynamic environmental field conducive to excessive precipitation in Jiangsu and other places [16].

It can be seen that the 6 h precipitation in Jiangsu from 18:00 on July 2 is affected by two weather systems: the low-pressure trough with the low-pressure center in Shandong and the Meiyu front across Anhui, Zhejiang and Jiangsu. The rain belt 1 formed in the north of Jiangsu is controlled by the low-pressure trough in Shandong, and the rain belt 2 formed in the south of Jiangsu is affected by the Meiyu front across Anhui, Zhejiang and Jiangsu. The original rain belt 1 moves from west to East offshore, and the precipitation will gradually increase in the next few days, the original rain belt 2 will move to the Northeast [8].

Figure 4a shows the 925 hPa wind field and relative humidity field on July 2, 2020. It can be seen from the figure that the relative humidity gradient in Northern Jiangsu Province, Northern Zhejiang, central Anhui and southern Jiangsu is large, indicating that there is a rain belt in each of the two places. Jiangsu Province is located at the top and front of a cyclonic circulation. As the ground temperature is higher than the upper air, the temperature gradually decreases during the rise of the air, and the water vapor is easily to be supersaturated and cause rain. Therefore, the precipitation in Jiangsu Province on July 2 caused by the cyclonic circulation is mostly concentrated in northern Jiangsu, with the maximum accumulated precipitation more than 60 mm. In addition, there is a strong rainfall zone in southern Jiangsu Province, northern Zhejiang and central Anhui. The location of the western ridge of the Western Pacific subtropical high is west compared with previous years, which leads to the transformation from the southeast air flow to the southwest air flow in the South China Sea and the Pacific. In addition, the weak East Asian summer monsoon is the driving factor conducive to the source of precipitation water vapor in southern Jiangsu and other places. The warm and humid air flowing northward and southward meets the cold air in Jiangsu and Zhejiang, forming a quasi-stationary front, resulting in another rain belt distributed in southern Anhui, northern Zhejiang and a few areas in southern Jiangsu. Figure 4b is a longitude height profile of the relative humidity field along 119.5°E, in which there is an obvious baroclinic structure between 35° N and 38°N, and the frontal shape tilts northward with the increase of height. It can be seen that the formation of the rain belt here is due to the cyclonic circulation, which is consistent with the analysis of the precipitation in Northern Jiangsu in Fig. 4a. Due to the inertia of the heating air mass and the fast speed and high density of the cold air mass, the cold air invades from the low, the warm air is forced to rise to the high altitude, the temperature drops, the air is supersaturated, and the cloud condenses to rain, providing favorable conditions for precipitation. In addition, there is an obvious positive

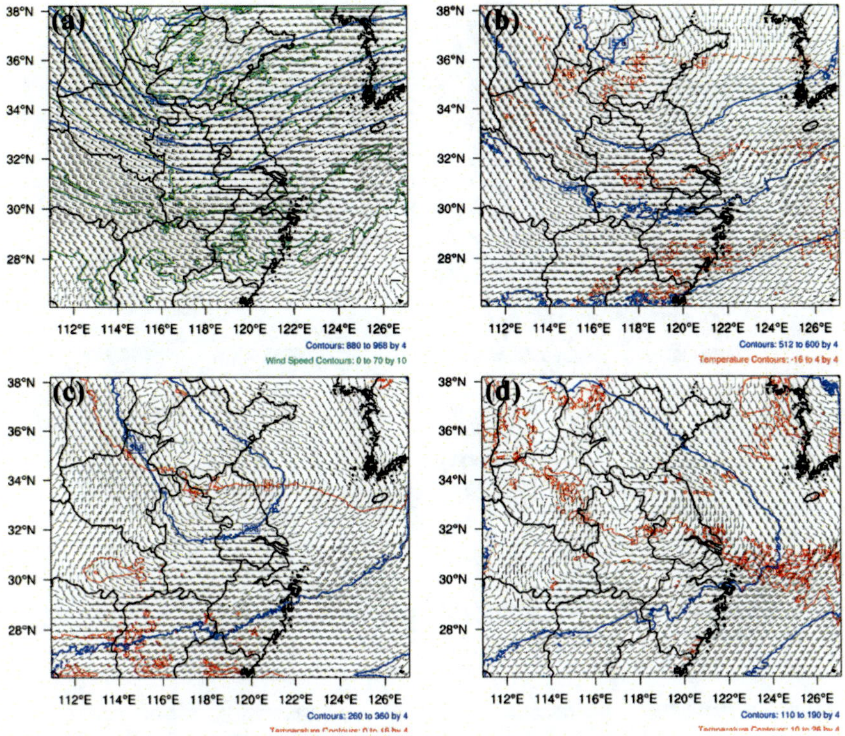

Fig. 3. (**a**)–(**d**) represent the circulation situation fields of 300 hPa, 500 hPa, 700 hPa and 850 hPa respectively. The blue line represents the wind field height, and the red line represents the wind field height.

pressure structure near 30.5°N ~ 32.5°N. The strength of the cold and warm air mass here is equal, the movement is slow, and the shape of the front rises almost vertically. It can be seen that the front controlling the precipitation here is a quasi-stationary front that stays in the Yangtze-Huai river basin for a long time. Figure 4c–d are latitude altitude profiles of relative humidity field along 30.1°N and 34.5°N, respectively. Figure 4c is the zonal profile of the rain belt 2 caused by the Meiyu front. From the figure, we can clearly see that the dry and cold air flow of 110°E ~ 112°E converges with the warm and humid air mass southward, forming a relatively stable Meiyu front near 112°E ~ 114°E, providing favorable environmental conditions for large-scale heavy precipitation weather. Figure 4d is the zonal profile of the rain belt 1 located in Northern Jiangsu. There is an obvious baroclinic structure here. It can be seen that the precipitation here is the result of the cyclonic circulation located at 110°E ~ 115°E.

Because vorticity can well correspond to the large-scale precipitation distribution area, vorticity is used as a variable to simulate the abnormal precipitation process during the Meiyu period [18]. The absolute vorticity is the sum of the relative vorticity and the planetary vorticity. As the Meiyu precipitation process is a small and medium scale movement, the planetary vorticity can be ignored. Therefore, the absolute vorticity

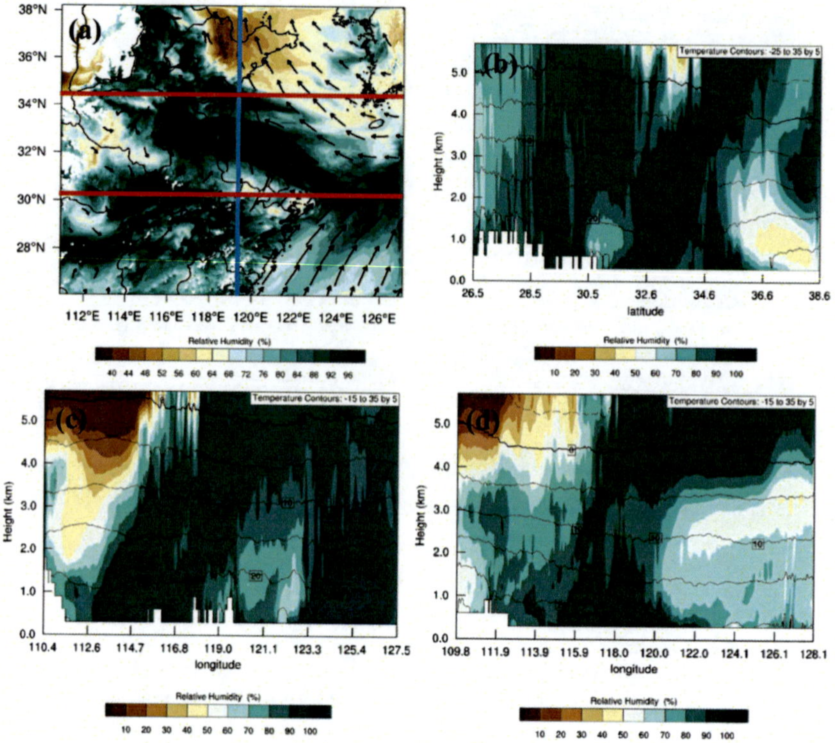

Fig. 4. Distribution of (**a**) 925 hPa wind field (vector, unit: m·s^{-1}) and relative humidity (shadow, unit:%) obtained by WRF simulation at 18:00 on July 2, 2020. (**b–d**) are the longitude altitude profile of relative humidity field along 119.5° E and the latitude altitude profile of relative humidity field along 30.1°N and 34.5°N (shadow, unit:%).

has a good correlation with large-scale precipitation, which can be used to judge the precipitation falling area and precipitation intensity. Figure 5a shows the distribution of 500 hPa absolute vorticity at 21:00 on July 2. It can be clearly seen from Fig. 5a that the concentrated distribution area of positive vorticity at 21:00 on July 2. Among them, in the high-level vorticity field, the distribution of convective system is reflected in two positive absolute vorticity zones. Part of the positive vorticity is concentrated in Anhui and Northern Jiangsu, which has multiple positive vorticity centers, roughly distributed in the northwest of Jiangsu, near 33.5°N, between 118°E and 119.5°E, corresponding to the rain belt 1 near 33.5°N. Another part of the positive vorticity is distributed in a small part of Eastern Hubei, Southern Anhui and Northern Zhejiang Province. There are also many scattered positive vorticity centers on this positive vorticity belt, which are distributed in a belt from east to west, which is consistent with the large-scale precipitation caused by the quasi-static stop front near 30°N and corresponds to the rain belt 2. There are multiple positive vorticity distributions at 500 hPa geopotential height, and their activities are relatively active and still developing, which hinders the northward transmission of negative vorticity at low latitudes to a certain extent and is conducive to the maintenance of Meiyu heavy precipitation. To sum up, the two large-scale precipitation in the north and south of Jiangsu Province are consistent with two positive vorticity zones respectively, and in the process of heavy precipitation in and around Jiangsu Province, the lower layer is a positive vorticity region, and its distribution and trend are very consistent with the rain zone. The maintenance of rain is closely related to large-scale circulation [25]. Positive vorticity in the northern hemisphere is the counterclockwise rotation of air, that is, cyclonic circulation, with upward movement. The positive vorticity zone is roughly consistent with the Meiyu front precipitation zone, which is conducive to the accelerated formation of the Central Asian low pressure and creates a favorable environment for promoting the precipitation in the Meiyu period. Figure 5b shows the distribution of 850 hPa potential vorticity. From the figure, it can be seen that there are two obvious positive potential vorticity bands. One positive potential vortex zone is distributed in Northern Jiangsu and extends to its offshore sea surface, and the other positive potential vortex zone is distributed in southern Jiangsu, Northern Zhejiang and central and southern Anhui. The maximum potential vorticity of the two positive potential vorticities is more than 5 PVU. The position distribution of the two potential vorticity zones is consistent with that of the positive vorticity zone and the two rain zones affecting heavy precipitation in Jiangsu Province. Therefore, the positive vorticity zone and potential vorticity zone can well reflect the large-scale heavy precipitation area. The mesoscale convective system is the direct influence system of large-scale precipitation, and the potential vorticity disturbance in the lower layer is the fundamental factor to stimulate the mesoscale convective system [19]. The change of potential vorticity may be related to the latent heat release effect of water vapor condensation in the process of large-scale precipitation. The potential vorticity on the low-level rain belt is likely caused by diabatic heating [20]. There is a positive feedback mechanism between water vapor vertical transport and condensation latent heat and potential vorticity, so as to maintain the mesoscale convective system [28].

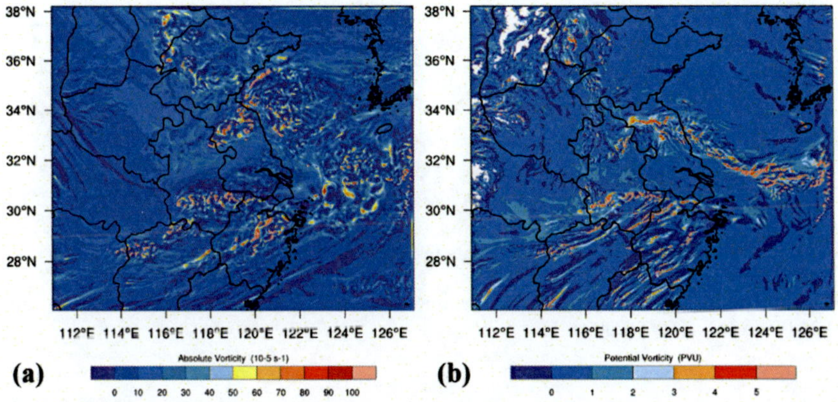

Fig. 5. Distribution of (**a**) absolute vorticity (shadow, unit: $10^5\ s^{-1}$) at 500 hPa and (**b**) potential vorticity at 850 hPa (shadow, unit: PVU) at 21:00 on July 2.

Figure 6 shows the echo distribution of the maximum radar intensity. From Fig. 6a, we can see that there is a strong radar echo zone in Northern Jiangsu at 18:00 on July 2, extending from northern Jiangsu to the offshore sea surface. The echo in some areas exceeds 45 dBZ, which corresponds to the precipitation zone 1 in Northern Jiangsu. In addition, there are strong radar echo zones in southern Jiangsu, central and southern Anhui and Northern Zhejiang. The distribution shape is banded, which corresponds to the distribution and trend of the rain belt 2. It can be seen that there is a good correlation between radar echo intensity and precipitation. The strong radar distribution area observed from the radar echo intensity distribution map corresponds well to the rain belt, which further confirms that the precipitation in Jiangsu is produced by different rain belts, and the influence mechanism of precipitation is also different. The variation of maximum radar echo intensity distribution can reflect the development trend of rain belt. It can be seen from Fig. 6b, c that within six hours from 18:00, the maximum radar intensity echo in Anhui and Northern Jiangsu moves eastward and extends to the offshore sea of Jiangsu. The maximum radar intensity echo center basically moves out of Jiangsu Province, and the intensity is significantly enhanced. The echo areas exceeding 45 dBZ are significantly increased, scattered from northern Jiangsu to the offshore sea, and the maximum radar intensity echo zone tends to be banded. The maximum radar echo intensity band distributed in a small part of Eastern Hubei, Southern Anhui and Northern Zhejiang Province moves to the northeast, the maximum radar echo intensity weakens, the scattered echo center becomes concentrated, and the influence range of the echo band is significantly reduced. Therefore, we can conclude that in the next 6 h at 18:00, the rain belt 1 located in Anhui and Northern Jiangsu under the influence of the cyclonic circulation will move slowly eastward and extend to the offshore sea of Jiangsu, and the precipitation will increase. The rain belt 2 located in a small part of Eastern Hubei, Southern Anhui and Northern Zhejiang Province will move to the Northeast in the future, reducing the scope of influence.

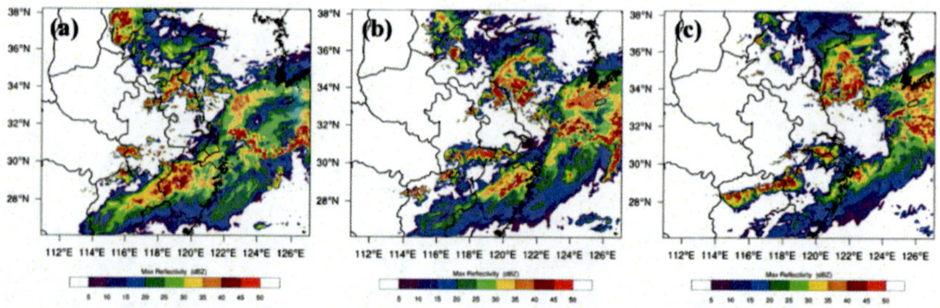

Fig. 6. Distribution of maximum radar echo intensity (shadow, unit: dBZ) at (**a**) 18:00 on July 2 (**b**) 21:00 on July 2 (**c**) 00:00 on July 3

5 Conclusions and Discussion

In 2020, the Meiyu weather in Jiangsu is abnormal, with a longtime Meiyu period, heavy rainfall and various influencing factors. In the six hours from 18:00 on July 2, there was large-scale precipitation in the north and south of Jiangsu Province, which was caused by different reasons. The rain belt 1 is located in the north of Jiangsu Province and is the result of the action of cyclonic circulation. Because Jiangsu Province is located in front of the low-pressure trough, the cold air flows to the front of the trough with the northerly air flow behind the trough, the air temperature rises and gradually decreases, and the water vapor is supersaturated, causing rain. The rain belt 2 is located in the south of Jiangsu Province and other places, and the precipitation is mainly affected by the Meiyu front. The West Pacific subtropical high intensity is strong, the west ridge is west, and the ridge position remains abnormal for a long time. In addition, the East Asian subtropical summer monsoon is abnormally weak, which is conducive to the continuous transmission of water vapor from the Bay of Bengal, the South China Sea and the Western Pacific to Jiangsu and Zhejiang. The prevailing warm and humid air flow provides sufficient water vapor conditions for precipitation in Jiangsu Province. The West Pacific subtropical high caused the warm air to lift northward and meet the cold air going southward in southern Jiangsu, Northern Zhejiang and Anhui. The Meiyu front formed by the equal cold and warm air flow remained near Jiangsu and Zhejiang for a long time, resulting in the rainstorm weather in the Meiyu period. The low-level jet, which has been maintained in Jiangsu Province for a long time, transports sufficient water vapor for the occurrence of precipitation in Meiyu period. With the strong upward movement, the water vapor can reach the high level, affecting the occurrence of convection in Meiyu rainstorm and the coupling of high and low-level jet, providing favorable environmental conditions for precipitation weather in Meiyu period.

Acknowledgments. This work is supported by the National Science Foundation of China (NSFC) under Grant 41905089.

References

1. Li, X., Zhang, L., Wang, J.: Sensitivity of Meiyu front rainstorm forecast to resolution and cumulus parameterization. Rainstorm Disaster **39**(06), 637–646 (2020)
2. Hu, X., et al.: Introduction to parameterization scheme of physical process in WRF mode. Gansu Sci. Technol. **24**(20), 73–75 (2008)
3. Wu, Z., Li, Y., Li, X., Hu, X., Zhou, G., Deng, C.: Influence of boundary layer parameterization scheme of WRF model on precipitation simulation of Southwest Vortex in Sichuan Chongqing Basin. Atmosp. Sci. **45**(01), 58–72 (2021)
4. Li, X., Zhang, L., Wang, J.: Sensitivity of Meiyu front rainstorm forecast to resolution and cumulus parameterization. Rainstorm Disaster **39**(06), 637–646 (2020)
5. Zhao, S., Wang, L., Li, X., Yin, F., Pei, K.: Numerical simulation and Q-vector diagnosis of a severe rainstorm in Nanjing. J. Chengdu Univ. Inf. Eng. **35**(06), 678–688 (2020)
6. Xing, S.: Simulation of rainstorm process in Zhejiang, Fujian and Jiangxi in mid-June 2010. Zhejiang University (2020)

7. Sha, S., Shen, X.Y., Li, X.F., et al.: The study of multi-scale energy interactions during a Meiyu front rainstorm. Part II: practical application. Chin. J. Atmosp. Sci. (in Chinese) **42**(5), 1119–1132 (2018)
8. Zhang, P., Xu, D., Shen, F., Shu, A., Bian, H., Zhou, X., Li, C.: Numerical simulation and diagnostic analysis of Meiyu peak rainstorm in Jianghuai. J. Lanzhou Univ. (Nat. Sci. Ed.) **56**(04), 527–536 (2020)
9. Yang, M., Liu, M., Ke, D., Chen, S.: Cause analysis of Meiyu anomaly in Jianghuai area in 2020. Rainstorm Disaster **39**(06), 555–563 (2020)
10. Cai, X., Zong, Z., Ma, J., Li, Y., Zhang, B., Yin, S., Mei, S.: Characteristic analysis of Meiyu in 2020 and performance test of model medium-term forecast. Rainstorm Disaster **39**(06), 629–636 (2020)
11. Liu, Y., Wang, Y., Ke, Z.: Characteristics and causes of climate anomalies in summer 2020 in China. Meteorology **47**(01), 117–126 (2021)
12. Zhang, S., Tao, S.: Diagnostic and Numerical Study on the influence of snow cover on the Tibetan Plateau on the Asian summer monsoon[J. Atmosp. Sci. (03), 372–390 (2001)
13. Qiu, H., Xiong, Y., Xing, W., Wang, L.: Characteristics and causes of rainstorm in the Yangtze River Basin during the main flood season in 2020. People's Yangtze River **51**(12), 104–110 (2020)
14. Sasha.: Study on multi-scale energy interaction in a Meiyu front rainstorm. Nanjing University of information engineering (2018)
15. Zhang, S.: Comparative analysis and development maintenance mechanism of two rainstorms in Jianghuai Meiyu period. Nanjing University of information engineering (2015)
16. Jin, R., Li, W., Zhang, B., Yan, C.: Study on the relationship between the activity of the subtropical westerly jet in East Asia and the Meiyu anomaly in the middle and lower reaches of the Yangtze River. Atmosp. Sci. **36**(04), 722–732 (2012)
17. Zhao, J., Chen, L., Wang, D.: Abnormal characteristics and cause analysis of Meiyu in China in 2016. Atmosp. Sci. **42**(05), 1055–1066 (2018)
18. Wang, S., Ji, L., Pan, X., Li, Y.: Mesoscale comparative simulation analysis of a Meiyu front rainstorm process. Meteorol. Sci. (06), 569–578 (2005)
19. Zhang, S., Min, J:. Development and maintenance mechanism of Meiyu front rainstorm in Jianghuai area on June 23, 2013. Meteorol. Sci. **38**(06), 707–718 (2018)
20. Zhao, Y., Li, Z., Xiao, Z., Chen, Y.: Potential vorticity inversion diagnosis of a case of continuous rainstorm on quasi stationary Meiyu front. J. Meteorol. (03), 353–371 (2007)
21. Liu, D., He, J., Yao, Y.: Review and prospect of Mei Yu research. Res. Meteorol. Disaster Reduct. **32**(01), 1–9 (2009)
22. Gao, H., Xue, F., Wang, H.: Influence of interannual variation of antarctic oscillation on Jianghuai Meiyu and its prediction significance. Sci. Bull. (S2), 87–92 (2003)
23. Qu, H.: On the climatic characteristics, social influence and cultural significance of Meiyu. J. Xiangtan Univ. (Philos. Soc. Sci. Ed.) **38**(03), 157–161 (2014)
24. Ye, Z.: Study on flood response of water cycle variation in typical plain water network area in the lower reaches of the Yangtze Huaihe River. Nanjing University (2011)
25. Wei, J., Chen, P., Sun, Y., Cheng, S., Li, N.: Simulation analysis of a heavy precipitation process in Jiangsu Province by WRF model. J. Atmosp. Sci. **34**(2), 232–238 (2011)
26. Shen, Y., Pan, M., Yu, J.: Quality evaluation of regional hourly precipitation fusion products in China. J. Atmosp. Sci. **36**(1), 37–46 (2013)
27. Wang, Y., Lou, D., Liu, Y.: Characteristics and causes of precipitation anomalies in the middle and lower reaches of the Yangtze River during the plum flood season in 2020. Rainstorm Disaster **39**(06), 549–554 (2020)
28. Lau, K., Yang, G.J., Shen, S.H.: Seasonal and intraseasonal climatology of summer monsoon rainfall over Eeat Asia. Mon. Wea. Rev. **116**, 18–37 (1988)

29. Thompson, G.: Explicit forecasts of winter precipitation using an improved bulk microphysics scheme. Part II: implementation of a new snow parameterization. Monthly Weather Rev. **136**(12), 5095–5100, 5102–5115 (2008)
30. Niu, G.Y., Yang, Z.L., Mitchell, K.E.: The community Noah land surface model with multi-parameterization options (Noah-MP): 1. Model description and evaluation with local-scale measurements. J Geophys Res 116: D12109. J. Geophys. Res. Atmosp. **116**, D12109 (2011)
31. Hong, S.Y., Noh, Y., Dudhia, J.: A new vertical diffusion package with an explicit treatment of entrainment processes. Mon. Wea. Rev. **134**(9), 2318–2341 (2006)
32. lawer, E.J., Taubman, S.J., Brown, P.D., et al.: Radiative transfer for in homogeneous atmosphere: RRTM, a validated correlated-k model for the longwave. J. Geophys. Res. **102**(D14), 16664–16682 (1997)
33. Moinn, A.S., Obukhow, A.M.: Basic laws of turbulent mixing in the atmosphere near the ground. Tr. Akad. Nauk. SSSR Geofiz. Inst. **24**(151), 163–187 (1954)
34. Chou, M.D., Suarez, M.J.: A solar radiation parameterization for atmospheric studies. NASA Technical Memorandum NASA/TM-1999-104606
35. Shen, Y., Zhao, P., Pan, Y.: A high spatiotemporal gauge-satellite merged precipitation analysis over China. J. Geophys. Res. **119**(6), 3065–3075 (2014)

Hyperspectral Image Classification Based on 3D Calibrated Attention

Menglin Zhang and Feng Gao(✉)

School of Computer Science and Technology, Ocean University of China, Qingdao 266100, China
gaofeng@ouc.edu.cn

Abstract. Hyperspectral image (HSI) classification aiming at identifying the pixel-level attributes, is a challenging task in remote sensing image interpretation. Informative features play an important role in HSI classification. Existing attention-based convolutional neural networks treat each convolutional layer as a separate process that miss the correlation among different layers. To slove the problem, we introduce the 3D calibrated attention module (3DCAM) to model the holistic dependencies among channels and positions. In particular, The 3DCAM can be used for the correlation among channels to achieve more powerful feature representation and correlation learning. Furthermore, we design a 3D calibrated attention-based network, which integrates 3D-CNN and 2D-CNN layers. Both the spectral and spatial feature are employed for classification. Extensive experiments on the Indian Pines and Pavia University datasets demonstrate that the proposed method performs favorably against three closely related methods.

Keywords: Hyperspectral image · 3D calibrated attention · Attention mechanism · Image classification

1 Introduction

Hyperspectral image (HSI) is capable of discriminating various ground objects owing to its refined radiation information. Therefore, HSI has attracted numerous industry and agriculture applications, such as mineral exploration [1] and environment monitoring [2]. HSI classification aiming at identifying the pixel-level attributes, is a significant and challenging task in remote sensing image interpretation [3].

Nowadays, numerous deep CNN-based methods have been proposed to solve the HSI classification problem. Zhong et al. employed the shortcut connection

This work was supported in part by the National Key Research and Development Program of China under Grant 2018AAA0100602, and in part by the National Natural Science Foundation of Shandong Province under Grant ZR2019QD011.

L. Wang et al. (Eds.): CHREOC 2022, LNEE 969, pp. 216–225, 2023.
https://doi.org/10.1007/978-981-19-8202-6_20

into CNN and proposed a spatial-spectral residual network [4]. Paoletti et al. [5] proposed a capsule network which can effectively reduce the network design complexity. Alipour-Fard et al. [6] presented a multibranch selective kernel network to capture features from different receptive fields. In [7], a network combined of 3D and 2D convolutions is proposed for HSI classification.

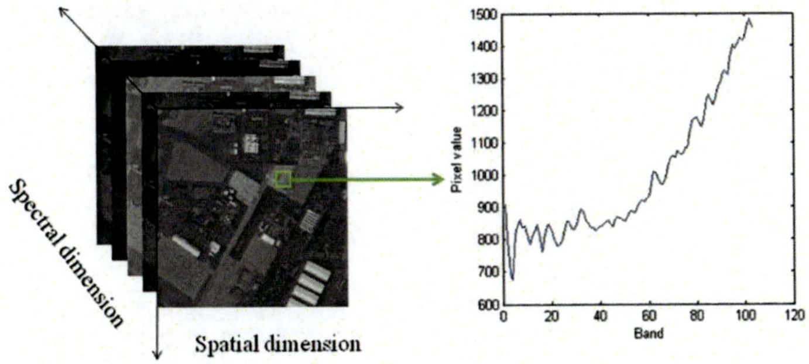

Fig. 1. Illustration of 3D calibrated attention module (3DCAM).

Recently, self-attention mechanism has been employed in computer vision to improve the feature representation [8–13]. The attention mechanism tells us where to focus and improves the intested feature representations. In particular, channel or spatial attentions are employed on the input data, therefore, the most important features are emphasized and the unnecessary ones are suppressed. Hu et al. [14] proposed SENet which introduced the channel attention to emphasize the critical channel-wise features. Following that, CBAM [15] and ECANet [16] follows the idea of SENet, presented more powerful channel attention. Inspired by these works, many self-attention based works have been proposed for remote sensing image interpretation [18–21]. Li et al. [22] proposed a dual-channel spatial, spectral and multiscale attention convolutional long short-term memory neural network for multi-modal data joint classification. Cai et al. [23] used the cross-attention for hyperspectral image feature representation. Attentions along the horizontal and vertical directions are exploited and the classification performance is improved. Zhu et al. [24] proposed an end-to-end residual spectral-spatial attention network for HSI classification. Sequential spectral-spatial attention module is designed to avoid overfitting and accelerate the training. Besides, shuffle attention [25], densely connected attention [26], and cooperative attention [27] are employed for remote sensing image classification.

Although the above-mentioned methods have made great progress, it is non-trivial to build a robust HSI classification model, due to the following reason: *Most existing CNN-based methods neglect the feature correlation of channels in feature maps.* The vast majority of networks deal with different layers of receptive fields are equal. Since the features in different channels are treated equally, some detailed information tends to get lost. Moreover, the long message information of the shallow network tends to get lost. This will lead to inaccuracy of predictions.

To solve this problem, inspired by holistic attention [28], we propose 3D calibrated attention module (3DCAM) to compute the correlation among channels to improve the discriminative ability of the hyperspectral representation network. The 3DCAM can be used for the correlation among channels to achieve more powerful feature representation and correlation learning. Furthermore, we design a **3D C**alibrated **A**ttention-based **Net**work (3DCANet), which integrates 3D-CNN and 2D-CNN layers. Both the spectral and spatial feature are employed for classification. Extensive experiments on the Indian Pines and Pavia University datasets demonstrate that the proposed method performs favorably against three closely related methods.

Table 1. Details of the proposed 3DCANet with window size of 25 × 25. "Class" in the last layer means the number of land-cover classes.

Layer name	Type	Kernel number@size	Shape
Input	–	–	(25, 25, 30, 1)
3D_Conv_1	3DConv	8 @ 5×5×7	(21, 21, 24, 8)
3D_Conv_2	3DConv	16 @ 3×3×7	(19, 19, 18, 16)
Reshape	Reshape	–	(19, 19, 288)
2D_Conv_1	2DConv	64 @ 3×3	(17, 17, 64)
Attention_1	3DCAM	–	(17, 17, 64)
2D_Conv_2	Conv2D	64 @ 3×3	(15, 15, 64)
Attention_2	3DCAM	–	(15, 15, 64)
Flatten	Flatten		(14400)
FC_1	Linear	–	(256)
FC_2	Linear	–	(128)
FC_3	Linear	–	(Class)

2 Methodology

2.1 Architecture of the Proposed 3DCANet

The hyperspectral pixels in the input HSI data cube exhibit the mixed land-cover classes, which introduces the high intra-class variability and inter-class similarity. Therefore, the principal component analysis (PCA) is employed to reduces the number of spectral bands to remove the spectral redundancy. Afterwards, we extract 3D neighboring patch $P \in \mathbb{R}^{r \times r \times b}$ for each pixel. We use similar hybrid network proposed by Roy [7] for feature extraction. 3D-CNN and 2D-CNN layers are assembled in the network to use both the spectral and spatial feature maps for classification.

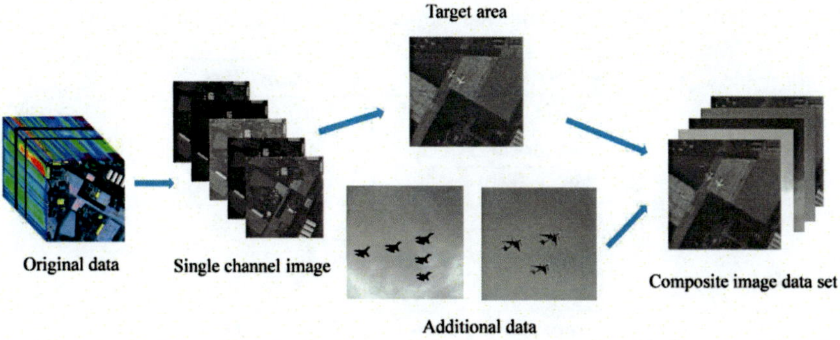

Fig. 2. Network architecture of the proposed 3DCANet.

As illustrated in Fig. 2, the proposed 3DCANet is comprised of two 3D convolutional layers, two 2D convolutional layers, and three fully connected layers. It should be noted that there is a 3DCAM after each 2D convolutional layer. Details of the proposed network are illustrated in Table 1.

2.2 3D Calibrated Attention Module

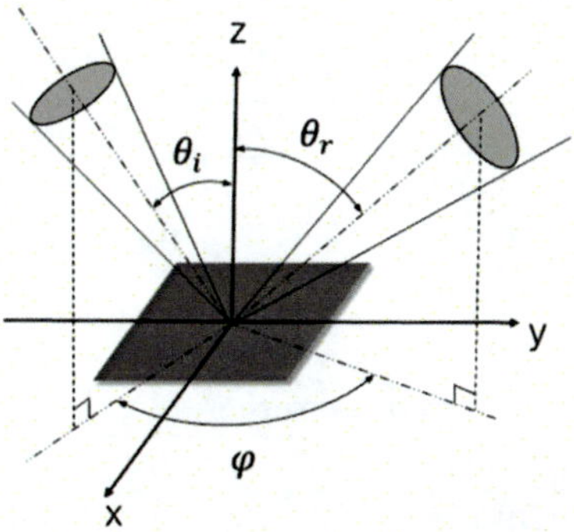

Fig. 3. Classification results of different methods on the Indian Pines dataset. **a** Result by 2D-CNN. **b** Result by 3D-CNN. **c** Result by DRCNN. **d** Result by the proposed 3DCANet.

Existing channel-wise attention mechanism ignore the position information, while the spatial-wise attention mechanism do not exploit the channel dimension formation. To alleviate the problem, we propose a 3D calibrated attention module (3DCAM) to compute the responses from all dimensions of the feature map, as illustrated in Fig. 1. Given an input feature $F \in \mathbb{R}^{H \times W \times C}$, we feed

Table 2. Train-test distribution of samples for the Indian Pines dataset.

#	Class	Train	Test
1	Alfalfa	5	41
2	Corn-notill	143	1285
3	Corn-mintill	83	747
4	Corn	24	213
5	Grass-pasture	48	435
6	Grass-trees	73	657
7	Grass-pasture-mowed	3	25
8	Hay-windrowed	48	430
9	Oats	2	18
10	Soybean-notill	97	975
11	Soybean-mintill	246	2209
12	Soybean-clean	59	534
13	Wheat	21	184
14	Woods	127	1138
15	Building-grass-trees-drives	39	347
16	Stone-steel-towers	9	84
	Total	1027	9322

Table 3. Train-test distribution of samples for the Pavia University dataset.

#	Class	Train	Test
1	Asphalt	66	6565
2	Meadows	186	18463
3	Gravel	21	2078
4	Trees	31	3033
5	Painted metal sheets	13	1332
6	Bare Soil	50	4979
7	Bitumen	13	1317
8	Self-blocking bricks	37	3645
9	Shadows	9	938
	Total	426	42350

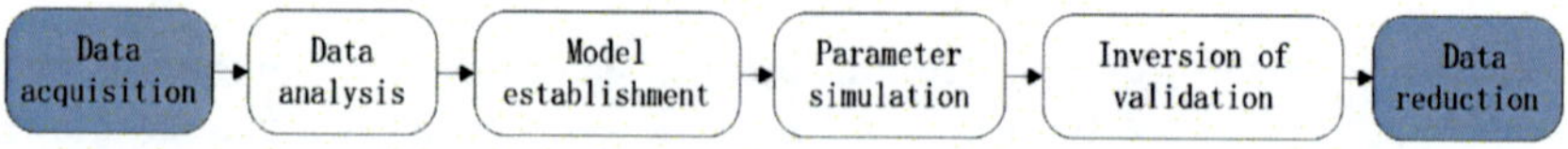

Fig. 4. Classification results of different methods on the Pavia University dataset. **a** Result by 2D-CNN. **b** Result by 3D-CNN. **c** Result by DRCNN. **d** Result by the proposed 3DCANet.

F to a 3D convolution layer to generate attention map by capturing joint channel and spatial features. The 3D convolution is performed with kernel size of $3 \times 3 \times 3$. Then, we obtain three groups of attention maps W_{cam}. Therefore, the module can extract powerful feature representations to describe inter-channel and intra-channel information in continuous channels.

Next, element-wise multiplication is performed between the attention map W_{cam} and input feature F. Afterwards, the output feature F_{out} is computed as follows:

$$F_{out} = \text{softmax}(W_{cam}) \odot \lambda F + F, \tag{1}$$

where λ is a scale factor which progressively improved in the follow iterations. $\odot$ is the element-wise product.

3 Experimental Results and Analysis

In this section, extensive experiments are conducted using three real SAR datasets. First, we describe the datasets used in the experiment and give the evaluation criteria. Then, the critical parameters that affect the change detection result are discussed.

3.1 Datasets and Evaluation Metric

Indian Pines and Pavia University datasets are selected as the benchmark data sets to verify the effectiveness of the proposed method. *Indian Pines dataset* contains a hyperspectral data cube of 145×145 pixels of 224 bands of spectral information. For Alfalfa, Grass-pasture-mowed, and Oats, we randomly select 15 samples for training. For the other classes, we randomly select 50 samples for training. *Pavia University dataset* comprises of 610×430 pixels with the spatial size of 1.3m. We randomly select 25% samples of each class for training.

We use the overall accuracy (OA) and Kapp coefficient (Kappa) to evaluate the HSI classification performance. The classification results of the proposed 3DCANet are compared with 2D-CNN [29], 3D-CNN [30], and DRCNN [31].

3.2 Ablation Study

In order to explore the effectiveness of 3DCAM module, ablation studies are conducted on both datasets. We designed two variants: 1) **3DCANet** refers to the proposed network with 3DCAM, 2) **w/o 3DCAM** refers to the proposed network removing the CAM module. From Table 4, it can be observed that removing the 3DCAM affects the feature representations and degrades the classification performance.

Table 4. Ablation Studies of the 3DCAM.

Method	Indian Pines	Pavia University
3DCANet	96.17	99.36
w/o 3DCAM	95.14	98.54

3.3 Performance Evaluation

Table 5 illustrates the classification results on the Indian Pines dataset. The corresponding classification maps are shown in Fig. 3. It can be seen that the proposed 3DCANet obtains the most smooth classification map. The OA value of 3DCANet is improved by 3.90%, 1.32%, and 1.55% over 2D-CNN, 3D-CNN, and DRCNN, respectively. It implies that the proposed 3DCANet can effectively model the holistic dependencies among channels and positions, which is valuable for HSI feature representation.

Table 5. Classification results of different methods on the Indian Pines dataset.

class	2D-CNN	3D-CNN	DRCNN	3DCANet
1	100.0	100.0	90.32	97.87
2	96.37	89.60	89.67	98.14
3	78.52	96.04	89.64	95.58
4	89.43	87.40	93.60	96.41
5	93.28	97.20	96.47	99.08
6	96.24	97.96	100.0	98.31
7	77.78	95.45	87.50	100.0
8	100.0	100.0	100.0	99.53
9	92.31	100.0	90.22	100.0
10	93.13	88.58	87.60	97.26
11	92.15	96.97	98.61	97.69
12	87.03	92.93	91.21	94.76
13	99.01	91.93	98.56	96.65
14	96.03	99.36	98.98	99.03
15	88.37	92.04	96.92	99.40
16	90.29	98.91	94.90	94.38
OA	92.27	94.85	94.62	96.17
Kappa	91.21	93.14	93.88	95.24

For the Pavia University dataset, the classification results are displayed in Fig. 4. The corresponding evaluation metrics are listed in Table 6. It obvious that the OA of 3DCANet is improved by 3.73%, 2.15%, 5.16% , and 1.95% over 2D-CNN, 3D-CNN, and DRCNN, respectively. It shows that compared with existing CNN-based methods, the proposed 3DCANet has better feature representations, and it is really an outstanding and stable HSI classification model.

Table 6. Classification results of different methods on the Pavia University dataset.

class	2D-CNN	3D-CNN	DRCNN	3DCANet
1	97.12	98.22	96.03	99.77
2	99.45	99.34	99.03	99.98
3	89.84	90.46	90.42	99.68
4	98.48	99.28	91.99	98.99
5	100.0	99.48	97.80	99.92
6	86.85	90.57	89.61	99.98
7	86.12	95.09	71.75	99.83
8	94.46	96.32	87.17	99.34
9	98.85	98.44	92.68	97.95
OA	95.63	97.21	94.20	99.36
Kappa	94.71	96.32	92.15	98.15

4 Conclusion

In this paper, we propose the 3DCANet which integrates 3D-CNN and 2D-CNN layers. Therefore, both the spectral and spatial feature are employed for classification. Furthermore, we design 3DCAM to model the holistic dependencies among channels and positions. Extensive experiments on the Indian Pines and Pavia University datasets demonstrate that the proposed 3DCANet outperforms the prevalent CNN-based HSI classification methods. In the future, we plan to work on large scale dataset to verify the proposed 3DCANet.

References

1. Audebert, N., Le Saux, B., Lefevre, S.: Deep learning for classification of hyperspectral data: a comparative review. IEEE Geosci. Remote Sens. Mag. **7**(2), 159–173 (2019)
2. Sun, W., Ren, K., Meng, X., Xiao, C., Yang, G., Peng, J.: A band divide-and-conquer multispectral and hyperspectral image fusion method. IEEE Trans. Geosci. Remote Sens. **60**, 1–13, Art. no. 5502113 (2022)
3. Song, M., Yu, C., Xie, H., Chang, C.-I.: Progressive band selection processing of hyperspectral image classification. IEEE Geosci. Remote Sens. Lett. **17**(10), 1762–1766 (2020)

4. Zhong, Z., Li, J., Luo, Z., Chapman, M.: Spectral-spatial residual network for hyperspectral image classification: A 3-D deep learning framework. IEEE Trans. Geosci. Remote Sens. **56**(2), 847–858 (2018)
5. Paoletti, M.E., et al.: Capsule networks for hyperspectral image classification. IEEE Trans. Geosci. Remote Sens. **57**(4), 2145–2160 (2019)
6. Alipour-Fard, T., Paoletti, M.E., Haut, J.M., Arefi, H., Plaza, J., Plaza, A.: Multibranch selective kernel networks for hyperspectral image classification. IEEE Geosci. Remote Sens. Lett. **18**(6), 1089–1093 (2021)
7. Roy, S.K., Krishna, G., Dubey, S.R., Chaudhuri, B.B.: HybridSN: exploring 3-D-2-D CNN feature hierarchy for hyperspectral image classification. IEEE Geosci. Remote Sens. Lett. **17**(2), 277–281 (2020)
8. Lai, Q., Khan, S., Nie, Y., Sun, H., Shen, J., Shao, L.: Understanding more about human and machine attention in deep neural networks. IEEE Trans. Multimed. **23**, 2086–2099 (2021)
9. Zhu, Y., Zhao, C., Guo, H., Wang, J., Zhao, X., Lu, H.: Attention CoupleNet: fully convolutional attention coupling network for object detection. IEEE Trans. Image Process. **28**(1), 113–126 (2019)
10. Lian, S., Jiang, W., Hu, H.: Attention-aligned network for person re-identification. IEEE Trans. Circuits Syst. Video Technol. **31**(8), 3140–3153 (2021)
11. Du, W., Wang, Y., Qiao, Y.: Recurrent spatial-temporal attention network for action recognition in videos. IEEE Trans. Image Process. **27**(3), 1347–1360 (2018)
12. Wang, W., Shen, J.: Deep visual attention prediction. IEEE Trans. Image Process. **27**(5), 2368–2378 (2018)
13. Guo, W., Zhang, Y., Yang, J., Yuan, X.: Re-attention for visual question answering. IEEE Trans. Image Process. **30**, 6730–6743 (2021)
14. Hu, J., Shen, L., Albanie, S., Sun, G., Wu, E.: Squeeze-and-excitation networks. IEEE Trans. Pattern Anal. Mach. Intell. **42**(8), 2011–2023 (2020)
15. Woo, S., Park, J., Lee, J.-Y., Kweon, I.: CBAM: Convolutional block attention module. In: Proceedings of European Conference on Computer Vision (ECCV) (2018)
16. Wang, Q., Wu, B., Zhu, P., Li, P., Zuo, W., Hu, Q.: ECANet: Efficient channel attention for deep convolutional neural networks. In: Proceedings of the IEEE/CVF Conference on Computer Vision and Pattern Recognition (CVPR) (2020)
17. Wang, K., Du, S., Liu, C., Cao, Z.: Interior attention-aware network for infrared small target detection. IEEE Trans. Geosci. Remote Sens. **60**, 1–13, Art no. 5002013 (2022)
18. Xiong, W., Xiong, Z., Cui, Y.: An explainable attention network for fine-grained ship classification using remote sensing images. IEEE Trans. Geosci. Remote Sens. **60**, 1–14, Art no. 5620314 (2022)
19. Han, Z., Gao, Y., Jiang, X., Wang, J., Li, W.: Multisource remote sensing classification for coastal wetland using feature intersecting learning. IEEE Geosci. Remote Sens. Lett. **19**, 1–5, Art no. 6008405 (2022)
20. Bai, L., Liu, Q., Li, C., Ye, Z., Hui, M., Jia, X.: Remote sensing image scene classification using multiscale feature fusion covariance network with Octave convolution. IEEE Trans. Geosci. Remote Sens. **60**, 1–14, Art no. 5620214 (2022)
21. Lu, Z., Liang, S., Yang, Q., Du, B.: Evolving block-based convolutional neural network for hyperspectral image classification. IEEE Trans. Geosci. Remote Sens. **60**, 1–21, Art no. 5525921 (2022)
22. Li, H.-C., Hu, W.-S., Li, W., Li, J., Du, Q., Plaza, A.: A3 CLNN: Spatial, spectral and multiscale attention ConvLSTM neural network for multisource remote sensing data classification. IEEE Trans. Neural Netw. Learn. Syst. **33**(2), 747–761 (2022)

23. Cai, W., Wei, Z.: Remote sensing image classification based on a cross-attention mechanism and graph convolution. IEEE Geosci. Remote Sens. Lett. **19**, 1–5, Art no. 8002005 (2022)
24. Zhu, M., Jiao, L., Liu, F., Yang, S., Wang, J.: Residual spectral-spatial attention network for hyperspectral image classification. IEEE Trans. Geosci. Remote Sens. **59**(1), 449–462 (2021)
25. Wang, X., Tan, K., Du, P., Pan, C., Ding, J.: A unified multiscale learning framework for hyperspectral image classification. IEEE Trans. Geosci. Remote Sens. **60**, 1–19, Art no. 4508319 (2022)
26. Wang, X., Fan, Y.: Multiscale densely connected attention network for hyperspectral image classification. IEEE J. Selected Top. Appl. Earth Observ. Remote Sens. **15**, 1617–1628 (2022)
27. Dong, Z., Cai, Y., Cai, Z., Liu, X., Yang, Z., Zhuge, M.: Cooperative spectral-spatial attention dense network for hyperspectral image classification. IEEE Geosci. Remote Sens. Lett. **18**(5), 866–870 (2021)
28. Niu, B., Wen, W., Ren, W., Zhang, X., Yang, L., Wang, S., Zhang, K., Cao, X., Shen, H.: Single image super-resolution via a holistic attention network. In: Proceedings of European Conference on Computer Vision, pp. 191–207 (2020)
29. Chen, Y., Jiang, H., Li, C., Jia, X., Ghamisi, P.: Deep feature extraction and classification of hyperspectral images based on convolutional neural networks. IEEE Trans. Geosci. Remote Sens. **54**(10), 6232–6251 (2016)
30. Li, Y., Zhang, H., Shen, Q.: Spectral-spatial classification of hyperspectral imagery with 3D convolutional neural network. Remote Sens. **9**(1) (2017)
31. Zhang, M., Li, W., Du, Q.: Diverse region-based CNN for hyperspectral image classification. IEEE Trans. Image Process. **27**(6), 2623–2634 (2018)

协同融合应用技术

Fine Registration of SAR and Optical Image Based on Improved Hausdorff Distance

Yifan Xu, Aifang Liu(✉), Hui Xu, Long Huang, and Fan Wang

Nanjing Research Institute of Electronics Technology, Nanjing, China
njuxyf@126.com

Abstract. The registration of SAR and optical image is a core step of multi-source image fusion. The quality of registration restricts the effect of multi-source image fusion. Due to the importance of SAR and optical image registration, we analyze the difficulties of SAR and optical image registration first. And we propose a fine registration method for SAR and optical image based on improved Hausdorff distance. Firstly, we correct the nonlinear distortion of SAR image in range direction by interpolation. Secondly, we use the genetic algorithm to search the mapping parameters that minimize the Hausdorff distance between the images to be registered. These parameters are used for affine transformation of optical images to achieve coarse registration. Finally, we use the Fourier operator to calculate the residual translation between SAR and optical image. Experiments on real data show the effectiveness of the proposed method, and the offset between the corresponding images is less than 2 pixels.

Keywords: Image registration · SAR image · Optical image · Hausdorff distance

1 Introduction

With the development of remote sensing technology, we can get a large number of remote sensing images obtained by different imaging systems, such as infrared images, optical images, and SAR images. Remote sensing images acquired by different types of imaging systems have their advantages and disadvantages. SAR images are microwave images. Not only are they immune to weather, temperature, and nighttime, they can also reflect the target's dielectric constant, roughness, polarization, and other information. However, SAR images acquired in coherent systems are easily affected by speckles, which are easy to cover up the details. Optical remote sensing images are visible images with precise details and good visual effects. Though they are the most common remote sensing images, they are easily affected by weather and night.

The fusion of multi-source remote sensing images can overcome shortcomings of different images, and highlight their advantages. Further more, it provides more abundant information, and help to describe the observed targets entirely and objectively. To realize the fusion of multi-source remote sensing images, the registration of multi-source remote sensing images must be carried out. That is, the remote sensing images of various resolutions and various frequencys obtained by different systems are matched at pixels.

L. Wang et al. (Eds.): CHREOC 2022, LNEE 969, pp. 229–241, 2023.
https://doi.org/10.1007/978-981-19-8202-6_21

Generally, the imaging region to be registered is the same. But different remote sensing images have different imaging mechanisms, different resolutions, different observation angles, different frequencys, and the same or different imaging time. Therefore, pixel-level image registration is not an easy task. The quality of registration directly affects the effect of multi-source image fusion and other post-processings. So, high-quality remote sensing image registration must be carried out.

Among the multi-source remote sensing images, SAR images and optical images are the most typical remote sensing images. They have good complementarity. People have done a lot of researches on their registration. In [1], the closed region is extracted by image segmentation, and then the edge of the closed region is matched to achieve registration. In this method, the images must have a closed edge. If there is no closed edge, it is challenging to achieve registration. In [2], the method based on image correlation is used for registration. However, due to the significant difference in imaging mechanism, cross-correlation matching is not wholly effective. In [3], firstly, the edge of SAR images and optical images are extracted and then matched according to a certain similarity measure. This method is restricted by the accuracy of image edge extraction and complex edge screening.

2 Methodology

The registration of SAR and optical image is challenging. The difficulty lies in the following aspects.

- The range of SAR image shows the slant range resolution, not the ground range resolution. For SAR images, at different oblique viewing angles, its ground range resolution is different. In other words, there is nonlinear distortion in the range of SAR images.
- The azimuth and range resolutions of SAR image are inconsistent. According to the principle of SAR imaging, the range resolution of SAR image is related to the bandwidth. The azimuth resolution is associated with the antenna length. Generally, the two are not the same, which means different zoom ratios in range and azimuth between SAR image and optical image.
- There is a rotation between the SAR image and the optical image, and the rotation angle between them is unknown.

Under the combined effect of the above factors, it is not easy to realize the registration of SAR and optical image. This paper proposes a fine registration method of SAR image and optical image based on improved Hausdorff distance and Fourier operator to solve this problem. As shown in Fig. 1, the method includes the following steps. Firstly, correcting the nonlinear range distortion of the SAR image by range interpolation. Secondly, several feature points are selected manually and connected by lines to get feature point pairs. Thirdly, using the genetic algorithm to search the parameters that can minimize the improved Hausdorff distance, and remove the significant components of scaling, rotation, and translation, to realize a coarse registration. Finally, calculating the residual translation by the Fourier operator to achieve a fine registration of SAR image and optical image.

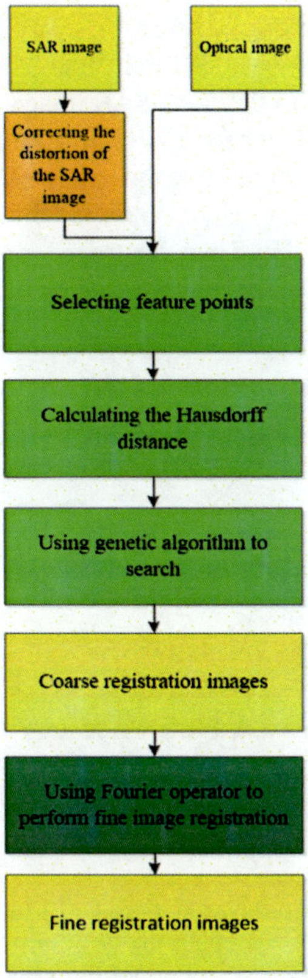

Fig. 1. The flow chart of the proposed algorithm

2.1 Correcting the Nonlinear Distortion of the SAR Images

SAR systems achieve high range resolution by transmitting large bandwidth LFM signals, and achieve high azimuth resolution by forming a synthetic aperture through the movement of the antenna. The range resolution of a SAR image can be expressed as (1).

$$\rho_r = \frac{c}{2B} \tag{1}$$

where B represents the bandwidth of the transmitted signal, and c represents the light speed. It can be seen from (1) that the range resolution of a SAR image is related to the transmission bandwidth B. The larger the transmission bandwidth, the higher the resolution, and vice versa. Figure 2 shows the relationship between ground range resolution and slant range resolution under radar imaging geometry. It can be seen that ρ_r represents the slant range resolution of the SAR image. Usually, we need to convert the slant range resolution ρ_r to the ground range resolution ρ_e, as shown in (2)

$$\rho_e = \frac{\rho_r}{\sin\theta} = \frac{c}{2B\sin\theta} \tag{2}$$

where θ is the radar incidence angle, it can be seen that the ground range resolution ρ_e becomes better with the increase of the incidence angle θ when the transmission bandwidth is fixed. Therefore, in the same SAR image, the ground range resolution of the near end is different from that of the far end. And it will change nonlinearly with varying angles of incident. Assuming that the height of the SAR system is 6000 m, the incidence angle of the radar line of sight center is 40°, the range beam width is 4°, the LFM signal bandwidth is 420MHz. Then the ground range resolution of the near end of the SAR image is 0.5801m, and that of the far end is 0.5337 m.

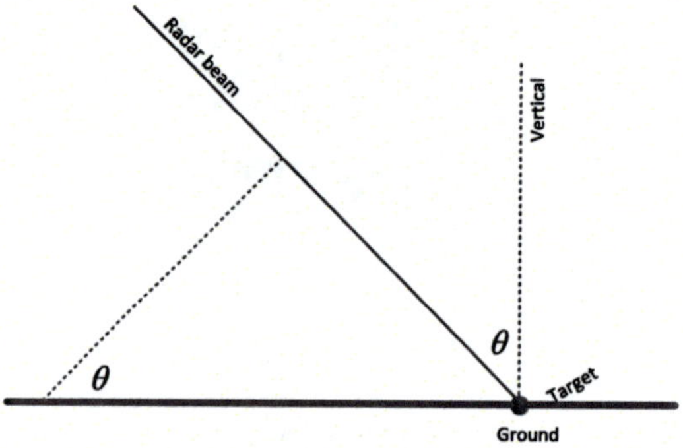

Fig. 2. The relationship between ground range resolution and slant range resolution

It can be seen that there is nonlinear distortion in the ground range direction of the SAR image, especially in the areas with large topographic relief. Therefore, it is impossible to directly use polynomial correction or affine transformation to register the SAR image with the optical image. It is necessary to resample the image in the ground range direction to remove the nonlinear distortion and get the SAR image with uniform resolution in the ground range direction.

As shown in Fig. 3, the actual SAR image obtained is uniform in the slant range direction, but we need to get a SAR image with a uniform ground range resolution. According to (2), the required image can be obtained by interpolating the actually acquired SAR image to correct the nonlinear distortion of the ground range direction.

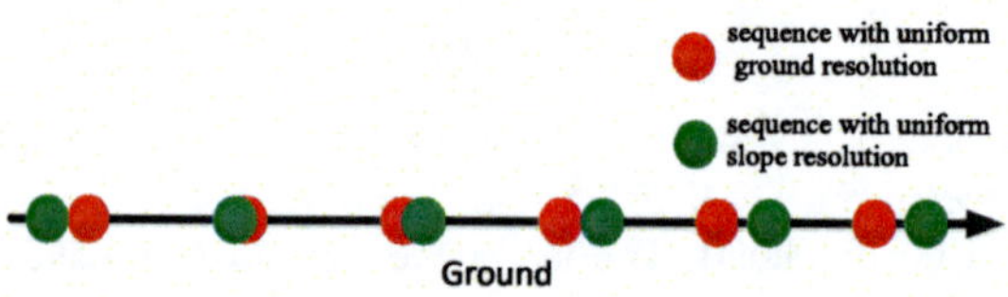

Fig. 3. Image sequence with uniform ground range resolution and image sequence with uniform slope range resolution

After correcting the nonlinear distortion of the SAR image in the ground range, it can be considered that the optical image can be registered with the SAR image through scaling, rotation, and translation. The geometric transformation between SAR image and optical image can be expressed as (3).

$$\begin{pmatrix} x'' \\ y'' \end{pmatrix} = \begin{pmatrix} s_x & 0 \\ 0 & S_y \end{pmatrix} \begin{pmatrix} \cos\varphi & -\sin\varphi \\ \sin\varphi & \cos\varphi \end{pmatrix} \begin{pmatrix} x' \\ y' \end{pmatrix} + \begin{pmatrix} x_0 \\ y_0 \end{pmatrix} \quad (3)$$

where x' and y' represent the coordinates of the optical image in ground range and azimuth,respectively. x'' and y'' represent the coordinates of the SAR image in ground range and azimuth,respectively. The transformation consists of two scales s_x and s_y, one rotation parameter φ, and two translation parameters x_0 and y_0. Where s_x and s_y indicate that SAR images have different scaling in azimuth and ground range. φ is the rotation angle of SAR image relative to optical image. x_0 and y_0 represent the translation of SAR image relative to optical image in ground range and azimuth. As long as the above five parameters are obtained, the registration of SAR and optical image can be realized.

2.2 Selecting Feature Points

To obtain the five parameters mentioned in (3) and realize the registration of SAR and optical image, we first select the corresponding points of SAR and optical image. Then the five parameters are solved. Finally, the coarse registration of SAR and optical image is realized by an affine transformation.

During image registration, common characteristics including intensity, points, texture, edges, and contours. The intensity of a optical image represents the scattering characteristics of the object in the visible light band. The intensity of a SAR image reflects the scattering characteristics of the object in the microwave band. Due to the difference in scattering characteristics and imaging mechanism, the intensity representation of SAR and optical image are diverse. It isn't easy to use the intensity for SAR and optical image registration.

The edges and contours of a image need to be obtained through segmentation. Due to the significant difference between SAR images and optical images, and the speckles in SAR images, it is difficult to obtain continuous edges and contours in the SAR image corresponding to the optical image. In contrast, based on structural information, the extraction of points is relatively simple, and a large number of corresponding points can always be found in SAR images and optical images.

In this paper, feature point pairs are extracted in the following ways:

Step1: Manually select multiple pairs of feature points in the SAR image and the optical image. The feature points in the SAR image are marked as A_{I}, and the feature points in the optical image are marked as B_{I}. Collectively referred to as type I feature point pairs.

Step2: For the feature points A_{I} in the SAR image and the feature points B_{I} in the optical image, they are connected by straight lines from left to right and top to bottom. And the pixels passed by the straight lines are also extracted as feature points, which are named as feature points A_{II} and B_{II}. They are called class II feature point pairs. As shown in Fig. 4, The points surrounded by a circle are manually extracted points A_{I}, and the pixels passed by the straight lines are feature points A_{II}.

Through the above operations, two types of feature point pairs can be obtained, namely, type I feature point pairs and type II feature point pairs.

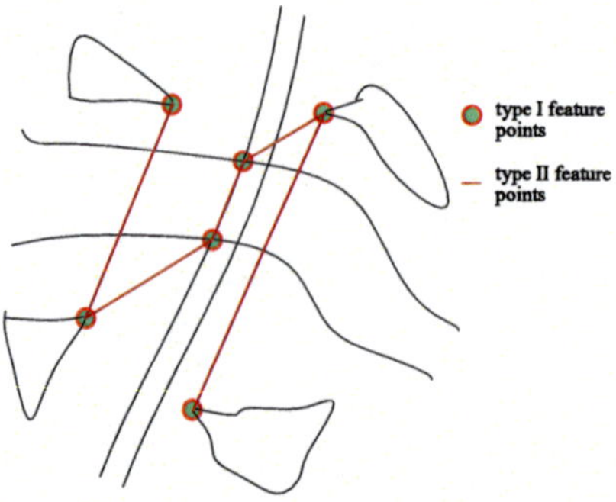

Fig. 4. Selected type I feature points and type II feature points

2.3 Using the Genetic Algorithm and the Improved Hausdorff Distance to Realize a Coarse Registration

In Sect. 2.2, based on the structural information, we can obtain two types of feature point pairs: type I feature point pairs and type II feature point pairs. In this section, based on the extracted feature point pairs, the genetic algorithm is used to search for the five parameters that can minimize the improved Hausdorff distance, to realize the coarse registration of the SAR and the optical image.

The Hausdorff distance [4] is a measure of the similarity between two sets of points, and it is a definition of the distance between two sets of points. Suppose there are two point sets, A and B, denoted by (4) and (5), respectively.

$$A = \{a_1, a_2, ...a_i, ..., a_p\} \tag{4}$$

$$B = \{b_1, b_2, ...b_j, ..., b_q\} \tag{5}$$

where, a_i and b_j represent the elements in A and B respectively, and the subscripts i and j represent the sequence numbers of the elements,respectively. Then the Hausdorff distance $H(A, B)$ between these two point sets is defined as (6).

$$H(A, B) = \max(h(A, B), h(B, A)) \tag{6}$$

where, $h(A, B)$ and $h(B, A)$ can be expressed as (7) and (8), respectively.

$$h(A, B) = \max_{a_i \in A} \min_{b_j \in B} ||a_i - b_j|| \tag{7}$$

$$h(B, A) = \max_{b_j \in B} \min_{a_i \in A} ||b_j - a_i|| \tag{8}$$

where, $|| \cdot ||$ represents the norm, $\min(\cdot)$ represents the minimum value, and $\max(\cdot)$ represents the maximum value. $H(A, B)$ is called two-way Hausdorff distance, which is the basic Hausdorff distance. The $h(A, B)$ and $h(B, A)$ in (6) are called one-way Hausdorff distances from A to B and from B to A, respectively. To calculate $h(B, A)$, first calculate the norm of all points in B to a certain point a_i in A, and find the smallest norm, denoted as S. Then calculate the distance S for all points in set A, where the maximum value of S is $h(A, B)$. $h(B, A)$ can be obtained in the same way.

It can be seen from (6) that the two-way Hausdorff distance $H(A, B)$ is the larger of the one-way Hausdorff distance $h(A, B)$ and $h(B, A)$, which measures the position mismatch between the two point sets. It can be seen that when the two-way Hausdorff distance is the smallest, the positions of the two sets are the closest,and the position similarity is the highest.

In the previous section, we obtained two types of feature point sets: type I feature point set and type II feature point set. Among them, the type I point set is determined by manual selection of structural information. The type II feature point set is obtained by connecting the type I feature points. And the type I feature point set has higher accuracy. To comprehensively utilize the type I feature point set and the type II feature point set, an improved Hausdorff distance $H_{modified}(A, B)$ is proposed, which is defined as (9).

$$H_{modified}(A, B) = \alpha H(A_{,}B_{)} + (1 - \alpha)H(A_{,}B_{)} \tag{9}$$

where $H(A_{\mathrm{I}}, B_{\mathrm{I}})$ is the Hausdorff distance of type I point sets; $H(A_{\mathrm{II}}, B_{\mathrm{II}})$ is the Hausdorff distance of type II point sets; α is a coefficient between 0 and 1, which is used to adjust the weight between type I point sets and type II point sets.

From (9), it can be seen that the improved Hausdorff distance $H_{modified}(A, B)$ measures the mismatch between two point sets. If the improved Hausdorff distance $H_{modified}(A, B)$ reaches the minimum, these point pairs in the images have achieved coarse registration.Therefore, the following steps can be used to achieve a coarse registration of SAR and optical image.

Firstly, change the five parameters in (3) and perform projection transformation on the feature point sets A_{I} and A_{II} in the SAR image according to (3) to obtain new feature point sets A_{I} and A_{II} . Secondly, calculate the improved Hausdorff distance $H_{modified}(A, B)$ based on the new feature point pairs. Thirdly, change the five parameters and calculate the improved Hausdorff distance $H_{modified}(A, B)$ cyclically until the smallest $H_{modified}(A, B)$ is obtained. At this time, the corresponding five parameters are the required mapping parameters. Finally, according to (3), the coarse registration image can be obtained by mapping the original optical image with these five parameters.

In order to speed up the search speed, the genetic algorithm can be used to search until the smallest Hausdorff distance is obtained. The genetic algorithm is a method to imitate natural selection and evolution in high-dimensional space. It provides a general framework for solving optimization problems of complex systems, especially in the optimization of multi-extremum problems, it may temporarily stay in some sub advantages. Still the mutation may make it jump to the best. The specific information can be referred to [5], and this paper will not give a detailed description.

Since the genetic algorithm search results may temporarily stay on some sub-advantages, the smallest $H_{modified}(A, B)$ may not be obtained in one optimization. The optimal mapping parameters can be obtained through multiple iterative searches.

2.4 Using Fourier Operator to Perform a Fine Image Registration

Through the above operations, a coarse registration is achieved between SAR and the optical image. In order to achieve the pixel-level registration, further processing is required, a fine registration based on Fourier operator is proposed.

As shown in (10), the reference image is $f_1(x, y)$, the image to be registered is $f_2(x, y)$, where $f_2(x, y)$ is $f_1(x, y)$ translation (x_0, y_0).

$$f_2(x, y) = f_1(x - x_0, y - y_0) \tag{10}$$

where, x and y are the coordinates of the image in the ground range and azimuth, respectively. x_0 and y_0 represent the offset of the two images in the ground range and azimuth. Two-dimensional Fourier transform is performed on the reference image $f_1(x, y)$ and the image to be registered $f_2(x, y)$, and their two-dimensional Fourier transform results are respectively $F_1(u, v)$ and $F_2(u, v)$. According to the Fourier transform principle, the two have the relationship as (11).

$$F_2(u, v) = e^{-j(ux_0 + vy_0)} F_1(u, v) \tag{11}$$

where, u and v are the frequency axis of ground range and azimuth, respectively. It can be seen from (11) that the displacement in the time domain of the image is equivalent to the phase shift in the frequency domain. Similarly, the phase shift in the frequency domain can also realize the displacement in the image domain.

The cross power spectrum (energy spectrum) of $f_1(x, y)$ and $f_2(x, y)$ in the frequency domain can be expressed as (12).

$$R = \frac{F_1(u, v)F_2^*(u, v)}{|F_1(u, v)||F_2^*(u, v)|} = e^{-j(ux_0 + vy_0)} \tag{12}$$

It can be seen from (12) that the phase difference between the two images is equivalent to the phase of the cross power spectrum of the two images. The inverse Fourier transform of (12) will form an impulse function $\delta(x - x_0, y - y_0)$ at (x_0, y_0) in the corresponding image domain. Therefore, the required translation amount can be determined by the peak position in the image.

Combining the above analysis, in order to obtain a pixel-level registration, the Fourier operator is used to further process the coarse registration image. The specific steps are as follows.

Step1: Interpolate the feature point image of the reference image and the feature point image of the coarse registration image. The interpolation multiple can be set according to the required registration accuracy. The larger the interpolation multiple, the higher the registration accuracy, but the corresponding operation will also become more extensive.

Step2: Perform two-dimensional Fourier transform on the feature point image of the interpolated reference image and the feature point image of the interpolated coarse registration image to obtain $F_1(u, v)$ and $F_2(u, v)$.

Step3: Obtain the cross power spectrum of the feature point image of the reference image and the feature point image of the image to be registered through (12) in the frequency domain.

Step4: Perform a two-dimensional inverse Fourier transform on the complex power spectrum image, and find the point with the highest power and its corresponding coordinates (x_0, y_0), through which the frequency shift of the image to be registered relative to the reference image can be obtained the amount.

Step5: Pixel-level registration can be realized by shifting and resampling the coarse registration image.

3 Registration Result and Evaluation

In order to verify the effectiveness of the method proposed in this paper, a registration test is carried out using a SAR image captured by the N-SAR airborne SAR system [6] produced by the 14th Research Institute of China Electronics Technology Group Corporation and a Google Earth optical image. The SAR image was acquired on November 3, 2017. The azimuth width is 2.71km, the range width is 2.1km, the range resolution is 0.71m, and the azimuth resolution is 0.65m. Figure 5 shows the SAR image and optical image to be registered.

3.1 Registration Result

According to the registration process, first, select feature points. As shown in Fig. 5, the points chosen by the red circle are type I feature point pairs, and most of these feature points are some apparent intersections or corners of ground objects. Connect the type I feature points in the order from left to right and top to bottom, and the points passed by the lines are the type II feature point pairs.

The improved Hausdorff distance method based on genetic algorithm is used for coarse registration of SAR and optical image. Figure 6a–d show the spatial positions of feature point pairs at different iteration times. Figure 6a is the spatial position of the feature points in the original SAR image and the optical image. Figure 6b is the spatial position of the feature points after one iteration. Figure 6c is the spatial position of the feature points after two iterations. Figure 6d is the spatial position of the feature points after three iterations. It can be seen from Fig. 6 that when the number of iterations is less, there are some significant gaps between feature points. With the increase of the iteration, the gap between feature points gradually decreases.

After using the Fourier operator to perform the fine registration processing on the rough registration result, the spatial position of the feature points is shown in Fig. 6e. It can be seen that compared to Fig. 6d, the gap between the feature point pairs is smaller.

In order to better display the registration effect, as shown in Fig. 7, we select different sub-blocks from the registered optical image and SAR image, respectively,and stitch them into one image. It can be seen from Fig. 7 that the transition between the SAR image and the optical image at adjacent sub-blocks is very smooth and natural. The continuity of features (roads, bridges, and farmland) is excellent. This shows that the SAR image and the optical image are well registered.

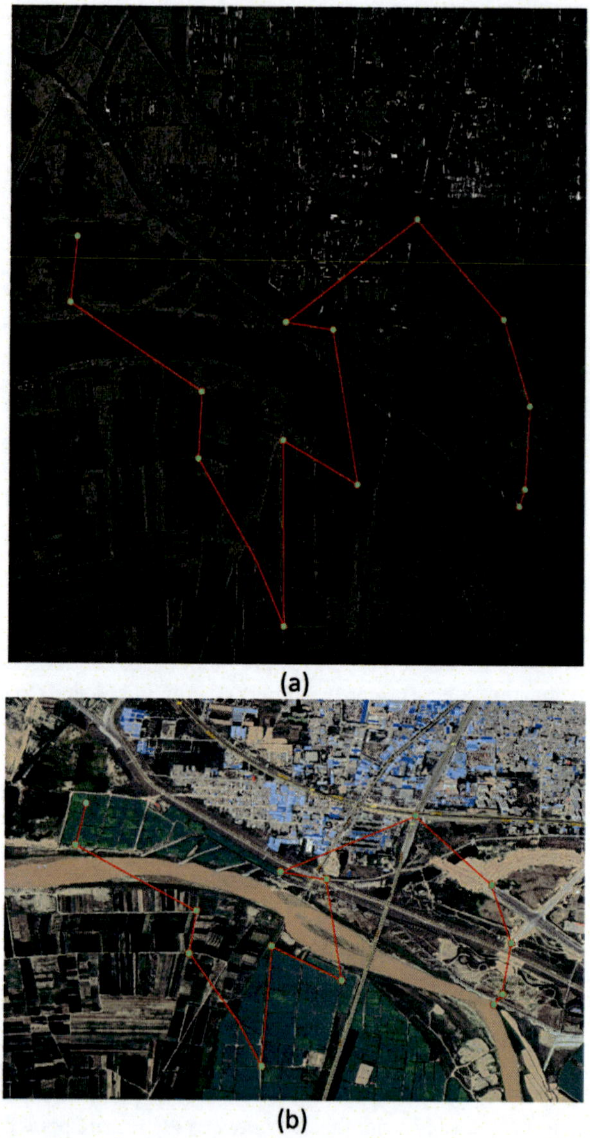

Fig. 5. Selected SAR image and optical image. **a** The selected SAR image which was captured by the N-SAR airborne SAR system, **b** the selected optical image is a Google Earth optical image

3.2 Evaluation of Registration Result

The root mean square distance method is used to quantitatively analyze the registration accuracy. The root mean square distance method judges whether the registration is successful by comparing the root mean square distance between the test point sets of

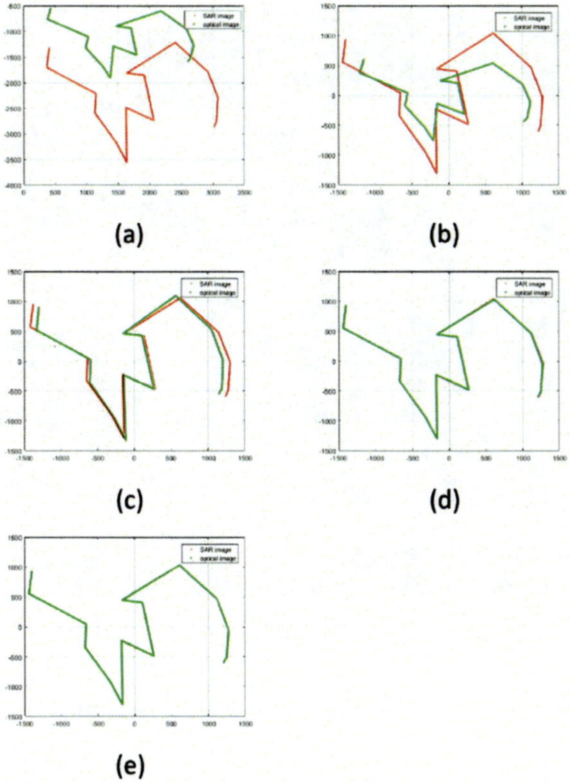

Fig. 6. The spatial positions of feature point pairs at different iteration times. (a) The spatial position of the feature points in the original SAR image and the optical image, (b) The spatial position of the feature points after one iteration, (c) The spatial position of the feature points after two iterations, (d) The spatial position of the feature points after three iterations, (e) The spatial position of the feature points after using the Fourier operator

two images. Suppose there are a total of n pairs of test points, where $x_k{}'$ represents the ground range coordinate of the *kth* test point in the optical image. $y_k{}'$ represents the azimuth coordinate of the *kth* test point in the optical image. $x_k{}''$ represents the ground range coordinate of the *kth* test point in the SAR image. $y_k^{''}$ represents the azimuth coordinate of the *kth* test point in the SAR image. The root mean square distance error $RMSE_x$ and $RMSE_y$ in the distance and azimuth between the registered images are (13) and (14), respectively.

$$RMSE_x = \sqrt{\frac{\sum_{k=1}^{n} (x'_k - x_k{}''^2}{n}} \tag{13}$$

$$RMSE_y = \sqrt{\frac{\sum_{k=1}^{n} (y'_k - y_k{}''^2}{n}} \tag{14}$$

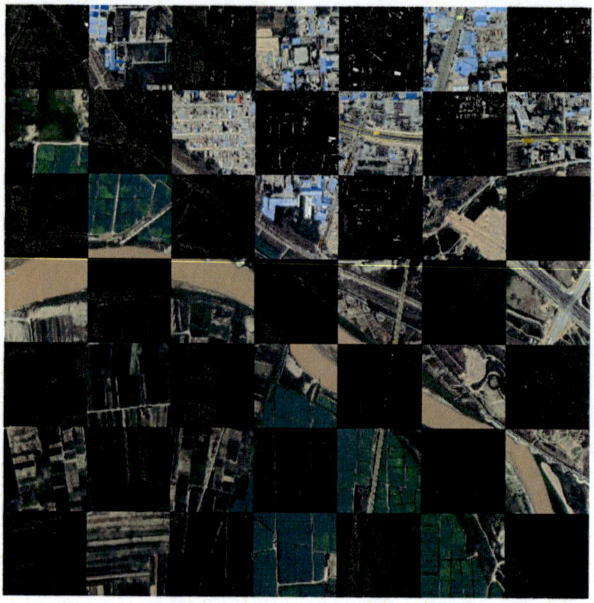

Fig. 7. The registration result of SAR and optical image

The total mean square error $RMSE_T$ in ground range and azimuth can be expressed as (15).

$$RMSE_T = \sqrt{\frac{\sum_{k=1}^{n} [(x_k' - x_k{''}^2 + (y_k' - y_k{''}^2]}{n}} \tag{15}$$

The better the two images are registered, the smaller the corresponding range root mean square distance error $RMSE_x$, the azimuth root mean square distance error $RMSE_y$, and the total root mean square distance error $RMSE_T$.

Table 1. The registration accuracy of SAR and optical image

No. of region	No. of points tested	$RMSE_x$	$RMSE_y$	$RMSE_T$
1	20	0.86	0.72	1.13
2	20	1.29	1.22	1.77
3	20	0.88	1.05	1.37
4	20	0.80	0.81	1.14

Select 20 sets of test points in four different regions of the registered optical image and SAR image to calculate the root mean square distance error. The root mean square error table of the test points is shown in Table 1. It can be seen from Table 1 that the root mean square distance errors of the 4 regions are all less than 2 pixels, indicating that the registration accuracy has reached the pixel level.

4 Conclusion

This paper first analyzes the reasons for the difficulty of registration between SAR and optical image. It points out that the nonlinear distortion of SAR image in range direction, the difference of resolution between range direction and azimuth direction, and the rotation between SAR and optical image cause the difficulty of registration. Then the range nonlinear distortion is corrected by interpolation. The improved Hausdorff distance is minimized by the genetic algorithm and the Fourier operator is used to achieve a fine registration at pixel level. Finally, the actual image is used for registration test. The registration result is good, and the registration accuracy is within 2 pixels, which proves the effectiveness of the method in this paper and shows that the proposed method has certain application value.

References

1. Hong, T., Schowengerdt, R.: A robust technique for precise registration of radar and optical satellite images. J. Photogramm. Eng. Remote Sens. **71**(5), 585–593 (2005)
2. You, H., Hu, Y.: Investigation on fine registration for SAR and optical image. J. Radars **79**(2), 269–274 (2014)
3. Suri, S., Reinartz, P.: Mutual-information-based registration of TerraSAR-X and Ikonos imagery in urban areas. J. IEEE Trans. Geosci. Remote Sens. **48**(2), 939–949 (2010)
4. Dubuisson, M., Jain, A.: A modified Hausdorff distance for object matching. In: International Conference on Pattern Recognition, vol. 25, no. 1, pp. 566–568. IEEE (1994)
5. Mao, H., Qiuze, Y.: Matching SAR image to optical image using modified hausdorff distance and genetic algorithms. In: Conference on Pattern Recognition and Computer Vision, vol. 11, no. 3, pp. 15–17 (2007)
6. Liu, A., Wang, F.: N-SAR: a new multi-channel multi-mode polarimetric airborne SAR. J. Sel. Top. Appl. Earth Obs. Remote Sens. **11**(9), 3155–3166 (2018)

High Resolution Imagery-Based Statistical Analysis for Urban Rivers Extraction and Quality Classification—Case Study in Beijing District of China

Dongsheng Liu[1,2] and Ling Han[1](✉)

[1] School of Geology Engineering and Geomatics, Chang'an University, Xi'an, China
hanling@chd.edu.cn
[2] PIESAT Information Technology Co., Ltd, Beijing, China

Abstract. The previous work on monitoring the urban water quality using remote sensing focused mainly on the specific spectral/band math and the regression relationship between the spectral signature and water surface's Chlorophyll-a content, while conducting urban wastewater monitoring on the basis of indirect interpretation indicators is still a gap to be bridged. Here we report a novel hybrid method to extract and classify the urban wastewaters (caused mainly by eutrophication) from 4-band high resolution imagery using a multi-step statistical approach. First, the MTSUWI (Modified Two-Step Urban Water Index) algorithm is presented for extracting the urban water automatically, and the Kappa Coefficient comes up to 0.92. Second, using the Minimum Message Length Criterion-Expectation Maximization algorithm (MML-EM), the histogram of the water-masked 1st principal component image was screened into two subpopulations, which are mainly a re-flection of the existing background of the waters, rather than the pollution. Third, the water floating matters (most of them are green algae) is selected as an indirect interpretation key of the polluted wastewaters, and by gradually reducing the interpretation target area, pixels containing green alga were enhanced and then extracted in the brightness image. Fourth, any river reach containing at least one patch of green alga is labeled as "polluted", otherwise they are labeled as "clean/fresh", and finally, eight black-and-odorous wastewaters in the study area were selected out, which are consistent with literatures and observations from the field. This research is one of the first to apply indirect interpretation indicators to 4-band high resolution imagery for the classification of urban polluted rivers.

Keywords: Water quality monitoring · Heavily polluted water · GF-2 · Algorithm

L. Wang et al. (Eds.): CHREOC 2022, LNEE 969, pp. 242–263, 2023.
https://doi.org/10.1007/978-981-19-8202-6_22

1 Introduction

The water quality urgency problem lies primarily in a constant deterioration of ecological state of inland water reservoirs, urban rivers, coastal zones and oceanic bays [1]. Pollutants can be found in waters everywhere, including plastic, oil, toxic chemicals, radioactive waste, domestic and industrial sewage, etc.,. The contamination is also caused by the excessive nutrients and sewage discharge from agricultural pesticides and fertilizers flowing into water bodies [2]. These contamination have adverse influences on marine/inland ecosystem, underground water, death of organisms mass in water, air pollution, aquaculture included. In particular, the deterioration of water quality may be caused by the frequent activities of human, towards being turbid, stink, or black (the black-odorous-water). Remote sensing is utilized in a variety of application domains, and it already has played a major role in water interpretation and water quality analysis [3–5]. In this article we will focus on high resolution imagery-based urban wastewater analysis, for map and monitor environmental pollutants are cretical to make sure a sustainable urban ecosystem in China. Actually, only a few researchers have stepped in this area [6].

Before solving the urban water-quality problem, it is necessary to review the water-extraction problem first. Many algorithms have been proposed for identifying water using remote sensing imagery. Among these algorithms, the spectral water index (WI) is used most widely for it is not difficult to obtain a relative good accuracy [7]. In these methods two or more spectral bands are combined, which use different algebraic operations to enhance the discrepancy between water and land. Generally, the principle of most WIs is similar to that of the normalized-difference vegetation index (NDVI) [8]. Considering the available WIs for high resolution images, NDWI (normalized-difference water index) is still the only option for ex-tract water in urban area for that most high-resolution images only have visible and NIR bands, such as ZY-3, GF-2, IKO-NOS, and RapidEye. However, there are problems when NDWI is applied to high-resolution imagery. For example, it is difficult to map narrow, shallow or turbid water bodies using the NDWI image, and water bodies and dark shadows cannot be easily separated by their spectra [7]. To address these problems, new water indices are proposed with additional shadow detection methods combined. For example, Yao et al. [9] proposed a new WI named High Resolution Water index (HRWI), as well as a method for building shadow detection, to extract urban waters. Wu et al.[10] proposed a novel urban water extracting method, Two-Step Urban Water Index (TS-UWI), which adds Urban Shadow Index (USI) and Urban Water Index (UWI). However, one existing gap of these methods is the little consideration of the thresholding algorithm, and the empirical/default threshold or threshold determined by the OTSU method can be unstable in practice. In this paper we will make some modifications.

With the extracted water bodies at hand, several methods have been proposed to monitor the water quality or water pollution status using remote sensing (RS) imagery. For example, one of the earliest attempts refers to [11], a set of algorithms are proposed for remote assessment of water quality from the perspective of Chlorophyll-a content of phytoplankton, the content of Bacteriochlorophyll-a, together with total suspended matter concentration in wastewater reservoir and oxidation ponds using aircraft imaging spectrometer. Pulliainen et al. [12] also tried to retrieve Chlorophyll-a concentration (which is a sensitive indicator reflecting water quality) from airborne/space-borne spectrometer, and the channel ratio of the used spectrometer seemed to be correlated with the chla concentration, regardless of turbidity level. With the focus on the urban "black-odorous" waters in several heavily contaminated river sections [13–15], etc., attempted to establish a linear regression model between specific spectral reflectance and water quality parameters (e.g. Chlorophyll-a, suspend-ed matter, turbidity, etc.). However, due to the uncertainties of the chemistry of urban wastewater, the related studies have not been promoted. Similar researches can be found in, but not limited to [16, 17], etc. With spatial and spectral resolution of recently launched satellites increased, there are new opportunities for remotely sensed data, and a series of RS image-ry-derived spectral characteristics was developed to detect/monitor the polluted waters. For example, Keith et al. [18] employed hyperspectral imagery to create a three-band approach for Chlorophyll-a inversion. Vignolo et al. [19] constructed a water-quality-index using blue, green and red bands of the Landsat TM imagery. Using the SPOT 5 imagery [20], and coworkers have detected and classified surface water pollution along the coastline by conducting spectral enhancements (e.g. Band blue/green and Principal Component Analysis bands). Besides, Ji et al. [21] constructed a discriminant index for urban "black-odorous" waters (), where R represents reflectance, and VIS is any visible spectral band or band combination. Wen et al. [22] used a ratioing method to observe urban wastewaters by (), where N1 and N2 are empirical thresholds. Based on these, Yao et al. [6] proposed a new black-odorous-water index.

However, in order to map and analyze the polluted waters, previous studies focus mainly on the spectral/band math or the statistical relationship between imagery spectra and water quality parameters, and conducting wastewater interpretation from the perspective of (non-linear) statistical analysis is still a gap to be bridged. In fact, the frequency distribution of image pixel values is an effective way to reveal the distributing characteristics and spatial variation of land covers, and the statistical methods are widely used in this realm [23]. The rationality behind is not only to reduce the uncertainties during the image acquisition, pre-processing, and analysis, but also to approximate the heterogeneity and regional randomness of remote sensing imagery. Pixel-based statistical methods enable interpreters to get new insights into the pattern recognition and image classification, but it is still a challenging task due to the complex mixture and distribution of geographic objects. For instance, (1) how the fresh waters can be separated from the polluted ones in a statistical way? (2) How many types of contaminations such as organic compounds, agricultural, domestic discharge and industrial, can be found from their statistical behaviors in the enhanced RS image? (3) How to improve the generalization and robustness of the known spectral/band math. It should be noted that the statistical algorithm may lead to inferior results if the contaminated waters are spectrally inseparable in essence, and in this case, the search for new interpretation keys or visual cues becomes crucial.

Taken urban black-odorous water as case study, this article is to present a novel method to extract and classify the wastewaters from high resolution images using a multi-step statistical approach. To this end, the article is structured as follows. In Sect. 2, the experimental materials, as well as the proposed method/workflow for detection and analysis of urban "black-odorous" waters are introduced. Section 3 shows and compares the result of model processing and the accuracy assessment is also conducted. Section 4 discusses the proposed method and shifts the research efforts from direct interpretation signs or visual cues to indirect interpretation signs-green algae. The last section illustrates the conclusion of this research.

2 Materials and Methods

2.1 Subsection

One GF-2 scene (level L1A) with no cloud coverage cover the study area (Scene number: GF2_PMS2_E116.6_N40.0_20170609_L1A0002417596) acquired on 06/09/2017 is chosen to explore the algorithms to map the distribution of contaminated/waste waters automatically. Its spatial limits are N 39.9052°–N 40.1561° and E 116.441°–E 116.777°.

As for GF-2 imagery (sensor), there are four multispectral bands including blue (0.45–0.52 μm), green (0.52–0.59 μm) and red (0.63–0.69 μm), as well as Near Infrared (NIR, 0.77–0.89 μm), of which the spatial resolution is 3.2 m. There is also one panchromatic band (0.45–0.90 μm), of which the spatial resolution is 0.8 m. The ancillary materials include, but not limited to, the vector basemap recording the spatial distribution of sewage draining exits, contaminated and uncontaminated riverways which were publicly released [24], as well as the monthly water quality monitoring data.

Based on ENVI v5.3, the pre-processing of images involved radiometric calibration, atmospheric correction, ortho-graphical correction, etc. Here the scenes are georeferenced to CGS-WGS-84 coordinates; and to map the wastewaters presenting in narrow channels, pan-sharpened GF-2 imagery was created by fusing the multispectral bands with the 0.8m pan band using the method of pan-sharpening. However for shallow and narrow waters, the land adjacency effect is inevitable, and as a result, these water pixels are quite sensitive to the changes of threshold and are easy to confuse with land.

2.2 Analysis Methods

The proposed method for extracting and classifying the urban wastewaters from 4-band high resolution imagery mainly consists of the following steps: (1) urban water mapping method MTSUWI; (2) post-processing and image enhancement; (3) rationality and feasibility of screening algorithm; (4) extraction of Green Alga. The technical flow chart is displayed in Fig. 1. The detailed procedures are presented below.

2.2.1 Urban Water Mapping Method MTSUWI (Modified Two-Step Urban Water Index)

With high-resolution imagery, there are great challenges for extracting water bodies when it comes to complex urban area. For example, having only visible and NIR bands, high-resolution images lack the bands which are necessary to obtain the MNDWI (Modified Normalized Difference Water Index) which has the ability of enhancing open water features when suppressing efficiently and even removing built-up land noise and soil and vegetation noise [7]. Also, urban water body can be affected by human activities, the contamination inside can change its optical properties, making them quite different

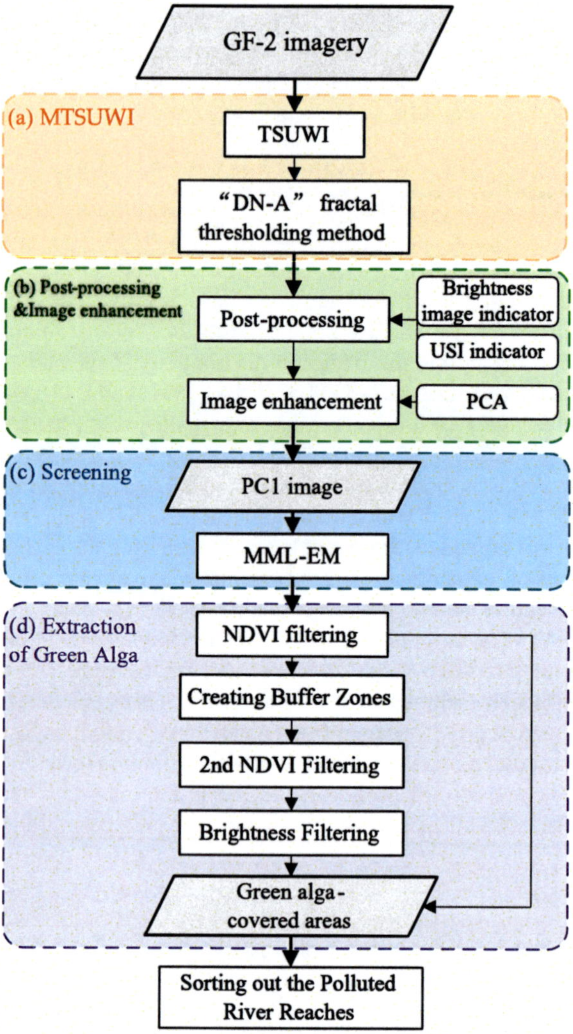

Fig. 1. Technical flow chart. **a** Urban water mapping method MTSUWI; **b** Post-processing and image enhancement; **c** Rationality and feasibility of screening algorithm; **d** Extraction of Green Alga.

from those of fresh water. Additionally, removing shadow noise from water maps in urban area is also difficult because of the spectral similarities. To address these problems [10], proposed a two-step urban water index (TSUWI), combining the Urban Shadow Index (USI) and Urban Water Index (UWI):

$$\mathrm{UWI} = \frac{\mathrm{Green} - 1.1 \times \mathrm{Re}d - 5.2 \times NIR + 0.4}{|Green - 1.1 \times \mathrm{Re}d - 5.2 \times NIR|} \quad (1)$$

$$\mathrm{USI} = 0.25 \times \frac{Green}{Red} - 0.57 \times \frac{NIR}{Green} - 0.83 \times \frac{Blue}{Green} + 1.0 \tag{2}$$

$$\mathrm{TSUWI} = (\mathrm{UWI} > T_1) \wedge (\mathrm{USI} > T_2) \tag{3}$$

where Green, Red, Blue, and NIR denotes respectively the reflectance of green, red, blue, and near infrared bands; T1 and T2 refers to the optical thresholds of the UWI and USI.

For urban water extraction, TSUWI is a simple yet robust method [10], however, currently there are no automatic approaches for determining the thresholds T1 and T2, and they were often decided based on multiple trials through visual examination. In order to determine the thresholds dynamically, here the "DN-A" fractal thresholding algorithm proposed by [25] is introduced in the following:

$$\mathrm{A(DN} \geq s) \propto \mathrm{DN}^{-D} \tag{4}$$

where A(DN≥s) denotes the area which is covered by pixels with DN or pixel values which is equal to or greater than a given observed DN value (s), and the D denotes a fractal dimension. According to the suggestions of [26], the variable "area" (A) is the product of the frequency (denoted by N) of a given s as well as the pixel size, that is to say A ∝ N, and "A" is better to be substituted by "N" because the frequency table of s (N) is easily obtained using PIE v5.0 or other commercial RS analysis software.

As shown in Fig. 2, a multifractal form is usually appeared in the ln(N) versus ln(DN) plot, which consists of a series of straight line segments (in this study it will be divided into three segments by default), namely scale-free intervals. It has upper and lower thresholds which are used to recognize different real-world objects [27]. To avoid missing those atypical water-bodies present within D2, here T' is selected as the threshold T1 or T2 in the above formula, which is adaptive.

Note that we cannot simply classify an image into water bodies and non-water bodies, especially when there are narrow, shallow or turbid water bodies and when there are pixels having similar spectral characteristics as waters. Intuitively, it is better to compulsively split the fractal schema into three segments, base on that the image can be classified into discrete classes or zones—these zones, according to [28] suggestion, often correspond to the features of real-world on the ground.

Through extensive experimentation we have found that most, if not all, of the pixels falling within D1 of Fig. 2 represent the non-water background, and D2 usually corresponds to the atypical water and noise (e.g. narrow, shallow or turbid water bodies), and D3 has the greatest chance of being detected as pure free-standing waters. Here we select T' (namely the cutoff value between D1 and D2) as the threshold delineating the entire water bodies.

2.2.2 Post-processing and Image Enhancement

MTSUWI works are effective when it comes to the surface water extraction in complex urban surfaces, but inevitably, the results may be mixed with noise objects such as buildings, roads, bridges and bare lands. Here the brightness image is used as an indicator

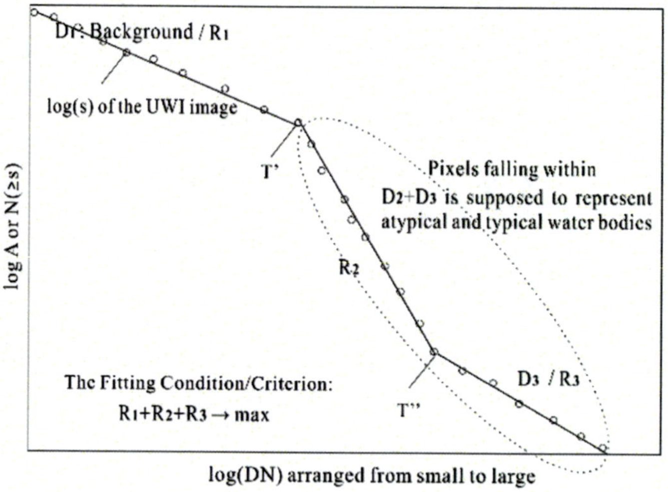

Fig. 2. The ln DN versus ln N(s) schema. Ri represents the goodness of fit, Di represents the scale-free interval, where $i = 1, 2, 3\ldots$.

to separate waters and other land covers which are usually characterized by lighter color than waters [29]. It is generated according to the following formula, and the relevant discriminative threshold is determined by the Otsu algorithm [30].

$$b(i) = \max_{1 \le k \le K} (band_k(i)) \tag{5}$$

where bandk(i) denotes the digital number (DN) of pixel i for the k-th visible spectral band; K refers to the total number of the visible bands; and b(i) is the brightness of pixel i. Obviously, the maximum value of the visible bands for each pixel i is denoted as the brightness b(i).

As for dark objects with similar spectral characteristics as waters (like dark vegetation, building shadows, etc.) in the RGB imagery, the USI indicator can make a clear discrimination. In addition, based on PIE V5.0's magic-stick algorithm [31], visual interpretation and limited manual editing was conducted to outline the omitted waters (the omission error is less than 10%).

On this basis, the PCA (principal component analysis) image was developed to exaggerate the difference between waters and pollutants, and/or between various water qualities. For analyzing correlated multivariable datasets, PCA is a widely used method and is common for multi-spectral images interpretation on the basis of linear algebraic matrix operations [32].

2.2.3 Rationality and Feasibility of Screening Algorithm

For the PC1 imagery, the DN range for each band, of the whole scene or a large enough subscene, can be obtained and then displayed as a histogram, and commonly, the distribution of DNs (pixel values) is unimodal and may be Gaussian or skewing Gaussian [33]. Multimodal distributions are most frequently bimodal, whose result if a scene or an object contains two or more dominant classes with distinctly various (often narrow) ranges of reflectance. The relevant threshold is chosen between two/multiple peaks in the grayscale histogram of the image data to separate different classes from the background [24, 34]. That is to say, the bimodal or multi-modal splitting method can provide a feasible way to find the threshold for water quality classification or detection of polluted water bodies. Here the MML-EM (Minimum Message Length Criterion-Expectation Maximization) [35] is taken into account. From it, a component-wise EM algorithm is proposed, which is robust regarding initialization and capable of selecting the number of components used for generating the data. By the way, Sun et al. [36] suggested that lack of self-similarity is not a limitation to the fractal technique, and new insights can be processed and yielded for "non-fractal" data [37]. But in practice, it might not perform well in approximating the background-class distributions in the image, for they are usually assumed to show a mixture normal distribution according to the literature [38]. The reason why the fractal algorithm was used in water extracting is because waters are indicated by brightest pixels (outliers) in the UWI and USI images, while particular sensitivity can be observed in fractal to outliers (anomalies) in a data set [27].

The basic principle of MML-EM is given as follows:

First, the general formula of a mixed distribution is:

$$\mathrm{p(x;}\,\theta) = \sum_{i=1}^{k} b_i \cdot f(x, \theta_i) \tag{6}$$

where $\theta = (k, \theta_1, \ldots, \theta_k, b_1, \ldots, b_k)$, θ_i is the i_{th} statistical parameters weight (b), mean (m) and covariance (Cov), b_i is the i_{th} weight (which is shown as percentage), and $\sum_{i=1}^{k} b_i = 1$ and $b_i > 0; f(x, \theta_i)$ is the i_{th} probability density function; k is the number of the mixture populations.

EM is a commonly used iterative algorithm which is designed for estimating the maximum probability of a whole set of parameters for a statistical model. EM can produce a sequence of estimates $\{\theta_i\}$ with two consecutive steps: Estimating the complete log-likelihood (the E step) conditional expectation and then updating the estimates of parameter according to the maximization (M) step; Until a given convergence criterion is satisfied, EM iterates between the 'E' step and the 'M' step can finish [39]. It is believed that with MML, an optimal model selection criterion can be obtained to decide one among a set of candidate modes.

Now consider a d-dimensional dataset y, which has been generated according to $\mathrm{p(y,}\,\theta)$, MML provide these following criterion (where the minimization with respect to θ can be understood as simultaneously in θ and d, namely the dimension of θ:

$$\widehat{\theta} = \arg\min_{\theta}\left\{-\log \mathrm{p}(\theta) - \log \mathrm{p(y|}\theta) + \frac{1}{2}\log|\mathrm{I}(\theta)| + \frac{d}{2}\left(1 + \log\frac{1}{12}\right)\right\} \tag{7}$$

where $I(\theta) = -E[D_{\theta}^{2}\log \mathrm{p(y|}\theta)]$ is the expected Fisher information matrix and $\mathrm{I}(\theta)|$ denotes its determinant; $\widehat{\theta}$ is a finite precision version of θ [40].

Finally, thresholding is also a basic but critical issue in image classification. Below or equal to the selected threshold, the grey-level values in an enhanced image are usually classified as a class or a real-world feature. While above this threshold, the values are classified as another class. For a bimodal distribution, the threshold (T) in between is calculated in the following way:

$$T = \frac{n_2\partial_1^2 - n_1\partial_2^2 + \partial_1\partial_2\sqrt{(n_1 - n_2)^2 + 2A(\partial_1^2 - \partial_2^2)}}{(\partial_1^2 - \partial_2^2)} \text{ and } A = \log\left(\frac{\partial_1}{\partial_2} \times \frac{n_2}{n_1}\right) \quad (8)$$

where ∂_1 and ∂_2 are their standard deviations; n1 and n2 refers to the mean values of two adjacent populations (classes), respectively, and n1 < n2 and they are fixed values. Through the MML-EM analysis, these parameters are by-products. As shown in Fig. 5, the (unimodal) normal probability-density curve cannot be separated and diagrammed without them.

In addition, output of the MML-EM analysis is a series of unlabeled classes, and expert knowledge is essential to assign each class a label (see Sect. 3.2).

2.2.4 Extraction of Green Alga

There are five steps for the extraction of Green Alga: (1) the NDVI filtering; (2) creating buffer zones around the (extracted) river banks; (3) the 2nd NDVI filtering within the buffer zones; (4) the brightness filtering within the NDVI-anomalous zones and (5) union of the results of step 1 and step 4. More details see Sect. 3.3.1.

3 Results

3.1 Water Extraction Maps

The output of the "TSUWI + thresholding" method is given in Fig. 3. Visual inspection of it indicates that we successfully discriminated urban water bodies from artificial facilities such as roads, bridges, buildings, etc. Figure 4 gives the post-processed image which is obtained by ruling out the light-colored patches in Fig. 3 (corresponding to, e.g., buildings and roads) using the brightness feature. Figure 4a is also known as the water-masked image based on which the subsequent analysis were performed.

The next step is to manually digitalize the omitted water bodies which are hard to be identified because of the high spectral variation within waters and problematic shadows. The objective of this step is to quantify the validation result of the new method and its limitations. Figure 5 gives the final water mapping result, and spatial analysis indicates that the rate of water-area in the whole scene reaches about 1.70% (namely 8.89 km^2).

Finally, an accuracy assessment was performed based on Fig. 3, calculating the relevant confusion matrix (Figure 4 is the reference map). According to the obtained confusion matrix, the Kappa Coefficient is 0.92, the commission error is 0.17%, and the omission error is 13.83%, which can be considered a good result of the proposed method.

3.2 Separation of Different Water Quality

Ju et al. [41] and its references believed that in an enhanced image, pixel-values which reflect multiple land cover types can be modelled as a subclasses mixture, represented by

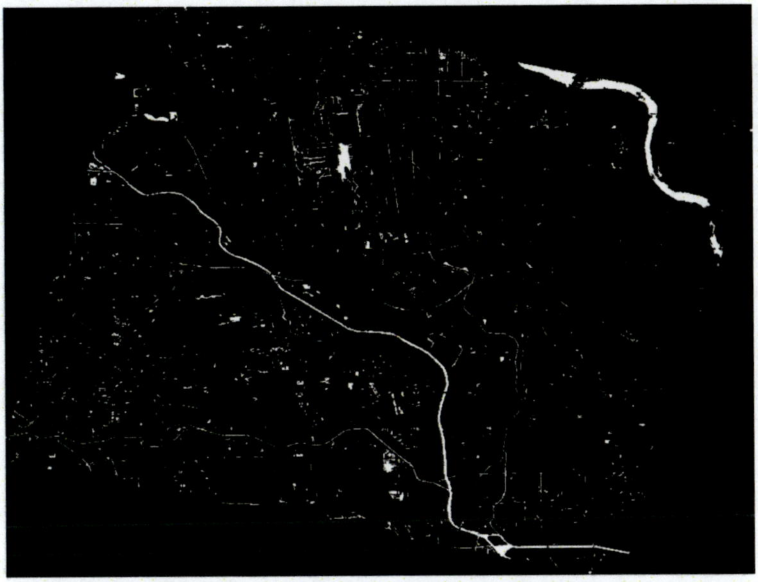

Fig. 3. The raw water-extracting result (binary image).

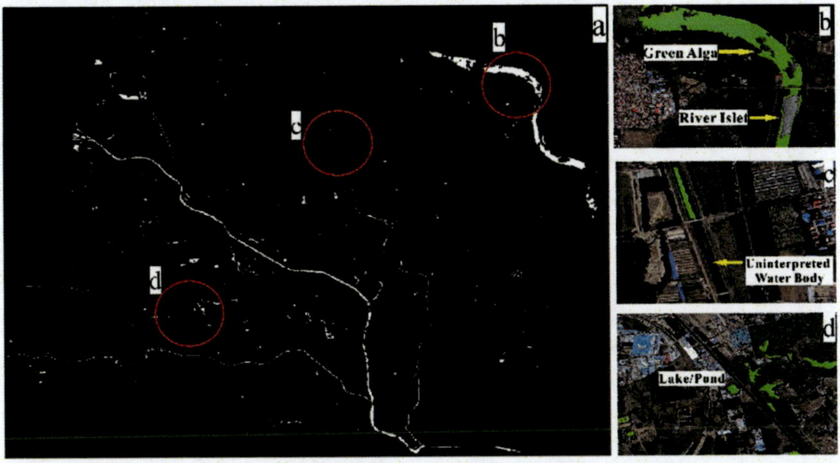

Fig. 4. **a** The post-processed water-extracting result (binary image); **b–d** are the highlighted regions b/c/d in a, which is enlarged (the green-colored areas represent automatically extracted water bodies).

an (approximately) Gaussian/normal distribution, and naturally the thresholds amongst them are useful and effective for image classification [42]. In the water-masked PC1 image, fresh waters and waters that are potentially contaminated can be intuitively

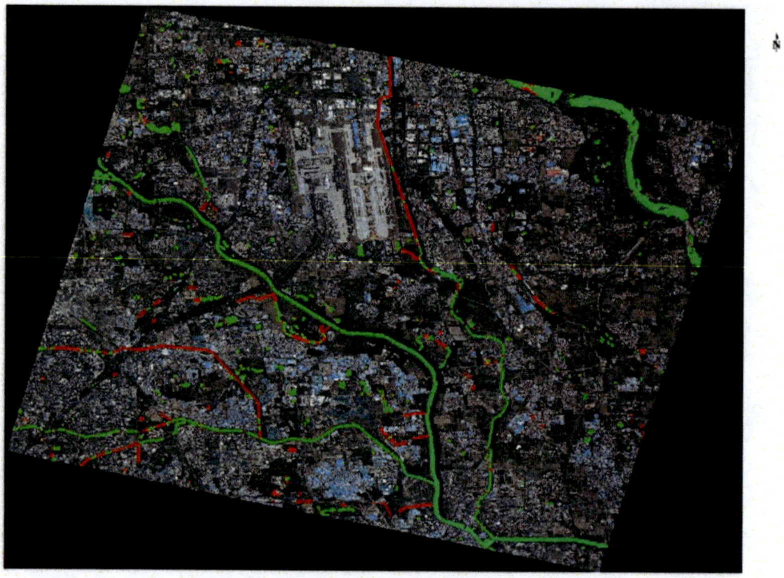

Fig. 5. The final result after manual editing (red line segments represent manually digitalized water bodies, and green line segments represent automatically extracted water bodies, which are consistent with Fig. 3a. Note that the river boundaries are exaggerated for illustration).

regarded as a two-class classification problem. Therefore, the number k in the MML-EM scheme is 2.

Using all 4 bands in the GF-2 imagery as input to PCA. However, PCA itself cannot decide which component is most fitting to enhance the contaminated water information. Here, only the first component (PC1) is taken into account because it is assumed that having usually the highest factor loading of each input component, PC1 can capture the essential qualities of a feature [43]. In this case, as listed in Table1, PC1 has positive loadings on band2 (green) and negative loadings on band4 (NIR), which is consistent with water's spectral characteristics. PC1 contains more than 97% of variance of the input data. This result implies that the contaminated waters might be concealed in the PC1 image. The histogram of the PC1 band, or say the MML-EM schema, is given in Fig. 6.

Table 1. The Eigenvectors matrix values of PCA for the 1,2,3 and 4 bands of GF-2 image.

Eigenvectors	PCA1	PCA2	PCA3	PCA4
Blue band	0.681512	0.557117	0.366096	0.301886
Green band	0.350937	0.097192	−0.033709	−0.930731
Red band	0.552817	−0.292772	−0752727	0.205132
Band	−0.326767	0.771011	−0.546113	−0.022917

It has delivered several important observations as follows: (1) the grey levels showed at least 2 populations distributed normally and independently. Moreover, each of them may correspond to a real-world feature on the ground. (2) As seen in Fig. 6, Population (Class) I and II turn out to reflect the differences among water bodies, for example, clean and deep water pixels exhibit stable spectral profiles. Contrarily, narrow, shallow or polluted water pixels might generate unstable spectral profiles or features, due to mixed reflectance caused by land, sediment, or phytoplankton. (3) According to the expert knowledge inferred from Table 1, PC1 is a reflection of water bodies, so Population I with the smallest mean value (m1) reflects the fresh waters (fresh waters show low reflectance in the visible and NIR range or the PC1 band), and Population II reflects waters affected by the land adjacency effect (barren lands usually have higher reflectance). (4) Due to lack of the required spectral information [44], the polluted/black-odorous water bodies themselves cannot form an independent population, and it is thus inseparable. We also tested the performance of water-masked Normalized Difference Vegetation Index (NDVI), Normalized Difference Water Index (NDWI) and texture images under the MML-EM framework, and the findings are all disappointing.

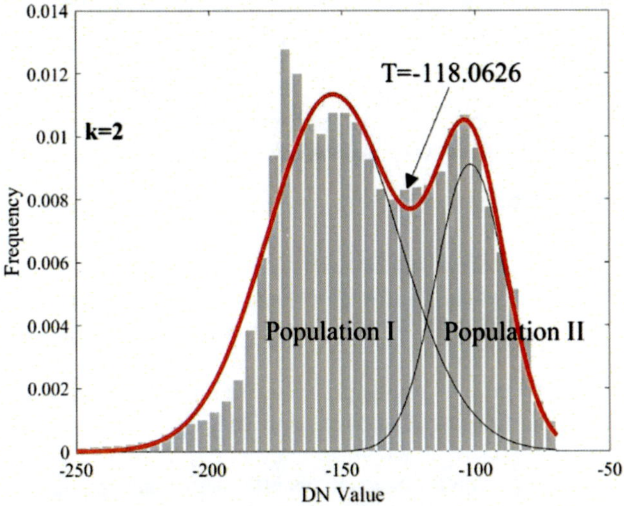

Fig. 6. The MML-EM schema of the water-masked PC1 band. (Population I: pixels with pixel values $< T = -118.0626$; population II: pixels with values $\geq T$)

The experimental results suggest that in our case there may be no direct interpretation signs or visual cues for separation of different water quality, and a search for indirect interpretation keys should be put on the agenda. It is reported that the urban black-and-odorous water bodies which is the main manifestation of heavily polluted water bodies in the studied area usually appear in closed or semi- closed riverways characterized by slow water circulation, and thus, household wastes and alga mixed with black sludge are always floating on the water surface [6]. Therefore the floating wastes, green algae, river islets, etc., can be considered as the indirect interpretation keys of wastewaters (Fig. 7).

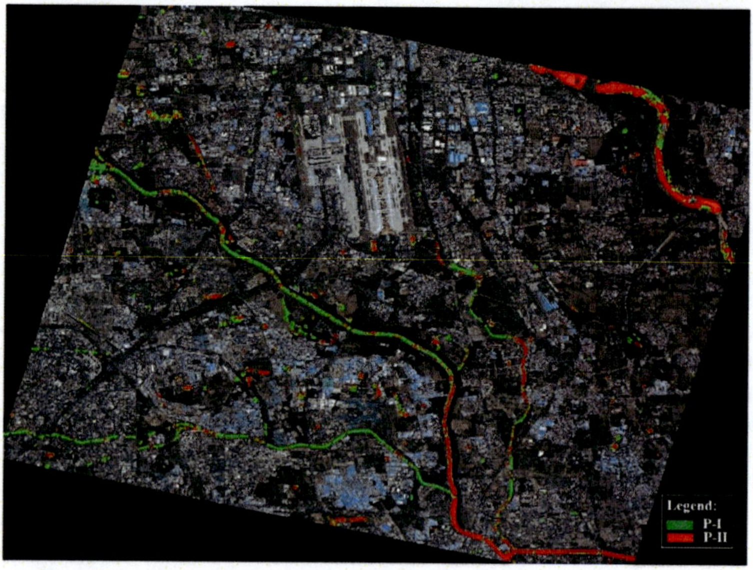

Fig. 7. Distribution of Population I (P-I) and Population II (P-II) waters. Note that the river boundaries are exaggerated for illustration.

3.3 A Search for Indirect Interpretation Keys

3.4 Extraction of Green Alga

Step 1: The NDVI Filtering

Fill the "gaps" in the river channels of Fig. 4 caused by green algae or other aquatic floaters, and the purpose of the NDVI filtering is to sort out the floating green algae on the river's surface, if any. To this end, the gray-level NDVI image must be converted into a binary image by selecting an appropriate decision threshold, and the resulting connected components (objects) that have fewer than 100 pixels will be removed as "salt and pepper" noise. Here an empirical threshold of 0.2 is recommended, and the gray-level values which are below or equal to this threshold are regarded as background, while the gray-level values which are above it are classified as objects (vegetated areas). As shown in Fig. 8, within the boundaries of the known water bodies, green algae and free standing waters are easily distinguished because of their significant spectral differences.

Step 2: Creating Buffer Zones around the (Extracted) River Banks

With a few exceptions, most of the green algae fall outside, but are close to, the extracted water bodies, so it is necessary to generate circular buffers around them to incorporate these green algae. Here, the buffering distance was set as 80 m (namely 100 pixels), and Fig. 9 gives the result.

Fig. 8. Result of the NDVI filtering (Only a clipped part of the study site is shown).

Fig. 9. The extracted water bodies with buffers (RGB: 321).

Step 3: The 2nd NDVI Filtering within the Buffer Zones

The gray-level NDVI image was generated within the buffered water bodies shown in Fig. 8, and the decision threshold 0.2 is used again to convert it into a binary image.

During this process, buildings, bare lands, sand lands, waters and other non-vegetated areas were removed, while grasslands, green algae, and forests were retained.

Step 4: The Brightness Filtering within the NDVI-Anomalous Zones

Within the vegetated areas generated by Step 3—a gradually reduced interpretation-target-area, the gray-level brightness image was generated. As shown in Fig. 10, relative to forests and grasslands, green algae are typically characterized by particular high brightness or reflectance in the clipped brightness imagery. A threshold of 500 determined by the MML-EM algorithm was found to successfully identify pixels containing green algae, and again, the connected components (in the resulting binary image) that have fewer than 100 pixels was eliminated as "salt and pepper" noise or small green-alga blocks.

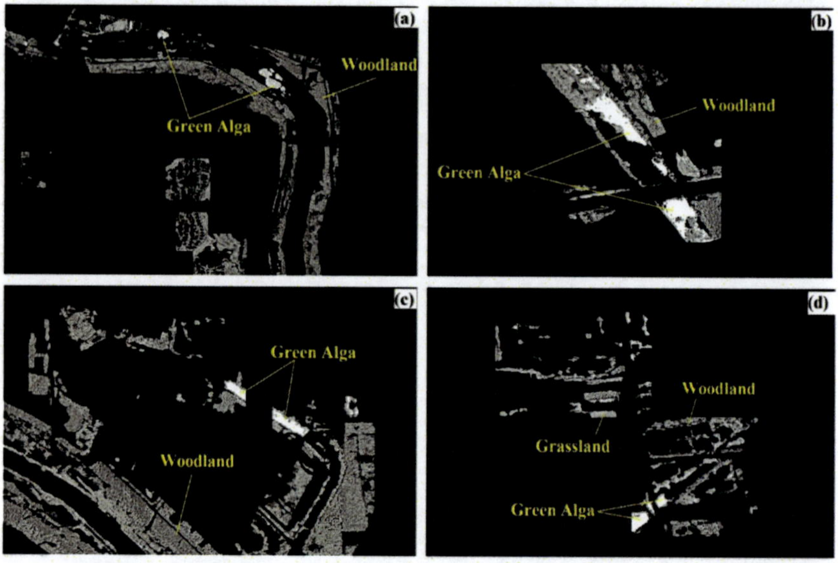

Fig. 10. The extremely high brightness of green alga in the clipped brightness image (Only four clipped parts of the study area are shown).

Step 5: Union of the Results of Step 1 and Step 4

Calculate the geometric union of the resulting shapefiles or binary images of Step 1 and Step 4, so that the green alga-covered areas can be identified as a whole (see Fig. 11). The overall accuracy of classification is over 97.0% (The main confusion occurred be-tween some crops and green algae.), indicating the good performance of the proposed method.

3.4.1 Sorting Out the Polluted River Reaches

According to the geometrical and hydrological features and polynomial water quality trend-lines, a river can be segmented into several reaches. To extract the local water pollution in the study area, the concept of "river reach" should be taken into account. According to the government documents, there are more than 50 river reaches in this area, and the water quality of 44 reaches is on the record. These were digitized to shapefile in GIS software for further use. However, 6 reaches cannot be detected with the GF-2 imagery (0.8 m resolution) because the river widths of them are less than 1 m.

Obviously, any river reach containing at least one patch of green alga can be labeled as "polluted", otherwise they are labeled as "clean/fresh". Following this principle, Fig. 11 maps eight black-and-odorous wastewaters in this scene, which are consistent with literatures [24] and observations from the field (Fig. 12 is an example). The green-alga patches falling outside the known river reaches are either alga-dominating lakes/ponds or, to a lesser extent, crops. In addition, the percentage of the contaminated water area against the total water area is 25.00%, which is about 2.175 km^2.

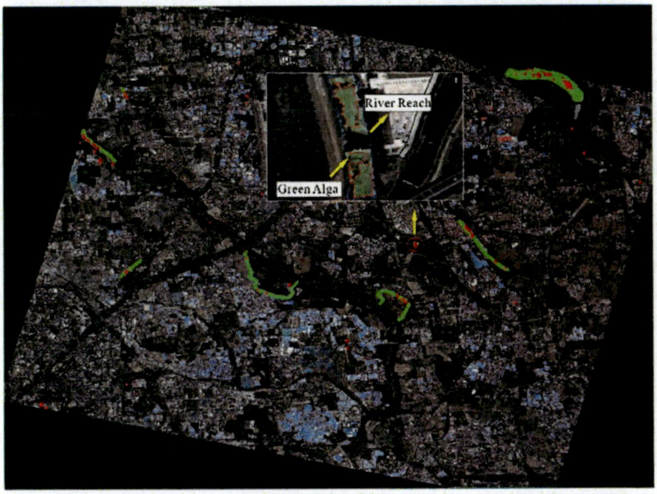

Fig. 11. The obtained black-and-odorous/polluted river reaches (green-colored) as well as the extracted green algae (red-colored). Note that both the river-reach's and the green-alga's boundaries are exaggerated for illustration.

4 Discussion

The proposed algorithm is a five-step procedure, which is discussed below:

(1) the first step is to extract urban waters under complex background. In this aspect, Ref. [10] provided us with the necessary rules to extract the shallow, narrow and impure waters, as well as the deep, broad and clean waters simultaneously using the high resolution imagery. On this basis, the fractal thresholding technique was introduced to determine the most appropriate cut-off value distinguishing water and non-water. Taken together, the modified TSUWI algorithm is proposed.

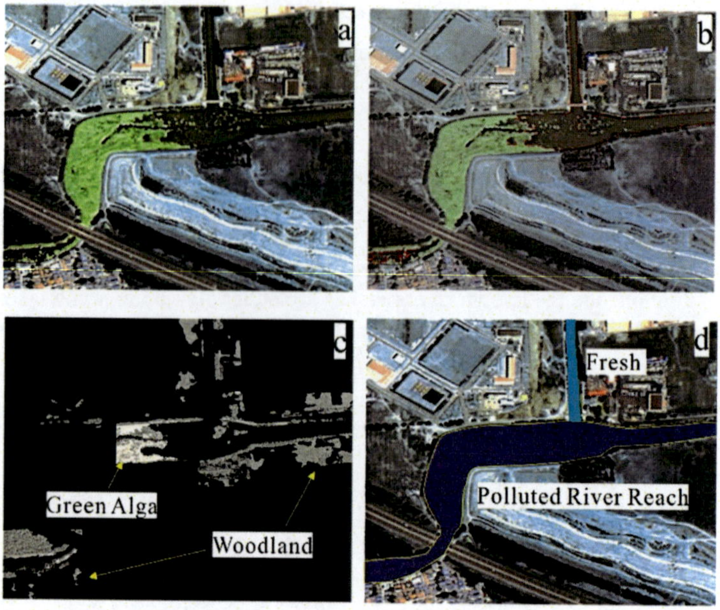

Fig. 12. The verification images/photos of the extracted polluted river reaches. (**a**) is the background image (true color). (**b–d**) Are the image screenshots of several representative polluted river reaches with floating green algae, and (**e**) is the reference image with fresh water. (**f–h**) are the on-site photographs of polluted river reaches (**b–d**), and (**i**) is the reference photograph with clean water. The shooting locations of them are shown in Fig. 11b–e. Note that during the field investigation of reaches (**b–d**), a distinct odor could be smelled in the air.

(2) However, due to the high spectral variation and similarities, there are many false alarms and omission alarms, and a simple method to address this issue is the visual interpretation and manual edition of the resulting shapefile of water, because human beings have the ability of comprehensively using shape, color, contextual information, spatial patterns, as well as knowledge and experience.
(3) The third step is the mixed-Gaussian-distribution screening, which is based on the first principal component band. However, the separation of different water quality is not possible in this paradigm, and the subpopulations are mainly a reflection of the existing background of the waters: for example by visual interpretation the subpopulation I only reflects pixels in deep and clean waters, and subpopulation II reflects pixels in shallow, narrow waters. The contaminated waters, on the other hand, cannot form an independent and separable subpopulation.
(4) Inasmuch, the research efforts in separating different water qualities have shifted from direct interpretation signs or visual cues to indirect interpretation signs—green algae (Other aquatic floaters e.g., floating wastes, river islets, etc., were not investigated in this study because of their variable spectral characteristics and small amounts.). By gradually reducing the interpretation target area, pixels containing green alga can be greatly enhanced and then extracted in the brightness image.

(5) In the GIS software, any river reach containing at least one patch of green alga is labeled as "polluted", otherwise they are labeled as "clean". Because the under-sized green-alga blocks were deleted in advance, this procedure will not result in misclassification errors. Finally, eight polluted river reaches in the study area were selected out.

Using 4-band high resolution image, this is the first time such a hybrid and semi-automatic method is used in detecting urban polluted waters. However, the extraction and analysis of urban polluted waters is still challenging, mainly due to two factors: (1) According to the internal data [24], the water quality of the river reaches can be generalized into three levels: standard, moderate and severely polluted. In this experiment, only the heavily polluted water bodies were discerned, while the proposed method cannot detect and separate the moderately polluted river reaches. This is because the limited spectral resolution of GF-2 imagery and the restriction of its bands wavelengths (visible-NIR) preclude a discrimination method of river reaches based on water-quality. (2) Field investigation around the polluted river reaches indicated that the green floats in the input image are mainly green algae, and to a lesser extent, duckweed which are a result of the eutrophication [45]. This is believed to be the main cause of Beijing black-and-odorous waters. However, if the pollution results from inorganic substances, a significant accumulation of phytoplankton may not appear and the proposed method becomes unavailable. (3) In Fig. 4, there are a number of water bodies uninterpreted using the proposed automated algorithm, and the spatial buffering analysis cannot cover them up completely. Therefore, the green algae distributing around these water bodies, if any, are always ignored, neither is the relevant water-pollution information. We accept that these challenges can hardly be overcome by using the GF-2 data.

To assess the generality or robustness of the proposed method, we further tested another two images with different acquisition times, sites and sensor types, see the annotations of Figs. 13 and 14. After the same treatment (the NDVI filtering → creating buffer zones → the 2nd NDVI filtering → the brightness filtering →…), the obtained overall accuracies of water and green-alga classification are all over 97%, and the black-and-odorous river reaches are also accurately labeled. These results are as good as the results derived in Chap. 5 and are thus satisfactory and reliable.

5 Conclusions

This study aimed at applying the MTSUWI (Modified Two-Step Urban Water Index) method and screening algorithm in extracting urban black-and-odorous water. The objective were satisfactorily met and several conclusions with regard to this study can be made from the results:

(1) With the MTSUWI method, the final Kappa Coefficient is 0.92. And to complete the water thematic map, the workload of visual interpretation and limited manual editing is less than 10%.
(2) The separation of different water quality is not possible based on the multispectral or texture image, for the contaminated waters (most, if not all, of them are the black-and-odorous waters) cannot form an independent and separable subpopulation.

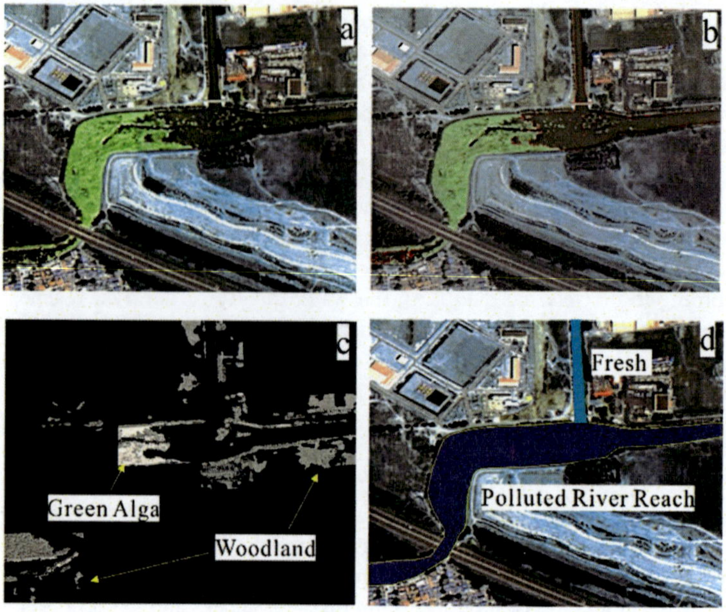

Fig. 13. The verification images/photos of the extracted polluted river reaches. (**a**) is the background image (the pre-processed GF-2 scene acquired on 06/17/2017 and its spatial limits are N 39.845°–N 39.853°and E 116.634°–E 116.643°, RGB:123). (**b**) is the raw water-extracting result using the MTSUWI method. (**c**) is the clipped brightness image. (**d**) shows the obtained black-and-odorous/polluted river reaches (the river-reach information was manually digitized to shapefiles in advance).

(3) The high resolution data are ideal for extraction of floating green alga–an important but indirect interpretation indicator for identification of the polluted water bodies, because the brightness image derived from these data is sensitive to algae, and they have red band and near-IR band, having sensitivity to vegetation.
(4) Eight black-and-odorous wastewaters in the study area were selected out, which are consistent with literatures and observations from the field.

This research, although not perfect, is the first to apply statistical algorithms to map urban polluted rivers using a 4-band high resolution imagery.

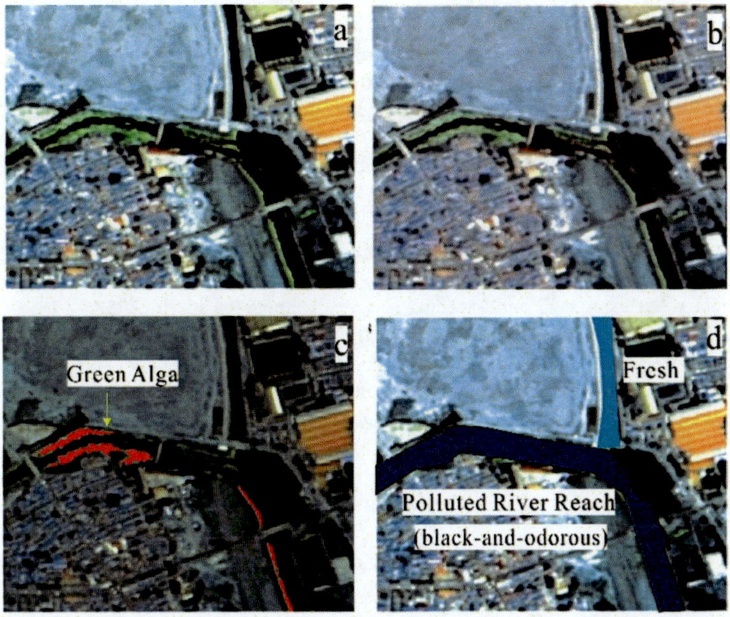

Fig. 14. The verification images/photos of the extracted polluted river reaches. (**a**) is the background image (the pre-processed ZY-3 scene acquired on 06/17/2017 and its spatial limits are N 39.486°–N 39.85°and E 116.702°–E 116.708°, RGB:123). (**b**) is the raw water-extracting result using the MTSUWI method. (**c**) is the extracted green-alga patches. (**d**) shows the obtained black-and-odorous/polluted river reaches (the river-reach information was digitized to shapefiles in advance).

Acknowledgments. Great thanks go to Dr. Wang Yuxiang, who is the president of Beijing PIESAT Information Technology Co., Ltd., for his help in providing the relevant remote sensing imagery. This work was financially supported by Beijing Water Authority's scientific research project: high resolution imagery-based urban black-odorous-water mapping and Fundamental Re-search Funds for the Central Universities of China (set for developing innovative R & D teams) (Project No. 30010226940). Helpful review comments on the manuscript were provided by several anonymous reviewers.

References

1. Kondratyev, K.Y., Pozdnyakov, D.V., Pettersson, L.H.: Water quality remote sensing in the visible spectrum. Int. J. Remote Sens. **19**(5), 957–979 (1998)
2. Hafeez, S., Wong, M.S., Abbas, S., Nichol, J.: Detection and monitoring of marine pollution using remote sensing technologies. monitoring of marine pollution. IntechOpen (2018)
3. Obade, V.D.P., Lal, R., Chen, J.: Remote sensing of soil and water quality in agroecosystems. Water Air Soil Pollut. **224**(9), 1658 (2013)
4. Vincent, D.P.O., Lal, R., Moore, R.: Assessing the accuracy of soil and water quality characterization using remote sensing. Water Resour. Manag. **28**(14), 5091–5109 (2014)

5. Loughland, R.A., Saji, B.: Remote sensing: a tool for managing marine pollution in the Gulf. In: Protecting the Gulf's Marine Ecosystems from Pollution, pp. 131–145 (2008)
6. Yao, Y., Shen, Q., Zhu, L., Gao, H.: Remote sensing identification of urban black-odor water bodies in Shenyang city based on GF-2 image. J. Remote Sens. **23**(2), 230–242 (2017). (in Chinese with English abstract)
7. Xu, H.Q.: Modification of normalised difference water index (NDWI) to enhance open water features in remotely sensed imagery. Int. J. Remote Sens. **27**(14), 3025–3033 (2006)
8. Rogers, A.S., Kearney, M.S.: Reducing signature variability in unmixing coastal marsh thematic mapper scenes using spectral indices. Int. J. Remote Sens. **25**(12), 2317–2335 (2004)
9. Yao, F., Wang, C., Dong, D., Luo, J., Shen, Z., Yang, K.: High-resolution mapping of urban surface water using zy-3 multi-spectral imagery. Remote Sens. **7**(9), 12336–12355 (2015)
10. Wu, W., Li, Q.Z., Zhang, Y., Du, X.: Two-step urban water index (TSUWI): a new technique for high-resolution mapping of urban surface water. Remote Sens. **10**(11), 1704 (2018)
11. Gitelson, A., Stark, R., Dor, I.: Quantitative near-surface remote sensing of wastewater quality in oxidation ponds and reservoirs: a case study of the naan system. Water Environ. Res. **69**(7), 1263–1271 (1997)
12. Pulliainen, J., Kallio, K., Eloheimo, K., Koponen, S., Hallikainen, M.: A semi-operative approach to lake water quality retrieval from remote sensing data. Sci. Total Environ. **268**(1–3), 79–93 (2001)
13. He, W., Shan, C., Liu, X., Chen, J.: Water quality monitoring in a slightly-polluted inland water body through remote sensing-case study of the guanting reservoir in Beijing. China. Front. Env. Sci. Eng. **2**(2), 163–171 (2008)
14. Wu, C.Q.: Establishing a remote sensing assessing method of lacustrine eutrophication. Environmental Monitoring in China (2011)
15. Jin, H.X., Pan, J.: Urban black-odor water body remote sensing monitoring based on GF-2Satellite data Fusion. Sci. Technol. Manag. Land Resour. **34**(4), 107–117 (2017). (in Chinese with English abstract)
16. Giardino, C., Candiani, G., Bresciani, M., Lee, Z., Gagliano, S., Pepe, M.: Bomber: a tool for estimating water quality and bottom properties from remote sensing images. Comput. Geosci. **45**(6), 313–318 (2012)
17. Tyler, A.N., Svab, E., Preston, T., Présing, M., Kovács, W.A.: Remote sensing of the water quality of shallow lakes: a mixture modelling approach to quantifying phytoplankton in water characterized by high-suspended sediment. Int. J. Remote Sens. **27**(8), 1521–1537 (2006)
18. Keith, D.J., Schaeffer, B.A., Lunetta, R.S., Gould, R.W., Rocha, K., Cobb, D.J.: Remote sensing of selected water-quality indicators with the hyperspectral imager for the coastal ocean (hico) sensor. Int. J. Remote Sens. **35**(9), 2927–2962 (2014)
19. Vignolo, A., Pochettino, A., Cicerone, D.: Water quality assessment using remote sensing techniques: medrano creek, argentina. J. Environ. Manag. **81**(4), 429–433 (2006)
20. Kaiser, M.F., Aboulela, H., El Serehy, H., Ezz Edin, H.: Spectral enhancement of spot imagery data to assess marine pollution near port said, Egypt. Int. J. Remote Sens. **31**(7), 1753–1764 (2010)
21. Ji, G.: Research and application on black and odorous water body by remote sensing. Diss. (in Chinese with English abstract) (2017)
22. Wen, S., et al.: Remote sensing identification of urban black-odor water bodies based on high-resolution images: a case study in Nanjing. Environ. Sci. **1**, 57–67 (2018). (in Chinese with English abstract)
23. Cheng, Q.: Multifractality and spatial statistics. Comput. Geosci. **25**(9), 949–961 (1999)
24. PIESAT Information Technology Co., Ltd. Report on urban black-and-odor water monitoring of Beijing district using remote sensing technology. Internal data (2017)

25. Shahriari, H., Ranjbar, H., Honarmand, M.: Image segmentation for hydrothermal alteration mapping using PCA and concentration-area fractal model. Nat. Resour. Res. **22**(3), 191–206 (2013)
26. Han, L., Zhao, B., Wu, J.J., Zhang, S.Y., Pilz, J., Yang, F.: An integrated approach for extraction of lithology information using the spot 6 imagery in a heavily quaternary-covered region-north baoji district of china. Geol. J. **53**, 352–363 (2017)
27. Wang, R., Lin, J., Zhao, B.: Integrated approach for lithological classification using ASTER imagery in a shallowly covered region-the Eastern Yanshan Mountain of China. IEEE J. Sel. Top. Appl. Earth Obs. **11**(12), 4791–4807 (2018)
28. Cheng, Q., Li, Q.: A fractal concentration-area method for assigning a color palette for image representation. Comput. Geosci. **28**(4), 567–575 (2002)
29. Yao, Y.Q., Zhang, Y.B., Liu, J.D., Shen, Z.J.: Model for evaluating urban water shortage risk: a case study in Beijing. Int. J. Digit. Content Technol. Appl. **6**, 68–79 (2012)
30. Liu, D., Yu, J.: Otsu method and K-means. In: Ninth International Conference on Hybrid Intelligent Systems IEEE Computer Society (2009)
31. Fan, J., Yau, D.K.Y., Elmagarmid, A.K., Aref, W.G.: Image segmentation by integrating color edge detection and seeded region growing. IEEE Trans. Image Process. **10**(10), 1454 (2001)
32. Liu, L., Zhou, J., Han, L., Xu, X.: Mineral mapping and ore prospecting using landsat tm and hyperion data, wushitala, Xinjiang, northwestern china. Ore Geol. Rev. **81**, 280–295 (2017)
33. Kerroum, M.A., Hammouch, A., Aboutajdine, D.: Textural feature selection by joint mutual information based on Gaussian mixture model for multispectral image classification. Pattern Recogn. Lett. **31**(10), 1168–1174 (2010)
34. Muñoz-Marí, J., Bruzzone, L., Camps-Valls, G.: A support vector domain description approach to supervised classification of remote sensing images. IEEE Trans. Geosci. Remote Sens. **45**(8), 2683–2692 (2007)
35. Figureueiredo, M.A.T., Jain, A.K.: Unsupervised learning of finite mixture models. IEEE Trans. Pattern Anal. Mach. Intell. **24**(3), 381–396 (2002)
36. Sun, W.G., Gong, P., Liang, S.: Fractal analysis of remotely sensed images: a review of methods and applications. Int. J. Remote Sens. **27**(22), 4963–4990 (2006)
37. Véhel, J.L., Legrand, P.: Signal and Image processing with FracLab. In: Novak, M.M. (ed.) Thinking in Patterns: Fractals and Related Phenomena in Nature, pp. 321–322. World Scientific Publishing Co. Pte. Ltd. ISBN #9789812702746 (2011)
38. Hastie, T., Tibshirani, R.: Discriminant analysis by gaussian mixtures. Appl. Stat. J. R. Stat. Soc. **58**(1), 155–176 (1996)
39. Zhang, S.F.: EM algorithm and its application in parameter estimation for Gaussian mixture. J. Space-Craft TT&C Technol. **23**(4), 47–52 (2008). (in Chinese with English abstract)
40. Christakos, G., Bogaert, P., Serre, M.: Temporal GIS: advanced functions for field-based applications. Appl. Stat. J. R. Stat. Soc. **52**(4), 690 (2010)
41. Ju, J.C., Kolaczyk, E.D., Gopal, S.: Gaussian mixture discriminant analysis and sub-pixel land cover characterization in remote sensing. Remote Sens. Environ. **84**(4), 550–560 (2003)
42. Tobias, O.J., Seara, R.: Image segmentation by histogram thresholding using fuzzy sets. IEEE Trans. Image Process. **11**(12), 1457–1465 (2002)
43. Mars, J.C., Rowan, L.C.: Spectral assessment of new aster swir surface reflectance data products for spectroscopic mapping of rocks and minerals. Remote Sens. Environ. **114**(9), 2011–2025 (2010)
44. Homma, K., Yamamoto, H., Shingu, H.: Water pollution monitoring using a hyperspectral imaging spectropolarimeter. In: Proceedings of Spie the International Society for Optical Engineering, vol. 5655 (2005)
45. Keating, K.I.: Blue-green algal inhibition of diatom growth: transition from mesotrophic to eutrophic community structure. Science **199**(4332), 971–973 (1978)

Design and Implementation of a Target Characteristics Recommender System for Image Interpretation

L. V. Shouye(✉), C. E. N. Pengrui, and Zheng Mei

Institute of Remote Sensing Information of Beijing, Beijing, China
lvshouye@126.com

Abstract. Aiming at the problem of target characteristics retrieval and image interpretation, a target characteristics recommender system framework based on the target characteristics database and scene template database is designed. The technical methods of target characteristics recommendation are proposed, including target characteristics retrieval using target directory information and word vector, target image recommendation using metric learning and collaborative filtering, and target image interpretation using target category statistics and scene template matching. A target characteristics information recommendation mechanism supporting image interpretation is established, and a target characteristics recommender system for image interpretation is constructed.

Keywords: Characteristics retrieval · Image recommendation · Image interpretation · Recommender system · Design and implementation

1 Introduction

At present, the rapidly developing spatial infrastructure provides a large amount of image data, and these data are still growing rapidly, which brings great challenges to the analysis, understanding and retrieval of image data. The interpretation of images has always been the focus of computer vision research. For the interpretation of image data obtained through spatial infrastructure, it is necessary to combine the description of images with natural language text in order to achieve better interpretation effect [1].

The main work of image interpretation is to generate the interpretation statement of the image according to the content of the image. It is necessary to identify the object in the image. In addition, it is also necessary to consider the grammatical correctness of the interpretation statement, so as to approach the interpretation result of human beings. Lei studied the problem of image recommendation based on the deep learning method. This study pointed out that the most important thing of image recommendation is to build a bridge between the semantic understanding of the image and the user's preference or intention for the image [2]. Bansal proposed a text recommendation method based on the cyclic neural network model. The method not only uses the scoring data, but also uses the text content information and text label data for recommendation [3]. Li proposed a new

L. Wang et al. (Eds.): CHREOC 2022, LNEE 969, pp. 264–273, 2023.
https://doi.org/10.1007/978-981-19-8202-6_23

replication mechanism. Using the pointer network, the model copies the new object label into the description statement, which improves the available vocabulary of the model [4].

The above models need a lot of manual interpretation data to train the model, but it is often impossible to achieve for image data without prior manual interpretation [5]. In this paper, we use the target characteristics database and scene template database to help interpret images. We find the target characteristics of the images to be interpreted from the target characteristics database and recommend the most similar images. We obtain the interpretation template from the scene template database, which can complete image interpretation with less manual interpretation data.

The work is supported by the Major Project of China High Resolution Earth Observation (GFZX0404120203), the National Defense Science and Technology Innovation Special Zone (2116301ZT00207901) and the National Defense Science and technology excellence Youth Science Foundation (2020JCJQZQ014).

2 Framework Design

Based on the labeled target of the image to be interpreted and the target characteristic information obtained by mining, the most similar target and the list of images are retrieved in the target characteristic database. The most similar images and their corresponding reference scenes in the target characteristic database are recommended by comparing the characteristics between the targets in the image by using the collaborative filtering technology. On this basis, based on the scene template generated from the scene template database, we apply the neural network to model and interpret the position relationship between the targets and the background, and generate meaningful description sentences to realize the understanding of the target scene. The target characteristics recommender system follows the service-oriented design idea and the modular design method to realize the loose coupling between the infrastructure module and the service module, so as to ensure the dynamic expansion of services and the rapid launch of new services. The framework design chart is shown in Fig. 1.

3 Technology and Method

3.1 Target Characteristics Retrieval

Based on the target characteristics database, the target characteristics retrieval module can quickly and accurately retrieve the qualified target elements and their corresponding images through the hierarchical directory information of the target and relevant query conditions [6]. When the hierarchical directory information of the target cannot be accurately matched, it is converted into a word vector to vaguely find the qualified target characteristics set or target image set, as shown in Fig. 2.

3.1.1 Target Element Retrieval

The input of the single target element retrieval module is a target element in the database to be retrieved. The information is analyzed to generate the corresponding query statement. The output of the module is the target element set or target image set that meets the query conditions of the target element to be retrieved.

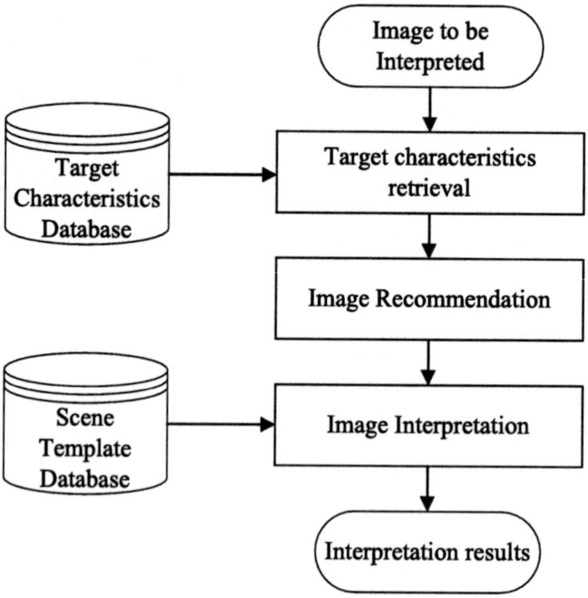

Fig. 1. Framework design chart

When retrieving the target characteristics, the directory information is analyzed first. If the name of the target to be retrieved can be accurately matched in the directory structure tree, then use the routing information from the root node to the leaf node, and use multithreading to search the database in parallel in the corresponding sub table. If the name of the target cannot be accurately matched, the trained word vector is used to express the target name, and then the similarity between the target name and each directory is calculated to find the first k similar directory names. Then, according to the directory structure, the routing information of the table is constructed, and finally it is searched in parallel in each sub table. Thus, the fuzzy retrieval of target characteristics can be realized, that is, when the target cannot match the directory item, it can also represent the semantic characteristics in its directory.

3.1.2 Target Image Retrieval

The input of the multi-target element retrieval module is the information of a target image in the database to be retrieved. The target image is analyzed to obtain multiple target elements, and the corresponding query statements are generated at the same time. The output of the module is the set of target images that meet the query conditions of the target image to be retrieved.

For the retrieval of target image, there is an N-to-N relationship between the target in the image to be retrieved and the image in the target characteristics database. It is necessary to aggregate the image. When all threads return results, the results are unrepeated in memory through HashSet to form an image set.

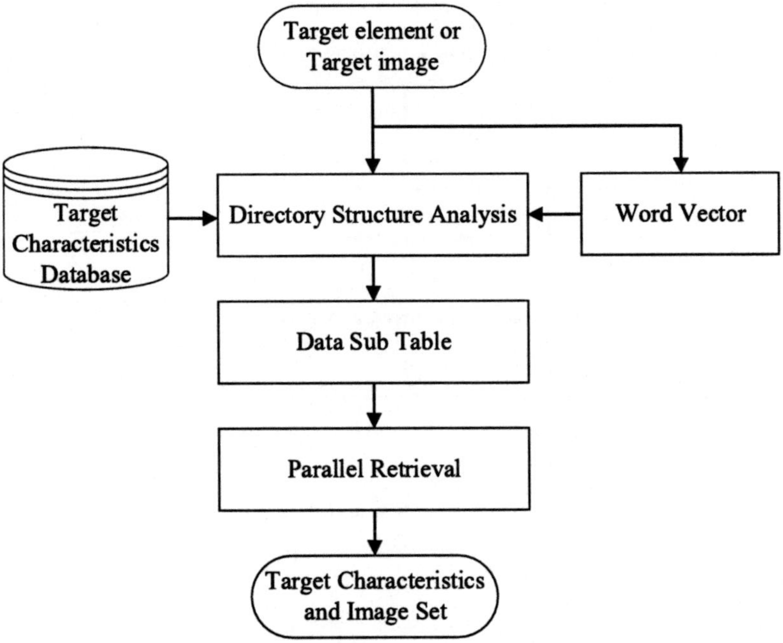

Fig. 2. Target characteristics retrieval chart

3.2 Target Image Recommendation

According to the target and its characteristic information in the image, we construct the target characteristic vector and apply metric learning to optimize the weight of different target characteristics in calculating the distance between the characteristic vectors. We train the similarity between the targets by using collaborative filtering technology and SLIM algorithm, and then use Hungarian algorithm to construct the perfect match between the targets of the images [7]. At the end, based on this perfect match, the similarity between the image to be recommended and the recommended image is calculated, as shown in Fig. 3.

3.2.1 Target Element Recommendation

For the distance measurement between characteristics vectors, metric learning is used to select the appropriate metric [8]. Most methods are based on constraints or triples, and learn metrics in a weakly supervised way. We use S as the two tuples of similar samples, D as the two tuples of dissimilar samples, and R as the three tuples of three samples. Supervised learning is adopted for the measurement learning training of targets in the image, and auxiliary information is provided between targets in the form of S, D and R constraint sets. The measurement form adopts Mahalanobis Distance linear measurement, uses measurement learning to calculate the distance between target elements, and recommends target elements from small to large according to the distance.

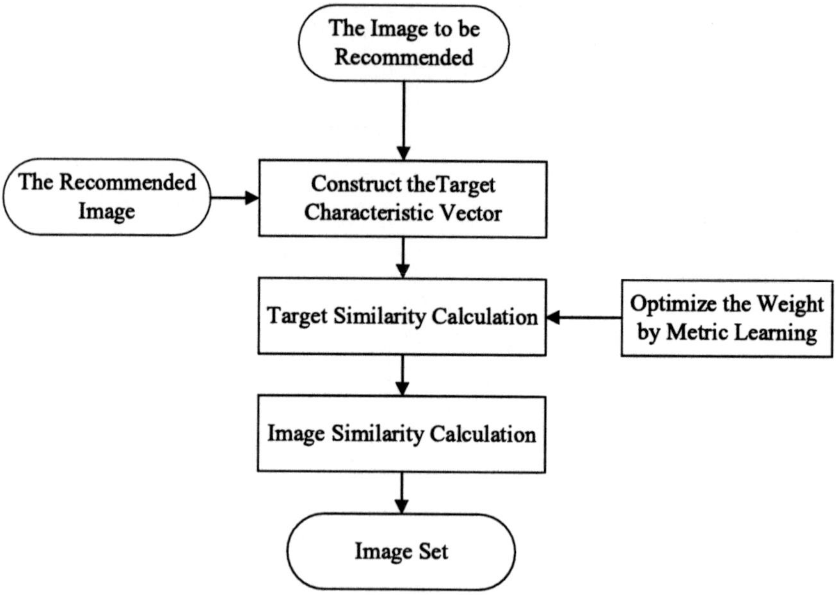

Fig. 3. Image recommendation chart

3.2.2 Target Image Recommendation

Taking the target in the recommended image and the image to be recommended as the node of the graph, the model can be regarded as a bipartite graph. If the similarity between the targets is greater than 95%, there is an edge between the two targets. Consider a bipartite graph $B = (U, V, e)$, U is the row vertex set, V is the column vertex set, $|U| = |V| = n$, and the edge set $E \subseteq U \times V$. The input of the algorithm is the graph of the target between the image to be recommended and the recommended image and a matching M. The output is a perfect match different from M. If the perfect match does not exist, M is returned.

By obtaining the perfect matching of the target between the image to be recommended and the recommended image, and then averaging the similarity between the target pairs in the perfect matching, the similarity of the image is obtained. Complete the target image recommendation from large to small according to the similarity.

3.3 Target Image Interpretation

Count the element distribution information of the target image to generate the description of different element types and quantities. Then structurally output the interpretation results of the element distribution information of the target image through the sequential dictionary. To interpret the target image, the target relationship pairs in the image are constructed [9]. For each relationship pair, the optimal template is matched in the scene template database to complete the description information interpretation, as shown in Fig. 4.

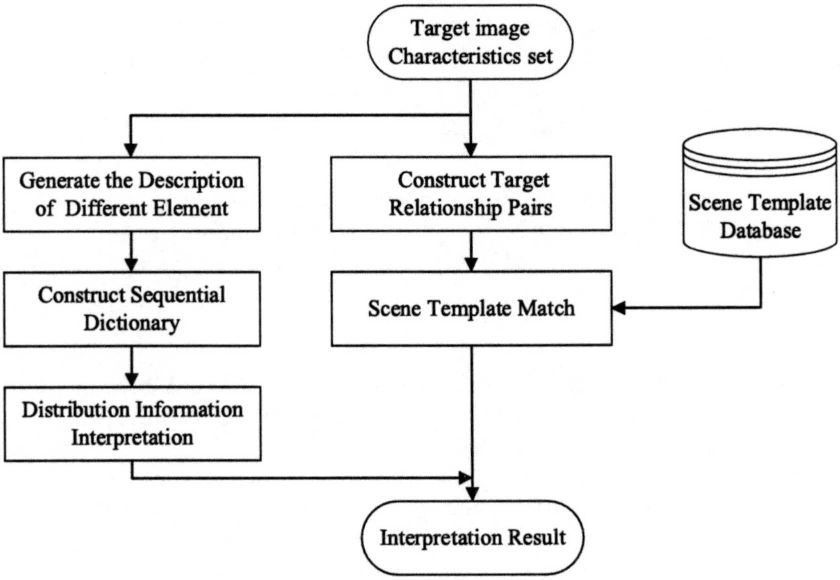

Fig. 4. Target interpretation chart

3.3.1 Target Element Distribution Information Interpretation

The target element distribution information is obtained by statistics of all target categories identified in the image. We get the category set and the target number of each category. Through the multi-level directory information, a multi-level list sequence dictionary for identifying the category subordinate relationship between entities is constructed. The distribution information interpretation result is generated for the category set through the sequence dictionary.

3.3.2 Target Scene Information Interpretation

By constructing the scene template database and matching the relationship pair to be interpreted with the scene in the scene template database, the interpretation of description information between target pairs is obtained. Replacing the entity information with the most similar scene template in the scene template database, we generate the interpretation result of the relationship pair in the image to be interpreted. The scene template database stores the target characteristics, the state relationship between targets, the scene information and the scene interpretation description information. The scene template database is used to construct the description information of all target pairs in the image, which is integrated into the description text of the target relationship pair of the image. The construction of scene template database adopts the mechanism based on incremental learning. When the current relationship pair can be interpreted through the scene template database, the interpretation results will not be put into the template database. When the current interpretation results cannot meet the needs of users, the user feedback information is used as the increment of the template database, which is dynamically updated to the scene template database through incremental learning.

4 Recommendation Mechanism

Through the user feedback, the labeled images uploaded to the target characteristics database are graded, so as to determine the recommended priority and the elimination of invalid characteristics, and finally realize the upgrading and updating of the characteristics database.

1. Image validity assurance
   Image validity has a great impact on the interpretation effect and recommendation accuracy. The design ensures the effectiveness of the image from two aspects: image content and user level. The timeliness and accuracy of the characteristics database are ensured by filtering the image content input into the characteristics database and regularly deleting the outdated content in the characteristics library. The content provided by high-quality users is pushed first.
2. New data filtering
   The system adopts the submit-feedback mechanism to filter the images uploaded by users. When the user uploads the image through the system, the system first reviews it through automatic recognition to ensure that the definition meets the requirements and the characteristics can be distinguished. Otherwise, the user fails to upload. After marking the target, the system will submit it to the administrator for the second review, and then add it to the target characteristics database. Otherwise, it will be fed back to the user that the upload does not meet the requirements.
3. Old data deletion
   When users upload data, the system not only automatically marks, but also scores the image. The clearer the characteristics of the image, the higher the score given by the system. The timeliness of data is marked by time weight. The system automatically updates the content in the target characteristics database at a fixed time, and automatically eliminates the images with low score after updating.
4. High-quality users evaluation
   The system can increase high-quality users by scoring users. The scoring mechanism is as follows. Set corresponding points for different users according to user category. Every time a user successfully uploads an image, the system increases the user points of the user according to the user level. The system can select high-quality users according to user points. For users with high scores, their uploaded data can be preferentially received into the target characteristics database. In the recommendation process, select high-level content to recommend to users according to the score of the image. When the scores are the same, the content uploaded by high-level users will be pushed first.

5 System Construction

The logical composition of the target characteristics recommender system includes data source layer, data business layer and data service layer. The data source layer includes the form of access data, which is the sum of the data resources of the whole system. The data business layer is the core business architecture of the system and provides support for the upper application of the platform. The data service layer opens services to the outside world and takes the data business layer as the support to provide service support for specific top-level application development. Each logic block supports each other to form a target characteristics recommender system for image interpretation, as shown in Fig. 5.

The data source layer contains the user database, data sharing database and data warehousing. The user database stores user login key, permission level and other information. Data sharing library contains target characteristics database, target scene template database, etc. Data warehousing is mainly used to support big data analysis applications.

The data business layer includes big data platform, data management module, data analysis and mining module and data visualization module. In addition, it also includes an external independent data governance platform for standardized data governance [10].

The data service layer is the application layer of the target characteristics recommender system. By using the relevant functional modules of the business layer, it completes the practical applications, such as target characteristics retrieval, target image recommendation and target image interpretation.

The target characteristics recommender system for image interpretation can retrieve the target characteristics according to the retrieval information provided by the user. The system calculates the similarity between the recommended target and the target characteristics to be recommended according to the target characteristics. The similarity is not less than 95%, and the recommended recall rate is not less than 95%. It can recognize and count the number of different target types and the relationship between targets according to the image to be interpreted provided by the user. For the target characteristics database of no less than one million, the system response time is in the order of seconds.

6 Conclusions

This paper proposed multi-user characteristics recommendation technologies such as target characteristics retrieval, target image recommendation and target image interpretation. A target characteristics recommender system for image interpretation was designed and implemented, which could retrieve the target characteristics according to the retrieval information provided by the user and interpret the user's input image. A target characteristics information recommendation mechanism supporting image interpretation was established, which could support business applications.

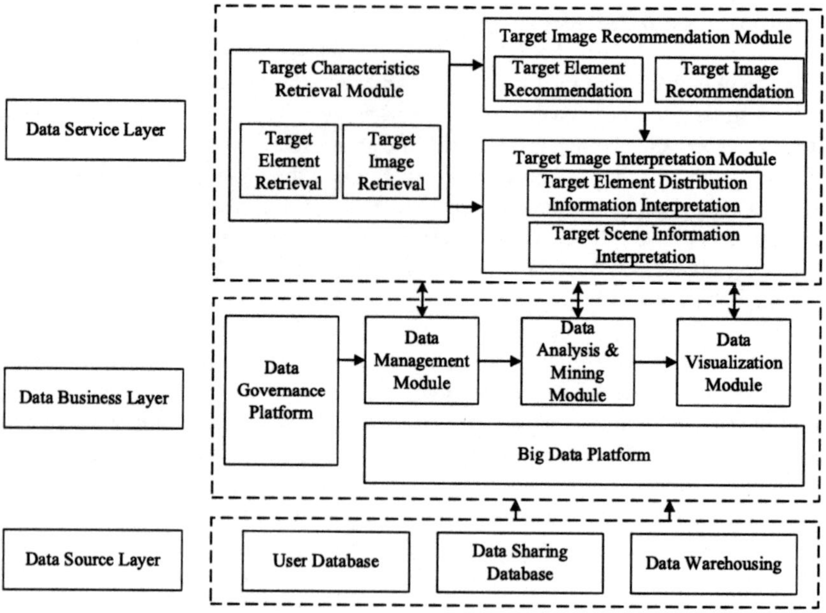

Fig. 5. System platform chart

Innovation. This paper designs a target characteristics recommender system framework based on the target characteristics database and scene template database. A target characteristics retrieval method based on the combination of target level directory information and word vector is proposed. A target image recommendation method based on metric learning and collaborative filtering is proposed. A target image interpretation method using target category statistics and scene template matching is designed. A target characteristics information recommendation mechanism supporting image interpretation is established. A target characteristics recommender system for image interpretation is constructed.

References

1. Adomavicius, G., Tuzhilin, A.: Toward the next generation of recommender systems: a survey of the state-of-art and possible extensions. IEEE Trans. Knowl. Data Eng. **17**(6), 734–749 (2005)
2. Lei, C., Liu, D., Li, W., et al.: Comparative deep learning of hybrid representations for image recommendations. In: Proceedings of the IEEE Conference on Computer Vision and Pattern Recognition. Las Vegas, USA, pp. 2545–2553 (2016)
3. Bansal, T., Belanger, D., McCallum, A.: Ask the GRU: multi-task learning for deep text recommendations. In: Proceedings of the 10th ACM Conference on Recommender System, pp. 107–114, Boston, USA (2016)
4. Li, Y., Liu, T., Jiang, J., et al.: Hashtag recommendation with topical attention-based LSTM. In: Proceedings of the 26th International Conference on Computer Linguistics, pp. 3019–3029. Osaka, Japan (2016)

5. Huang, L.W., Jiang, B.T., Lv, S.Y., et al.: Survey on deep learning based recommender systems. Chin. J. Comput. **41**(7), 1619–1647 (2018)
6. Gonde, A.B., Maheshwari, R.P., Balasubramanian, R.: Modified curvelet transform with vocabulary tree for content based image retrieval. Digit. Signal Process. (1) (2013)
7. Liu, Q., Xiong, Y., Huang, W.: Combining user-based and item-based models for collaborative filtering using stacked regression. Chin. J. Electron. (04) (2014)
8. Hsieh, C.K, Yang, L., Cui, Y., et al.: Collaborative metric learning. In: Proceedings of the 26th International Conference on World Wide Web, pp. 193–201. Perth, Australia
9. Li, Z., Fan, Y., Liu, W., Wang, F.: Image sentiment prediction based on textual descriptions with adjective noun pairs. Multimed. Tools Appl. **77**(1), 1115–1132 (2017). https://doi.org/10.1007/s11042-016-4310-5
10. Lin, H., Zhang, D.: The research of reader Portrait in University Library based on big data. In: Proceedings of 3rd International Conference on Mechatronics Engineering and Information Technology (ICMEIT 2019) (2019)

Application Research on Water Body Extraction of Gaofen-3 Polarimetric SAR Based on Deep Learning

Zhang Wei[1(✉)], Wang Guanghui[1,2], Qi Jianwei[1], and Zhang Tao[1]

[1] Land Satellite Remote Sensing Application Center, MNR, Beijing, China
dave6806@163.com

[2] China University of Mining and Technology, Xuzhou, China

Abstract. Quickly and accurately obtaining the area, volume and distribution of surface water is of great significance to the resource survey, urban planning, economic development, flood disaster monitoring, etc. In this paper, the domestic Gaofen-3 (GF-3) dual-polarization radar satellite data is used, combined with the semantic segmentation algorithm of the deep learning attention mechanism, to construct a polarized radar surface water information extraction model based on TransUnet, which has been carried out in the Poyang Lake area. The results show that the model can not only give full play to the advantages of radar water body information extraction, but also better ensure the accuracy of water body edges and reduces the interference of shadows and radar image speckle noise caused by terrain, overlay, foreshortening, etc. The experimental result achieves an overall classification accuracy of 95.8% and a Kappa coefficient value of 0.942, which are both better than traditional threshold segmentation algorithms and support vector machine classification algorithms.

Keywords: Gaofen-3 · Polarimetric SAR · Water body extraction · Deep learning

1 Introduction

Remote sensing surface water extraction has always been a research hotspot in the field of remote sensing image information extraction. With the emergence of radar remote sensing, it not only improves the accuracy of water extraction, but also has the advantages of being not affected by time and weather, and has been widely used in flood disasters [1], wetland detection [2], coastline extraction [3], resource survey [2] and other business scenarios.

With the continuous progress of radar technology innovation, water extraction technology has mainly experienced the transition from single-polarization feature extraction to multi-polarization information fusion. Single-polarization radar mainly completes water identification by setting a fixed threshold [4, 5]. Since the threshold is only suitable for local areas, this method is not suitable for large-area areas. With the continuous enrichment of multi-polarization features, the object-oriented water body information

L. Wang et al. (Eds.): CHREOC 2022, LNEE 969, pp. 274–283, 2023.
https://doi.org/10.1007/978-981-19-8202-6_24

extraction technology [6–8] can better integrate the characteristics of multi-type polarized water bodies and improve the accuracy of water body identification, but due to the complexity of the processing process and the pertinence of data use, it cannot be better used for business application.

The application of deep learning technology in remote sensing image information extraction is mainly based on semantic segmentation technology. Due to its strong feature learning ability, the classification accuracy of remote sensing objects is further improved. At the same time, the operation process is simplified and the universality of classification model is improved, which is gradually applied to radar water application. Xue [9] designed a dense network model based on coordinate recognition to integrate the water features of dual-polarization SAR images to overcome the influence of ground element interference and improve the accuracy and generalization of the water extraction model. Zhang [10] proposed a cascaded fully convolutional network to improve the performance of water body detection in high-resolution SAR images of Gaofen-3 data, which not only improved the water body pixel accuracy and boundary accuracy, but also improved the detection robustness, which speed performance is as well. The above methods are all based on convolutional neural network for semantic segmentation. With the development of attention mechanism, transformer [11] shines in the field of deep learning speech and image, but in remote sensing water extraction, especially based on polarimetric SAR, few studies on water extraction.

As Chinese first civil high-resolution C-band radar satellite, Gaofen-3 was successfully launched on August 10, 2016, which has 12 imaging modes, and the highest resolution can reach 1 meter, and has dual-polarization and full-polarization observations ability. However, its single-polarization mode is often used in water body extraction [12], and there are few studies on water body recognition based on dual-polarization combined with deep learning. In this paper, an algorithm is proposed based on the dual-polarization FSII data of Gaofen-3 and applied to the Poyang Lake area in July 2020, which uses polarization processing to generate false-color synthetic images to produce a semantic segmentation sample set, and uses TransUnet semantic segmentation network [13] integrated Transformer-structured for surface water extraction.

2 Main Research Contents and Results

2.1 Technical Methods

In a batch of dual-polarization data of Gaofen-3 in the demonstration area, firstly, through preprocessing such as radiometric calibration and polarization band synthesis, a polarization channel synthetic image with obvious features is obtained. Select a part of the data and its corresponding water body vector data to make training samples, and carry out the TransUnet semantic segmentation network for model training, and then complete the inference prediction and accuracy evaluation for the remaining part of the image. The specific process is shown in Fig. 1.

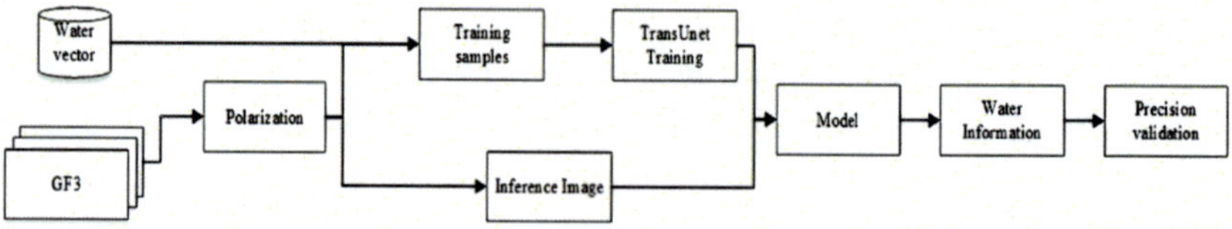

Fig. 1. The algorithm flow

2.2 GF3 Polarization Data and Preprocessing

2.2.1 The Study Area and Data

The fine strip pattern FSII data of GF-3 located in the Poyang Lake area of Jiangxi Province has been selected in this paper. The specific data information is shown in Table 1. Due to the continuous heavy rains in the Poyang Lake area of Jiangxi in July 2020, the water level of the lake has risen sharply. The 5-scenario GF-3 image range covering almost the entire Poyang Lake basin acquired on July 13 was selected as the study area, which is of great significance to extract research and assist emergency decision-making. The detailed data coverage is shown in Fig. 2.

Table 1. GF-3 FSII data information

Sensor	Gaofen-3
Polarization mode	HH + HV
Incident angle (°)	20 ~ 50
Strip imaging mode	Fine strip 2 (FSII)
Width (km)	100
Resolution (m)	10 × 10
Product type	SLC
Product level	Leavel-1

2.2.2 Preprocessing Method

For the original FSII data, it needs to go through the preprocessing process as shown in Fig.3, which mainly includes radiometric calibration, multi-look processing, speckle noise filter, DB conversion, geocoding, and band layer stacking. After the above steps, the polarization backscattering false color composite image of FSII can be obtained, in which the multi-look is processed by 3 × 3, and the speckle noise is processed by 7 × 7 refined Lee filter. DB transformation can better improve the contrast information of data images and improve the identification of water body information (as shown in Fig. 4b, image number IV, see the red box in Fig. 2). Combined with DEM for geocoding, the band combination of Red: HH, Green: HV, Blue: HH is used for false-color synthesis. Figure 5a, b show the false-color synthesis data results and the range of backscatter

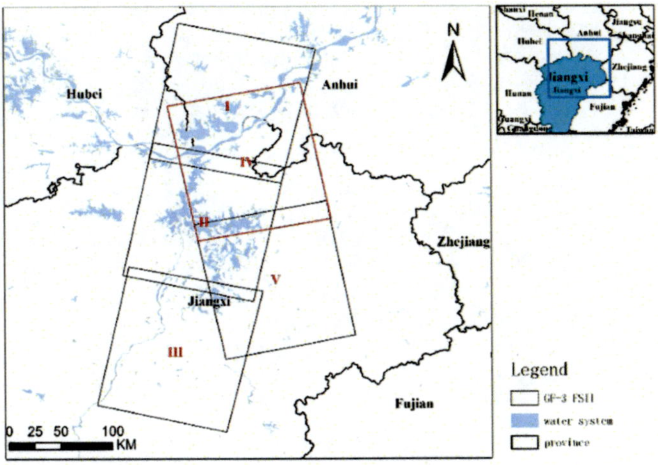

Fig. 2. The study area and GF-3 FSII data

values for each band of image numbers I and II respectively. It can be seen that after the preprocessing process, the water features of the FSII data are significantly enhanced on the false-color image, which is easier for target classification and extraction in subsequent steps.

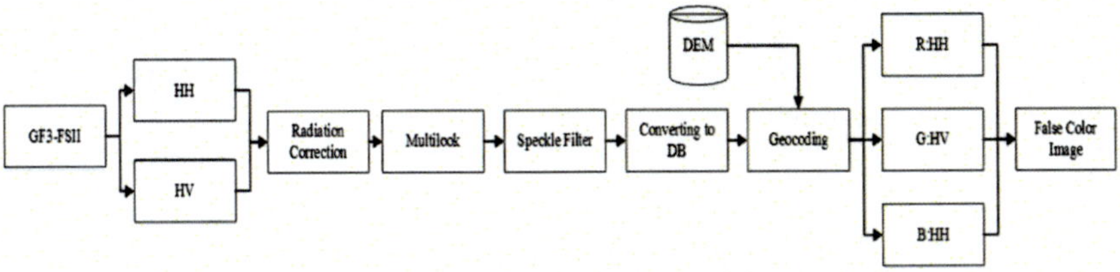

Fig. 3. Preprocessing process of GF-3 FSII data

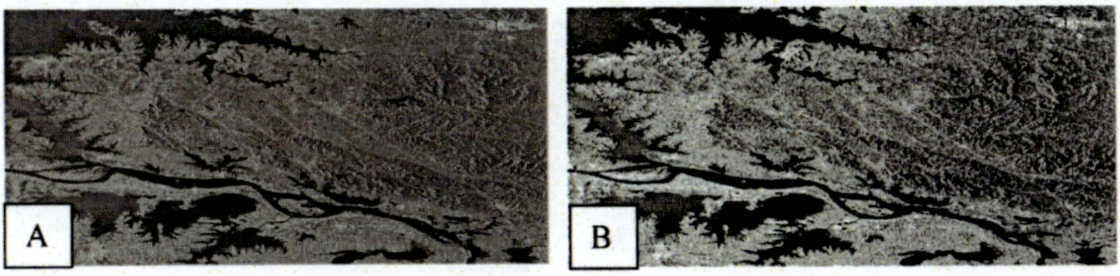

Fig. 4. The DB result of HH polarization

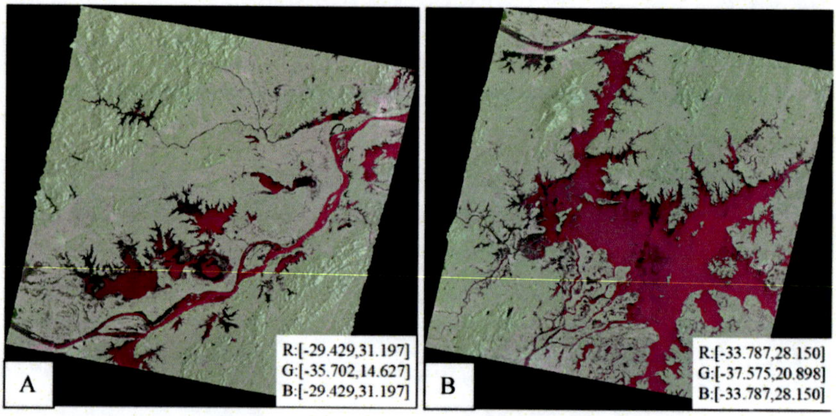

Fig. 5. Polarized false color composite image

2.3 Training Samples

Based on the false-color composite image data in the polarized band, the fourth scene image in the experimental area (red box in Fig. 2) and its corresponding water vector data were selected as training samples, and the frame-by-frame loop method was used to generate samples of size 512 * 512. For the label data, the target information in the vector data is marked as 255 and the background is marked as 0, generating a total of 540 samples. At the same time, sample enhancement algorithms are used, including random rotation, color enhancement, random noise addition, contrast enhancement, and sharpening processing to expand the total number of samples to 4860, which have been divided as the training sample set and the verification sample set according to the ratio of 8:2. Figure 6 shows the result of enhancement effect on the one sample.

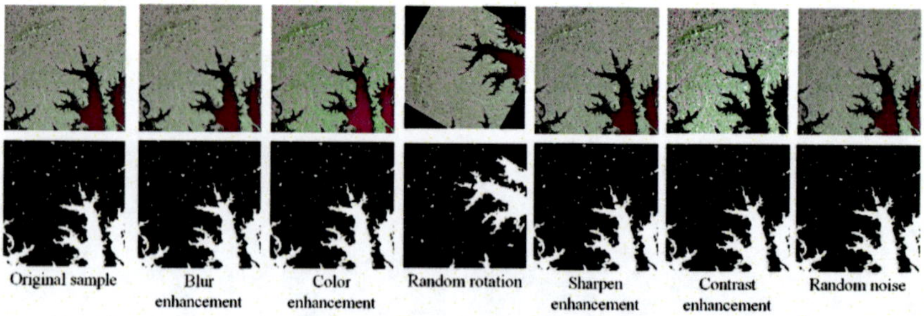

Fig. 6. Training sample enhancement

2.4 Water Extraction Network Model and Training

The TransUnet used in this paper is a semantic segmentation network that effectively combines Unet with the attention mechanism Transformer, which was first applied in the field of medical image segmentation. TransUnet has the dual advantages of Transformers and U-Net. On the one hand, the transformer is used to encode the tokenized image blocks from the convolutional neural network (CNN) feature map to improve the global target self-attention accuracy. On the other hand, the features are upsampled and then combined with high-resolution CNN feature maps for precise localization. In this way, the network model can not only improve the recognition accuracy of image target information, but also retain the boundary information of the target effectively. Its overall network architecture is shown in Fig. 7.

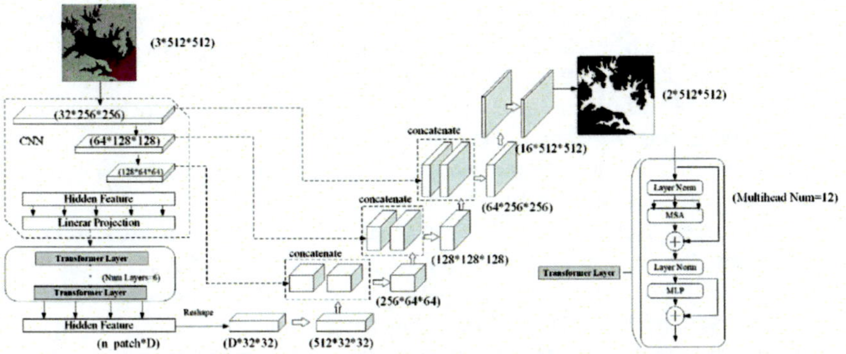

Fig. 7. Water extraction network structure based on TransUnet

Since water body extraction is a binary classification problem, the number of channels in the last layer of TransUnet is set to 2. At the same time, when using the ResNet50 network for feature extraction in the early stage, the last set of convolutional layers are discarded, so the minimum dimension of the feature layer is 1/8 of the original sample size. The Loss function is composed of Dice loss combined with Cross Entropy Loss, the training accuracy evaluation on the validation set uses the average IoU (Intersection-over-Union) and F1 Score, and the optimizer uses stochastic gradient descent (SGD). In the model training phase, the maximum number of epochs is set to 80, the batch size is set to 8, and the base learning rate is set to 0.01. The training is performed on the 4GPUs of NVIDIA TESLA V100 32G. The training error curve and the accuracy evaluation curve are shown in Fig. 8, from which it can be seen that on the validation set, a high training accuracy has been achieved, and the training model has converged. At the same time, it can be seen from the figure that using the ImageNet pre-training model can speed up the training efficiency, that the model converges at the early stage of training.

2.5 Model Inference and Accuracy Evaluation

The process of extracting water body information from the whole scene image is the inference process of the remote sensing deep learning model, that is, the network feed-forward propagation is completed after getting the model file from training and loading remote sensing images. Due to the large size of FSII data, data loading cannot be performed directly, and it needs to be processed in blocks. In the process of looping the

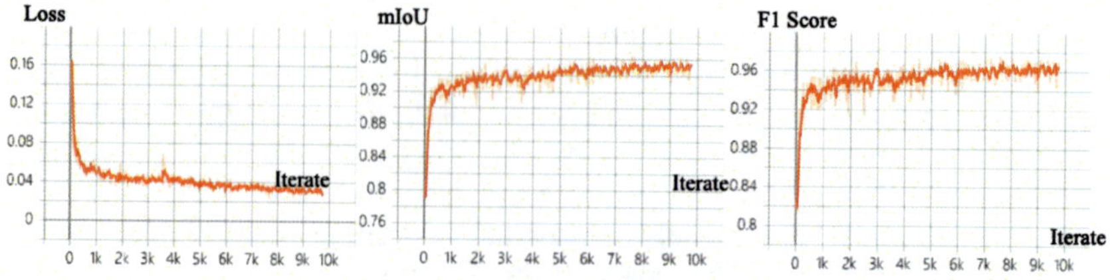

Fig. 8. Model training results

whole scene data frame by frame, there is no overlap between blocks, which will cause the inference results of each block to be discontinuous, thereby affecting the results of the panoramic data. Therefore, this paper proposes a model inference method that can smooth the blocks. When traversing the remote sensing image block by block, in addition to increasing the overlap between blocks, it is also necessary to complete the mosaic and smooth transition of the deduction results of each block, which ultimately ensures the integrity of the inference results of a scene image.

In order to objectively evaluate the effectiveness of the method proposed in this paper, 5 image data in the experimental area are mosaicked and model inference is completed, as shown in Fig. 9, which respectively show the results of threshold segmentation, SVM classification and deep learning method in this paper. From the results in the figure, it can be seen that the water extraction results of the method proposed in this paper are more continuous (as shown in the central area of Poyang Lake in Fig. 9), and are not disturbed by shadows. Whether it is threshold segmentation or SVM classification, a large body of water also mentioned in the shadowed area.

Secondly, the quantitative accuracy evaluation has been carried out. 500 verification points are randomly collected and the distribution is shown in Fig. 9. Due to the bad weather in the test area, it is impossible to obtain the real ground reference results. Therefore, this paper adopts a method based on visual interpretation to obtain the true value of the ground water information, counts the classification results and the true value results of each point, and calculates the confusion matrix and Kappa coefficient (as shown in Table 2), It can be seen that the classification accuracy of the method in this paper reaches 95.8%, and the Kappa coefficient is

0.942. Both indicators are higher than those based on threshold segmentation (PA: 90.5%, Kappa: 0.836) and support vector machine (PA: 85.3%, Kappa: 0.854). Combined with the chart, it can be seen that the accuracy of the deep learning method used in this paper is higher than the traditional classification results, and the edge accuracy of the water body is also guaranteed, avoiding the influence of speckle noise.

3 Conclusion

Remote sensing images can achieve high-precision acquisition of surface water information in a large-scale, non-contact, and high-efficiency manner, which is of great significance to flood control and disaster relief, resource surveys, and coastline mapping. In this paper, based on the dual polarization processing of the domestic radar satellite

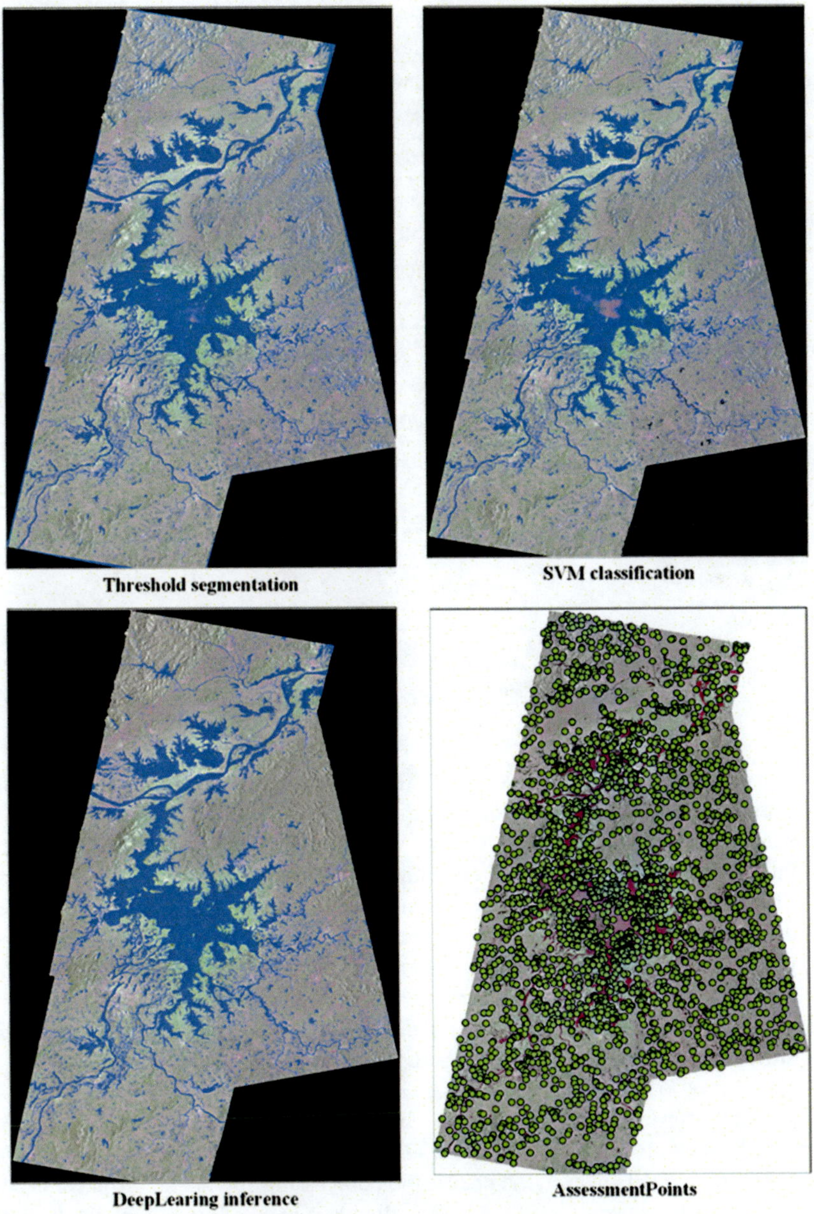

Fig. 9. Classification results of different methods

GF-3, a semantic segmentation network based on TransUnet is proposed to achieve high-precision extraction of surface water information. The experimental conclusions are as follows:

Table 2. The accuracy statistics table of the above methods

Method			Predicted class				
			Non-water	Water	Total	UA	Kappa
Deep learing	Actual class	Non-water	400	4	404	0.990	
		Water	5	91	96	0.948	
		Total	405	95	500		
		PA	0.988	0.958		0.982	
		Kappa					0.942
SVM	Actual class	Non-water	397	14	411	0.966	
		Water	8	81	89	0.910	
		Total	405	95	500		
		PA	0.980	0.853		0.956	
		Kappa					0.854
Threshold	Actual class	Non-water	388	9	397	0.977	
		Water	17	86	103	0.835	
		Total	405	95	500		
		PA	0.958	0.905		0.948	
		Kappa					0.836

(1) The polarization data of the domestic GF-3 radar satellite can be used to extract the surface water body information, and the false color synthesis of the polarization band can further improve the surface water body characteristics and reduce the interference between the water body information and other background information;
(2) The deep learning framework based on attention mechanism can better complete the task of surface water extraction from radar satellite remote sensing. Compared with traditional remote sensing water extraction algorithms, it not only has higher classification accuracy, but also retains water boundary information with high precision, which plays the effect of polarization treatment to better distinguish the difference between regional water bodies and shadows.

In a word, deep learning technology combined with radar polarization processing has a high application prospect, and it is a new idea for the extraction of surface information of polarization radar. However, the polarization processing applied in this paper is only simple band synthesis, and only uses polarization as a band of remote sensing images. In the future, the polarization feature decomposition method will be combined with the polarization feature decomposition theory to form the polarization feature combined with the deep learning classification technology, in order to further improve the accuracy and universality of deep learning polarimetric radar water extraction.

References

1. Bioresita, F., Puissant, A., Stumpf, A., et al.: A method for automatic and rapid mapping of water surfaces from Sentinel-1 imagery. Remote Sens. **10**(2) (2018)
2. Whyte, A., Ferentinos, K.P., Petropoulos, G.P.: A new synergistic approach for monitoring wetlands using Sentinels-1 and 2 data with object-based machine learning algorithms. Environ. Model. Softw. **104**, 40–54 (2018)
3. Fuse, T., Ohkura, T.: Development of shoreline extraction method based on spatial pattern analysis of Satellite SAR images. Remote Sens. **10**(9) (2018)
4. Otsu, N.: A threshold selection method from gray-level histograms. IEEE Trans. Syst. Man Cybern. **9**(1), 62–66 (1979)
5. Liang, J., Liu, D.: A local thresholding approach to flood water delineation using Sentinel-1 SAR imagery. ISPRS J. Photogramm. Remote Sens. **159**, 53–62 (2020)
6. Pekel, J., Cottam, A., Gorelick, N., et al.: High-resolution mapping of global surface water and its long-term changes. Nature **540**(7633), 418–422 (2016)
7. Sun, Z., Xu, R., Du, W., et al.: High-resolution urban land mapping in China from Sentinel 1A/2 imagery based on google earth engine. Remote Sens. **11**(7527) (2019)
8. Wan, J., Wang, J., Zhu, M.: Water extraction from fully polarized SAR based on combined polarization and texture features. Water **13**(333223) (2021)
9. Xue, W., Yang, H., Wu, Y., et al.: Water body automated extraction in polarization SAR images with dense-coordinate-feature-concatenate network. IEEE J. Sel. Top. Appl. Earth Obs. Remote Sens. **14**, 12073–12087 (2021)
10. Zhang, J., Xing, M., Sun, G., et al.: Water body detection in high-resolution SAR images with cascaded fully-convolutional network and variable focal loss. IEEE Trans. Geosci. Remote Sens. **59**(1), 316–332 (2021)
11. Vaswani, A., Shazeer, N., Parmar, N., et al.: Attention is all you need. Ad. Neural Inf. Process. Syst. **30** (2017)
12. Li, N., Niu, S.: High-precision water segmentation from synthetic aperture radar images based on local super-resolution restoration technology. J. Rad. **9**(1), 174–184 (2020)
13. Chen, J., Lu, Y., Yu, Q., et al.: Transunet: transformers make strong encoders for medical image segmentation (2021). arXiv:2102.04306

A Study on Buildings Liveability Index Based on High Resolution Land Cover Data

Jie Wang[1(✉)], Guanghui Wang[1,2], Jianwei Qi[1], Yu Liu[1], and Tao Zhang[1]

[1] Land Satellite Remote Sensing Application Center, The Ministry of Natural Resources, Beijing 100048, China
wangjie0039@163.com

[2] China University of Mining and Technology, Xuzhou 221116, China

Abstract. The increase of impervious surfaces and the decrease of urban green spaces are the important reasons to seriously affect the livability of the city. Using high resolution satellite data, urban land cover data could be extracted to effectively monitor the distribution of buildings and green spaces. This paper proposed a buildings livability index (BII) based on buildings greening rate and greening radiation factor, retrieved from high resolution land cover data. Taking Qingdao city as an example, this paper used the FROM-GLC10 dataset with a resolution of 10 m to extract the buildings and vegetation cover data in 2017, then calculated and analyzed the BII. The results showed that the BII of 10 districts in Qingdao was consistent with the distribution of buildings and vegetation, and was highly correlated with the urban heat island intensity. Among the 10 districts, Laoshan district had the highest BII in buildings pixels, reaching 1.33 due to more vegetation cover, and Huangdao and Pingdu districts had the highest BII in non-buildings pixels, pointing out the areas where development should be strengthened in urban planning. The results indicated that the BII could effectively reflect the rationality of the spatial layout of urban buildings and vegetation, and provide a scientific basis for urban spatial planning and governance.

Keywords: Land cover · Urban green spaces · Buildings greening rate · Buildings livability index

1 Introduction

In recent years, with the rapid development of society and economy, there is a series of urban ecological problems as reduction of increasing air pollution and climatic perturbations, which would greatly reduce the livability of the city [1]. Conversely, urban green spaces (UGSs) promote multiple effects such as absorbing CO_2, reducing noise, and alleviate the urban heat island (UHI) [2]. In response, there is a growing recognition that urban green spaces (UGSs) have a role in mitigating such environmental pressures [3].

The traditional method of obtaining UGSs is ground field survey, which are costly and time consuming [4], and is difficult to guarantee accuracy. In the past decades,

L. Wang et al. (Eds.): CHREOC 2022, LNEE 969, pp. 284–292, 2023.
https://doi.org/10.1007/978-981-19-8202-6_25

remote sensing technologies had occupied an important place in the study of UGSs as the characteristics of wide range, strong timeliness, short period and high precision. Nowak et al. [5] first used aerial photographs to determine urban tree cover. Ouma et al. [6] combined spectral and spatial information to significantly improve UGSs extraction accuracy. Hofmann [7] used GeoEve-1 imagery to obtain a vegetation map of the city of Bishkek in Austria. Pu and Landry [8] used IKONOS and WorldView-2 data for mapping urban tree species. Shojanoori and Shafri [9] used WorldView-2 data for mappings urban tree species. These studies focused on the extraction of UGSs and did not analyze the distribution of vegetation and buildings.

Schöpfer et al. [10] obtained the spatial distribution map of UGSs by grid division method. Based on the same idea, Gupta [11] used medium-resolution remote sensing images to measure the UGSs of Delhi, India. Lang [12] evaluated the green area of each street, and then obtained the spatial distribution map of UGSs in the whole area. Bao et al. [13] evaluated the distribution of UGSs in Baotou using Landsat-5 data and analyzed the quantitative relationship with UHI.

These studies provided ideas for the analysis of UGSs, but there was a lack of a university and standard evaluation index to quantitatively evaluate the rationality of urban buildings and vegetation distribution in a local area. This paper aimed to propose an index model based on buildings and vegetation cover data, considering the mechanism of how vegetation improve the urban problems, which could make full use of high-resolution satellite data to evaluate UGSs.

2 Methodology

2.1 Buildings Greening Rate

The buildings greening rate refers to the ratio of the total area of vegetation surfaces to the total area of urban impervious surfaces, which could reflect the balance between human activities and ecosystems, and is an important factor affecting UGSs and the livability of built-up areas.

Within a buffer zone, the formula for calculating the buildings greening rate at the central pixel is:

$$R_{vb} = \frac{s_V}{s_B} \tag{1}$$

where S_v and S_b are the sum of the vegetation and buildings area in the buffer zone, respectively. For block-scale analysis, the buffer zone can be taken as a square area of 110 m by 110 m. For the 10m resolution land cover dataset, the corresponding size of the buffer zone should be 11 rows and 11 columns. Through a moving window, size of which is 11 x 11, the buildings greening rate for each central pixel could be calculated.

2.2 Greening Radiation Factor

Vegetation has a positive impact on the surrounding environment, but the degree of influence varies with different distances. Therefore, for each vegetation pixel, within a certain radiation range, weight factors can be assigned to pixels with different distances. The closer to the vegetation pixel, the higher the weight factor. The formula for calculating the greening radiation factor of the central pixel is as follows:

$$P_v = \sum_{i=1}^{n} W_i \tag{2}$$

where n is the total number of vegetation pixels within the radiation range, and W_i is the weight factor at the *i*th vegetation pixel. The size of the radiation range is consistent with the buffer zone in Sect. 2.1. Considering the attenuation effect of vegetation influence, the weight factor matrix with 11 rows and 11 columns can be expressed as Fig. 1.

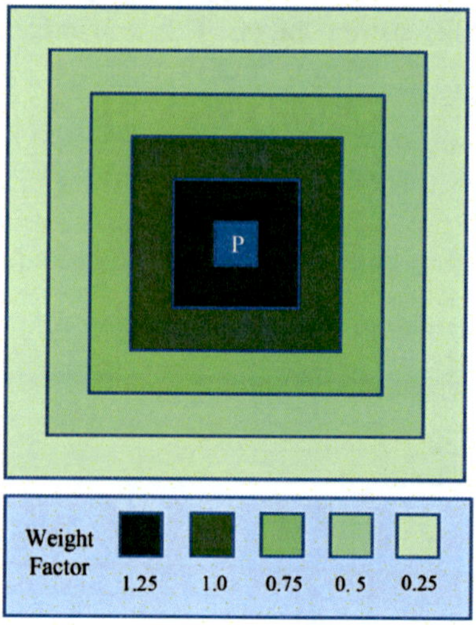

Fig. 1. The weight factor matrix of the buffer zone.

As shown in the Fig. 1, when all the pixels in the matrix are covered by vegetation, the green radiation factor calculation result at the center pixel P is 70; when the upper half of the 6 x 11 pixels are all covered by vegetation, and the lower half of the 5 x 11 pixels are all covered by non-vegetation, the greening factor calculation result at the center pixel P is 38.75.

2.3 Buildings Liveability Index

The buildings liveability index (BII) is determined by two factors, one is the area ratio of vegetation and buildings, the other is the distribution of UGSs. These two factors could be represented by buildings greening rate R_{vb} in Sect. 2.1 and greening radiation factor P_v in Sect. 2.2. The calculation formula of BII is:

$$BII = a_1 * R_{vb} + a_2 * P_v \tag{3}$$

where a_1 and a_2 are the weight coefficients of buildings greening rate and greening radiation factor respectively, which are related to the type of vegetation and height of buildings. The higher the branches and the denser the leaves of the vegetation, the larger the a_1 and a_2; Conversely, the higher the buildings, the smaller the a_1 and a_2. For urban built-up areas, the value of a_1 is usually from 0.05 to 0.1, and a_2 is from 0.5 to 0.8.

3 Dataset and Study Area

3.1 Land Cover Data

In order to extract urban buildings and vegetation cover data, this paper uses the 10m resolution global land cover product FROM-GLC10 in 2017, developed by Gong Peng et al. [14]. The world's first 10m resolution land cover dataset, the main types include crops, forests, grasslands, shrubs, wetlands, water bodies, tundra, impervious layers, bare land, and ice and snow, the overall accuracy of this dataset is 72.76%.

In this paper, three types of land cover, such as forest, grassland and shrubs, were extracted from the FROM-GLC10 dataset as vegetation cover data, and impervious surface was extracted as buildings cover data. Based on these two types of land cover data, urban green spaces and buildings spaces analysis was carried out.

3.2 Land Surface Temperature

In order to further verify the accuracy of the BII, this paper analyzed the relationship between the BII and UHI intensity. Therefore, the land surface temperature was retrieved using the 30 m resolution Landsat-8 data of Qingdao in 2017, and the UHI intensity was calculated according to the classification method proposed by Xu [15].

3.3 Study Area

This paper selected Qingdao as the study area. Qingdao has developed rapidly in recent years, and the urbanization process had accelerated, but the accompanying urban problems had become increasingly serious, and the urban and rural development had not been balanced. Through this method, the spatial distribution of buildings and vegetation in Qingdao could be monitored, and a basis for urban planning could be provided.

4 Results and Analysis

4.1 Buildings and Vegetation Cover in Qingdao

In this paper, the map of vegetation and buildings cover of Qingdao city extracted from FROM-GLC10 is shown in the Fig. 2.

The total area of buildings in Qingdao is 3210.76 square kilometers, and the total area of vegetation is 1105.51 square kilometers, accounting for 22.85% and 7.87% of the total city area of 14053.32 square kilometers. The ratio of the total area of buildings and vegetation is about 3:1.

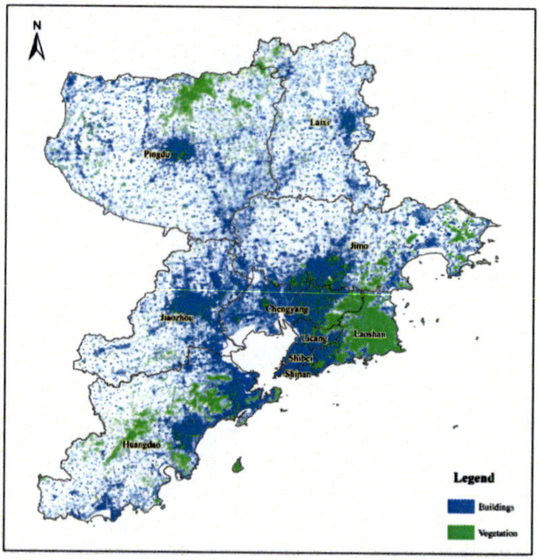

Fig. 2. The map of vegetation and buildings cover.

4.2 Buildings Livability Index in Qingdao

Based on the buildings and vegetation cover data in Qingdao, the buildings greening rate and the greening radiation factor were calculated through the moving window, and then the pixel-scale BII was calculated according to Eq. 3. Figure 3 is the distribution map of the BII of buildings pixels in Qingdao.

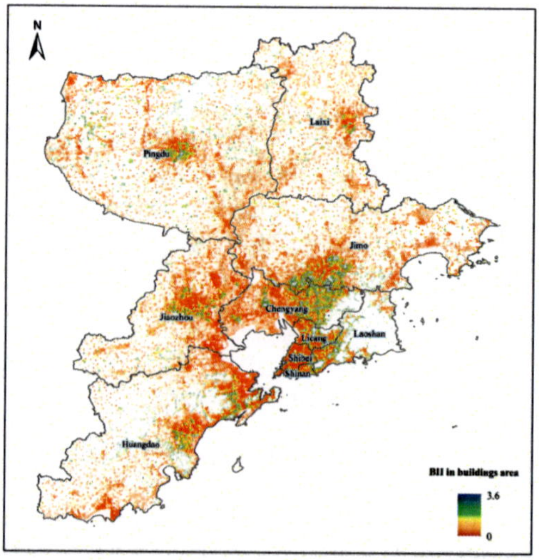

Fig. 3. The distribution map of the BII in buildings pixels.

In order to analyze the change law of the BII, and the rationality of the distribution of buildings and vegetation spaces in various regions of Qingdao, this paper analyzed the relationship between the BII, vegetation area and buildings area of 10 districts or counties in Qingdao, the results are shown in the Fig. 4.

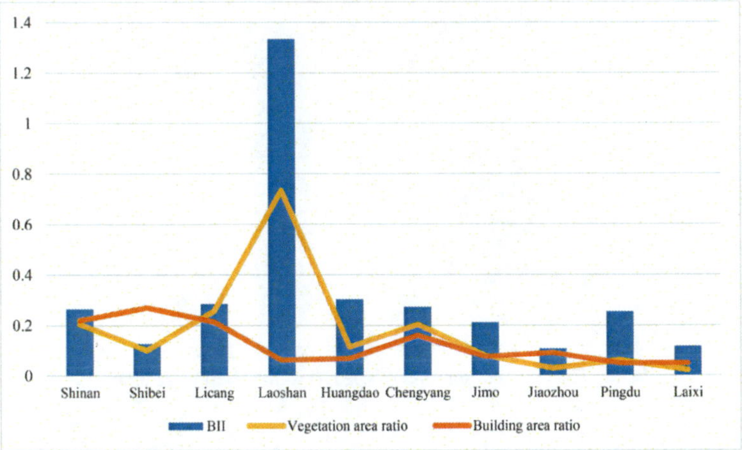

Fig. 4. The BII, vegetation area and buildings area of 10 districts in Qingdao.

Laoshan district has the highest BII, reaching 1.33 due to more vegetation cover. The distribution of buildings and vegetation spaces in Huangdao district is exceptionally reasonable, the vegetation cover of which is less than 12%, but the BII is 0.302; The distribution of that in Licang, Chengyang and Shinan districts is relatively reasonable, and the BII is 0.285, 0.272 and 0.264, respectively. Shibei district needs to improve the urban planning scheme and increase the greening degree of the urban areas, since the BII in which is as low as 0.249.

4.3 Analysis of Suitable for Building in Non-buildings Areas in Qingdao

In order to further tap the application potential of the BII, this paper also analyzed the distribution of the BII in the non-buildings cover areas, and pointed out the areas suitable for human habitation in the unused land, which would provide decision-making for land space planning and rural construction planning. Figure 5 is the distribution map of the BII in non-buildings pixels (crops and bare land types in the FROM-GLC10 dataset).

The Fig. 5 shows that there are fewer non-buildings pixels in urban areas, more non-buildings pixels in suburban counties. Huangdao and Pingdu districts have the highest BII in non-buildings pixels, urban development in these areas should be strengthened in urban planning.

4.4 Correlation Analysis Between BII and UHI Intensity

With the acceleration of urbanization, the problem of UHI has become increasingly prominent, which has seriously affected the urban ecological environment and people's

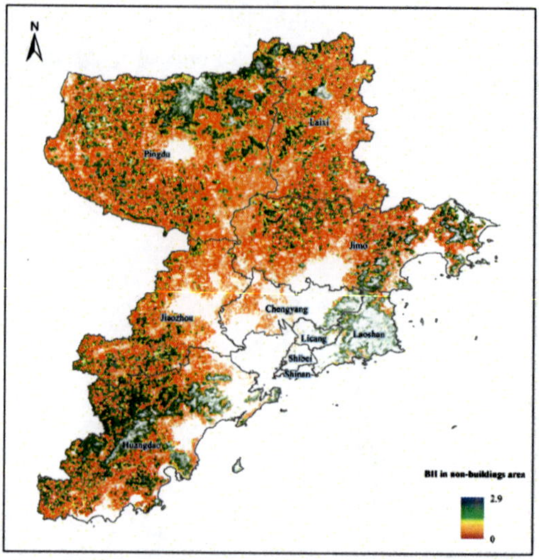

Fig. 5. The distribution map of the BII in non-buildings pixels.

living quality. The Fig. 6 is the UHI intensity of Qingdao in 2017 based on Landsat-8 data.

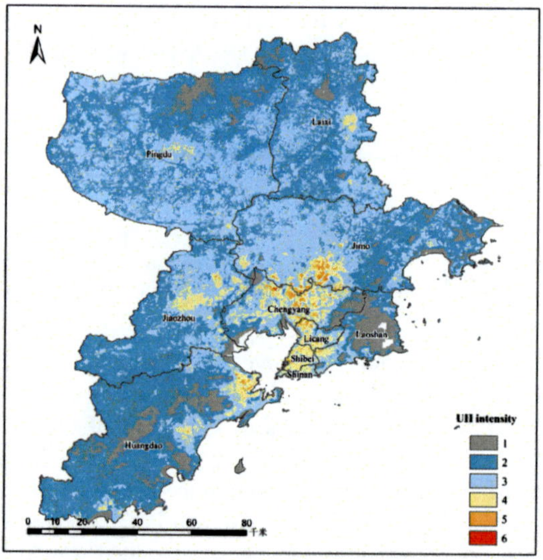

Fig. 6. The map of the UHI intensity.

The Fig. 6 shows that the UHI intensity is higher in the urban areas and lower in the suburban counties, and the distribution is very similar to BII. Actually, UHI intensity and BII are both affected by land cover changes and urban planning, and there is a certain relationship between the two parameters. Therefore, this paper further analyzed the quantitative relationship between BII and UHI intensity of Qingdao.

This paper randomly selected two areas in the urban and suburban, with a total of about 3000 pixels, to analyze the relationship between BII and UHI intensity. The results are shown in Fig. 7. It can be seen from the results that the BII is significantly negatively correlated with the UHI intensity, the correlation R^2 is 0.649. When the BII increases by 0.1, the UHI intensity decreases by 0.09. It shows that rationally planning the spatial layout of buildings and vegetation, and improving the livability of buildings would effectively reduce the UHI effect and improve the living environment.

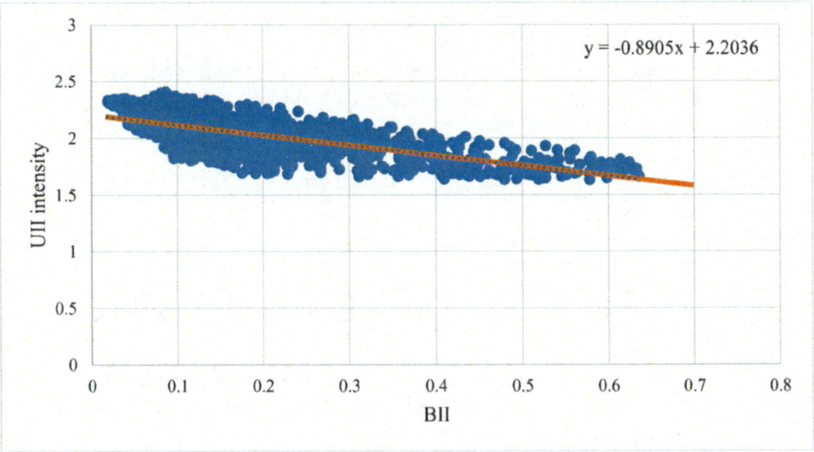

Fig. 7. The relationship between BII and UHI intensity.

5 Conclusion

Based on buildings and vegetation cover data, calculating buildings greening rate and greening radiation factor at the pixel-scale, this paper proposed a building livability index (BII) to characterize the rationality of the spatial layout of urban buildings and vegetation. The city of Qingdao was selected as the study area, and the FROM-GLC10 dataset was used to verify and analyze the BII proposed in this paper. Among the 10 districts or counties in Qingdao, Laoshan district has the highest BII, reaching 1.33; the distribution of buildings and vegetation in Huangdao district is reasonable. Huangdao and Pingdu districts have the highest BII in non-buildings pixels, suitable for strengthening urban development. The BII in Qingdao has a significant negative correlation with the UHI intensity, indicating that rational planning of the spatial layout of buildings and vegetation and improving the buildings livability could effectively reduce the UHI effect.

In general, the BII proposed in this paper could effectively reflect the rationality of the urban spatial layout of buildings and vegetation, would provide a scientific basis for urban spatial planning and governance, and have a significantly bright application prospect based on domestic high resolution satellite data.

References

1. Perini, K., Magliocco, A.: Effects of vegetation, urban density, building height, and atmospheric conditions on local temperatures and thermal comfort. Urban For. Urban Green. **13**(3), 495–506 (2014)
2. Conine, A., Xiang, W., Young, J., et al.: Planning for multi-purpose greenways in Concord. N. C. Landsc. Urban Plan. **68**(2–3), 271–287 (2004)
3. Shahtahmassebi, A., Li, C., Fan, Y., et al.: Remote sensing of urban green spaces: a review. Urban For. Urban Green. **57**, 126946 (2021)
4. Francis, M.: Urban open spaces. In: Zube, E.H., Moore, G.T. (eds.). Advances in Environment, Behavior, and Design. Plenum, vol. 1(7), pp. 71–106 (1987)
5. Nowak, D., Rowntree, R., McPherson, E., et al.: Measuring and analyzing urban tree cover. Landsc. Urban Plan. **36**(1), 49–57 (1996)
6. Ouma, Y., Josaphat, S., Tateishi, R.: Multiscale remote sensing data segmentation and post-segmentation change detection based on logical modeling: theoretical exposition and experimental results for forestland cover change analysis. Comput. Geosci. **34**(7), 715–737 (2008)
7. Hofmann, P., Strobl, J., Nazarkulova, A.: Mapping green spaces in Bishkek—how reliable can spatial analysis be? Remote Sens. **3**(6), 1088–1103 (2011)
8. Pu, R., Landry, S.: A comparative analysis of high spatial resolution IKONOS and WorldView-2 imagery for mapping urban tree species. Remote Sens. Environ. **124**, 516–533 (2012)
9. Shojanoori, R., Shafri, H., Mansor, S., et al.: The use of WorldView-2 satellite data in urban tree species mapping by object-based image analysis technique. Sains Malays. **45**(7), 1025–1034 (2016)
10. Schöpfer, E., Lang, S., Blaschke, T.:. A "Green Index" incorporating remote sensing and citizen's perception of green space. In: International Archives of Photogramm., Remote Sensing and spatial information sciences, pp. 1–6 (2005)
11. Gupta, K., Kumar, P., Pathan, S., et al.: Urban neighborhood green index–a measure of green spaces in urban areas. Landsc. Urban Plan. **105**(3), 325–335 (2012)
12. Lang, S., Schöpfer, E., Hölbling, D., et al. (2008). Quantifying and qualifying urban green by integrating remote sensing, GIS, and social science method. In: Use of Landscape Sciences for the Assessment of Environmental Security. Springer, Dordrecht, pp. 93–105
13. Bao, T., Li, X., Zhang, J., et al.: Assessing the distribution of urban green spaces and its anisotropic cooling distance on urban heat island pattern in Baotou, China. ISPRS Int. J. Geo-Inf. **5**(2), 12 (2016)
14. Chen, B., Xu, B., Zhu, Z., et al.: Stable classification with limited sample: transferring a 30-m resolution sample set collected in 2015 to mapping 10-m resolution global land cover in 2017. Sci. Bull **64**, 370–373 (2019)
15. Xu, H., Chen, B.: An image processing technique for the study of urban heat island changes using different seasonal remote sensing data. Remote Sens. Technol. Appl. **18**(3), 129–133 (2011)

Radiometric Cross-Calibration of GF1-B/C/D Satellite with Landsat-8 OLI

Hengyang Wang[1], Hongzhao Tang[1](✉), Honggeng Zhang[2], and Chen Hui[2]

[1] Fine Mechanics & Physics, Changchun Institute of Optics, Chinese Academy of Sciences, Changchun 130033, China
tanghz@lasac.cn

[2] Land Satellite Remote Sensing Application Center, Ministry of Natural Resources of P. R. China, Beijing 100048, China

Abstract. The Gaofen-1 (GF1) B/C/D is a Chinese civilian high-resolution optical satellite constellation of three satellites launched in March 2018. In this paper, radiometric cross-calibration of GF1-B/C/D Satellite was conducted at the homogeneous field test site in July 2018 using Landsat-8 OLI. The radiometric cross-calibration coefficients were compared and validated with the reflectance-based vicarious calibration result in Dunhuang at the same time. The validation results show the accuracy of the cross-calibration for GF1-B/C/D sensor using Landsat-8 OLI is less than 6.1%. This study can supply reference to the radiometric cross-calibration of Chinese satellite senor, and the results suggest that the radiometric cross-calibration coefficients can be useful for maintaining the quantitative application.

Keywords: Cross-calibration · Radiometric calibration · Validation · Landsat-8 · GF1-B/C/D

1 Introduction

With the deepening of remote sensing in various fields, quantification of remote sensing has gradually become the further development trend of remote sensing technology [1]. The study of comparison between multiple sensors, real-time evaluation of the radiation characteristics of sensors themselves and timely correction of the attenuation of sensors are all indispensable and important links of quantification of remote sensing. Radiometric calibration is to convert the DN value of the sensor into a characterization with physical parameters [2]. The equivalent radiance or equivalent reflectivity of the sensor at the pupil point can be obtained in the visible and near-infrared bands, which is the first step to achieve remote sensing quantification. High-precision radiometric calibration of domestic optical satellites is of great significance to the development of quantitative remote sensing, which can monitor the performance of satellite sensors in real time and timely find the attenuation generated after the sensors are in orbit, and promote the industrial application of domestic satellites and the comprehensive application of domestic and international satellites [3]. Absolute radiometric calibration, as

L. Wang et al. (Eds.): CHREOC 2022, LNEE 969, pp. 293–302, 2023.
https://doi.org/10.1007/978-981-19-8202-6_26

an important means of monitoring the performance of on-orbit sensors and the research basis of remote sensing quantification, directly determines the accuracy of remote sensing data products and can timely correct the radiation error caused by the attenuation of sensors during on-orbit, to ensure the accuracy requirements of quantitative remote sensing [4].

At present, the main radiometric calibration method for satellite-borne sensors in China is reflectance basis method, which synchronously measures atmospheric parameters and surface reflectance at the time of satellite transit, and then simulates the equivalent radiance of the sensor by radiative transfer model, thus completing the conversion of image DN value to the radiance of the sensor. Although this calibration method in a certain extent, can be accurate to obtain calibration coefficient, but it is often subject to weather conditions, Angle of satellite observations and surface observation instruments, but also need a lot of manpower and financial resources in the calibration experiment, this undoubtedly led to huge workload for the remote sensing sensor calibration work. In order to solve the difficulties of data acquisition and high experimental cost, many scholars are constantly studying and developing cross radiation calibration technology. Cross-radiation calibration is to update the calibration coefficient of the calibration sensor image based on the remote sensing image of the reference sensor [5]. This calibration method does not need to measure the field data, and can use historical data for calibration. Cross calibration needs to select the currently calibrated sensor as the reference sensor. Most Chinese scholars use Landsat-8 as the reference sensor for cross calibration. Landsat series of satellites has been known as the "gold standard" since its launch [6]. The accuracy of on-orbit radiometric calibration of LandSAT-8 OLI load is 2–3%. It can be used as a reference sensor for cross radiation calibration of high resolution optical remote sensing satellites in China. In cross calibration, not only the observation geometry and atmospheric conditions among different sensors should be considered, but also the difference of the spectral response function within the sensor, which is usually corrected by calculating the spectral matching factor.

On March 31, 2018, the Gaofen-1 02, 03 and 04 satellite constellations (GF1-B/C/D) were successfully launched and operated with 120° phase networking, forming the first operational satellite constellations in China's civil space infrastructure planning. The constellation can make repeated observations within two days in the global region between 80° north and south latitude, greatly improving China's satellite monitoring capability of total elements, full coverage, all-weather real-time investigation and monitoring, and can be widely used in quantitative remote sensing applications such as vegetation and water. The satellite has a design life of six years. It is equipped with a panchromatic multispectral camera, with a single satellite imaging width of 60 km and a spatial resolution of panchromatic images of 2 m and multispectral images of 8 m. Although the sensor parameters of the three satellites are the same and all have been calibrated before launch, the sensors may attenuate to different degrees under the influence of changes in the surrounding environment and motion state after the satellites operate in orbit [7], leading to different changes in their radiation performance. Therefore, based on the Landsat-8 OLI load, this paper carried out cross radiation calibration for GF1-B/C/D satellite, obtained the calibration coefficient of the satellite load, and evaluated the radiation capacity of the three satellites, so that the images of the three satellites could be used together, which not only improved the image efficiency, but also integrated remote sensing data resources to a certain extent. The time interval of remote sensing observation is further reduced, the observation range is further expanded, and more abundant data resources are provided for quantitative processing and analysis of domestic satellite remote sensing.

2 Methodology

Cross-calibration is a process in which a sensor with high radiation accuracy is selected as a reference, and the same ground object target is observed consistently with the sensor of the target to be determined, and finally the calibration is completed through geometric registration, time matching and spectral matching [8–10].The flow of cross radiation calibration is shown in Fig. 1.

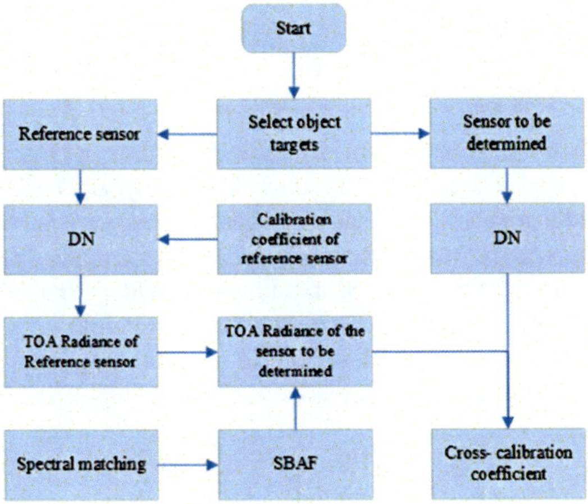

Fig. 1. Cross-calibration process.

Firstly, the reference sensor image is converted into radiance data by Eq. (1):

$$L_{ref} = DN_{ref} \times Gain + Bias \tag{1}$$

L_{ref} refers to the the top-of-atmosphere (TOA) radiance of the reference sensor image, expressed in units of $W/m^{-2}/sr^{-1}\mu m^{-1}$; DN_{ref} is the digital number from the reference satellite image. *Gain* and *Bias* are the official calibration coefficients in units of $W/m^{-2}/sr^{-1}\mu m^{-1}$. . Then, the spectral matching factor(SBAF) is calculated, and the observation geometry, observation time, spectral response function and atmospheric parameters of the sensor to be calibrated and the reference sensor are respectively input into the radiative transfer model to simulate their apparent radiance, as shown in Eq. (2).

$$SBAF = \frac{L_\lambda^A}{L_\lambda^{ref}} = \frac{\int L_\lambda \cdot RSR_\lambda^A d\lambda \Big/ \int RSR_\lambda^A d\lambda}{\int L_\lambda \cdot RSR_\lambda^{ref} d\lambda \Big/ \int RSR_\lambda^{ref} d\lambda} \tag{2}$$

Among them, RSR_λ is the relative spectral responses of sensor, L^A is the reflectance measure from SVC spectrometer, L_λ^A is the TOA reflectance of sensor A, and L_λ^B is the compensated TOA reflectance of sensor A using sensor B as reference sensor.

According to Eq. (3), the apparent radiance of the undetermined sensor modified by spectral matching factor is calculated.

$$\overline{L_\lambda^A}^* = SBAF \cdot L_\lambda^{ref} \quad (3)$$

Finally, the apparent radiance and image DN value of the sensor to be calibrated are obtained for fitting calculation, and the cross radiation calibration coefficient is obtained.

$$\overline{L^A}^* = DN^A \times Gain^A + Bias^A \quad (4)$$

where, $Gain^A$ and $Bias^A$ are the cross calibration coefficients of sensors to be calibrated.

3 The Data Processing

3.1 The Satellite is Introduced

The Landsat satellites is an outgrowth of NASA's Landsat program, which began in 1972. Nine satellites have been launched so far. Over the years, Landsat satellites have been called the "gold standard" for their rich data, short coverage periods, and global imaging, often compared to other sensors. Landsat-8, in particular, has been providing stable radiometric measurements since its launch, providing tracable and consistent apparent reflectance observations over a wide dynamic range (i.e. over land, snow/ice and water) with an absolute spectral irradiance uncertainty greater than 5% and an atmospheric top reflectance uncertainty greater than 3%[11]. The Landsat-8OLI land imager has nine bands with an imaging width of 185 × 185 km and a spatial resolution of 30 m. Therefore, selecting Landsat8 sensor as reference sensor can guarantee the feasibility and accuracy of its cross calibration.

GF1-B/C/D satellite is the first operational satellite constellation of China's civil space infrastructure program. It consists of three satellites with 120° phase network operation, which can repeatedly observe the global region between 80° north and south latitude within two days, greatly improving China's satellite monitoring capability of total elements, full coverage, all-weather real-time survey and monitoring. The spatial resolution of the satellite panchromatic image is 2 m, the spatial resolution of the multi-spectral image is 8 m, and the width of the single satellite image is 60 km. Specific band Settings are shown in Table 1.

3.2 Calibration Site

Dunhuang calibration field is an important remote sensing satellite radiometric calibration site in China. Many scholars have done a lot of research on the relatively normal site in Dunhuang, mainly including the uniformity of the site, the stability of the site reflectivity, the measurement of the bidirectional reflection distribution function and the vertical profile of the site aerosol. Dunhuang calibration site is located in Dunhuang City,

Table 1. Parameters of GAofen-1 B/C/D Satellite

The payload	Spectral coverage	Spectral region (nm)	Spatial resolution (m)	Breadth (km)	Revisiting ***period (day)***
Panchromatic multispectral camera	1	450–900	2	60	2
	2	450–520	8		
	3	520–590			
	4	630–690			
	5	770–890			

Fig. 2. GF1-B satellite multispectral image of Dunhuang site.

Gansu Province, as shown in Fig. 2. As an important remote sensing satellite radiometric calibration site in China, dunhuang calibration site is characterized by flat terrain, single ground object and good atmospheric observation conditions, and has been verified by many scholars to meet the requirements of remote sensing radiometric calibration.

3.3 Cross Calibration Data

In this paper, Landsat 8 OLI data from Dunhuang calibration field in July 2018 were selected for cross-calibration of GF1-B/C/D satellites. The parameters of cross-calibration satellite remote sensing data are shown in Table 2.Considering the influence of solar zenith Angle and atmospheric aerosol optical thickness, this paper selected Landsat 8 OLI on July 15, 2018 for cross calibration of GF1-B star on July 16 and GF1-C star on July 18.Cross-calibration of GF1-D star on July 18 was performed using Landsat 8 OLI on July 22, 2018 (Table 3).

4 Cross Radiation Calibration Results

Before cross calibration, images need to be preprocessed, including geometric correction of remote sensing image, radiometric calibration of reference load image and resampling

Table 2. Cross calibration satellite remote sensing data parameters

Satellite	Date	Time	Solar Zenith Angle (°)	Solar azimuth (°)	View Zenith Angle (°)	View azimuth (°)	AOD
Landsat 8	20180715	12:19	26.24	129.3	0	0	0.282
GF1-B	20180716	12:43	23.43	139.48	1.65	126.99	0.255
GF1-C	20180718	12:35	24.6	135.64	4.47	97.2	0.275
GF1-D	20180720	12:50	23.27	143.02	7.83	104.68	0.11
Landsat 8	20180722	12:25	27.281	130.853	0	0	0.07

of uncalibrated load image. The purpose of geometric correction is to eliminate the geometric deviation of the two sensors and control their accuracy within one pixel. Due to the different resolution of the two sensors, the number of pixels of the image in the calibration area is different, so it is necessary to resampling the image with higher resolution to make its resolution consistent with that of the image with lower resolution. Remote sensing image resampling methods mainly include nearest neighbor interpolation method, bilinear interpolation method, and cubic convolution interpolation method. In this paper, the cubic convolution interpolation method with high accuracy was selected to aggregate GF1-B/C/D image to Landsat 8 30 m resolution. The apparent radiance of the sensor to be calibrated was obtained by multiplying the atmospheric top radiance obtained by radiometric calibration of the reference sensor image with the spectral matching factor. The apparent radiance value and DN value were linearly fitted to obtain the cross calibration result. Figure 3 shows the fitting results of GF1-B cross radiometric calibration. It can be seen from the figure that the linear correlation coefficient between image DN value of each band under load and apparent radiance is better than 99%.

By comparing the cross-calibration coefficients obtained in this experiment with the official radiation calibration coefficients (site calibration) of GF1-B/C/D satellite updated in November 2018, the relative difference of GF1-B cross calibration coefficient and field calibration coefficient based on Landsat8 OLI are 2.58%, 2.29%, 2.13% and 3.24%, respectively. The relative errors of GF1-C cross calibration coefficient and field calibration coefficient are 4.92%, 4.36%, 2.82% and 4.93%, respectively. The relative errors of the cross-calibration coefficient of GF1-D satellite and the field calibration coefficient are 5.42%, 1.91%, 2.55% and 4.67%, respectively, and the errors are all within 6%. The cross-radiation calibration results of GF1-B/C/D satellite are good.

Table 3. Comparison between cross radiometric calibration coefficients and official radiometric calibration coefficients

Satellite	GF1-B			GF1-C			GF1-D		
sensor band	Radiometric calibration coefficient (Gain)								
	Cross-calibration	Official calibration	Relative error (%)	Cross-calibration	Official calibration	Relative error (%)	Cross-calibration	Official calibration	Relative error (%)
B1	0.03067	0.0299	2.58	0.03116	0.0297	4.92	0.0311	0.0295	5.42
B2	0.03674	0.0376	2.29	0.0383	0.0367	4.36	0.0373	0.0366	1.91
B3	0.04277	0.0437	2.13	0.0438	0.0426	2.82	0.042	0.0431	2.55
B4	0.03938	0.0407	3.24	0.0386	0.0406	4.93	0.0388	0.0407	4.67

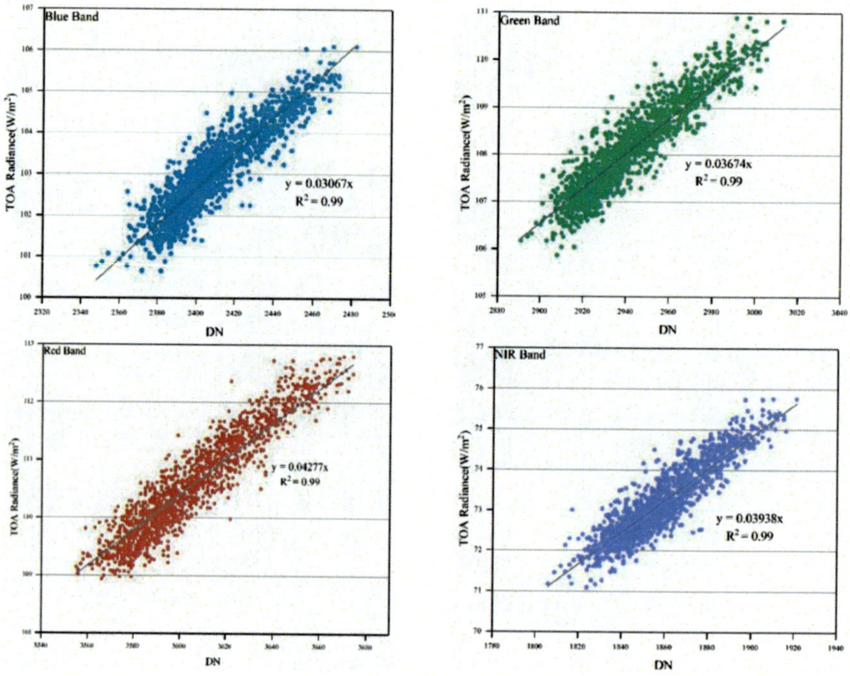

Fig. 3. GF1-B calibration results of cross-calibration

5 Cross Calibration Coefficient Accuracy Verification

To verify the accuracy of cross-calibration coefficient, the satellite and earth synchronous atmospheric and ground object spectral observation data of Dunhuang calibration field in July 2018 were selected to verify the radiation accuracy. On July 16, 18 and 20, 2018, the simultaneous observation of GF1-B/C/D satellite over Dunhuang calibration field was carried out, and the aerosol parameters and spectral reflectance data were obtained at the time of satellite transit. The parameters of satellite transit synchronous observation were put into the radiative transfer model to calculate and obtain the apparent radiance, denoted as L_{Band}. The apparent radiance L_{Cal} was obtained by cross calibration coefficient and SATELLITE image DN value, and the relative error %L they of L_{Band} and L_{Cal} were compared as the verification accuracy of cross radiation calibration coefficient.

$$\Delta L\% = \left| \frac{L_{Band} - L_{Cal}}{L_{Band}} \right| \times 100\% \tag{5}$$

It can be seen from Table 4 and Fig. 4 that the accuracy of cross radiative calibration coefficients of GF1-B/C/D stars is better than 6.1%. Combined with the calibration coefficient comparison results above, it can be seen that the cross calibration results of GF1-B/C/D satellite load with Landsat8OLI as reference load are authentic.

4. TableVerification results of GF 1-B / C / D star cross-radiation calibration accuracy

Satellite sensor band	Relative error $\Delta L\%$			
	B1 (%)	B2 (%)	B3 (%)	B4 (%)
GF1-B 20180716	5.30	5.16	4.98	5.50
GF1-C 20180718	4.35	5.99	6.03	4.75
GF1-D 20180720	5.73	5.89	5.22	5.19

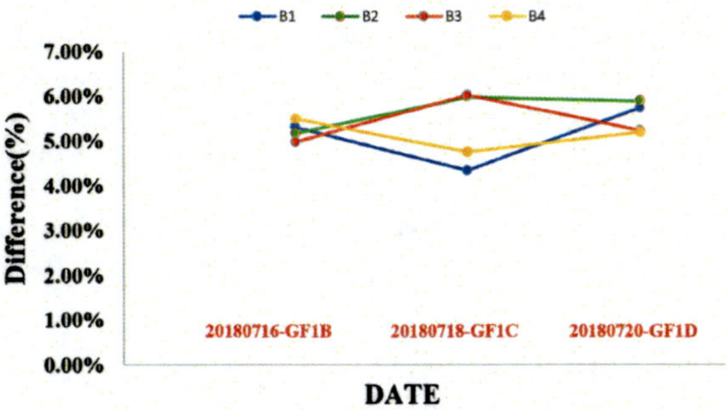

Fig. 4. GF1-B/C/D Calibration accuracy verification results of cross star radiation.

6 Conclusion

In this paper, Landsat-8 OLI was selected as a reference sensor to conduct cross radiation calibration for GF1-B/C/D satellite. The relative errors of the cross-calibration coefficients of the three stars and the official site calibration coefficients were all less than 5.5%, among which the average relative errors of coefficients of GF1-B were the smallest. All bands are less than 3.5%, and the cross-calibration results are good. In order to further verify the accuracy of the cross calibration coefficient, this paper uses the synchronous surface reflectance and atmospheric measurement data of GF1-B /C/D transit to simulate the radiation transmission, and compares the apparent radiance calculated by the cross calibration coefficient with that simulated by the measured data. It is found that the relative errors of each band of GF1-B/C/D stars are all less than 6.1%, which proves that it is feasible to use Landsat-8 OLI as the reference sensor for GF1-B/C/D satellite cross calibration. The obtained cross radiation calibration coefficients are of high accuracy, and the cross-calibration results are reliable. Without relying on the satellite transit synchronous radiometric calibration test, the in-orbit radiometric calibration and accuracy verification of domestic high-resolution satellites can be realized through high-precision cross radiometric calibration, which is beneficial to the quantitative application of domestic high resolution satellites.

Acknowledgment. This work was supported by the National Key R&D Program of China, 2018YFB0504800, 2018YFB0504805, 2016YFB0500600, 2016YFB0500602.

References

1. Dinguirard, M., Slater, P.N.: Calibration of space-multispectral imaging sensors: a review. Remote Sens. Environ. **68**(3), 194–205 (1999)
2. Thome, K.J.: Absolute radiometric calibration of Landsat 7 ETM+ using the reflectance-based method. Remote Sens. Environ. **78**(1–2), 27–38 (2001)
3. Gao, H.L., Gu, X.F., Yu, T., Li, X.Y., Gong, H., Li, J.G.: The research overview on visible and near-infrared channels radiometric calibration of space-borne optical remote sensors. Remote Sens. Inf. **4**, 12 (2010)
4. Biggar, S.F., et al.: Radiometric calibration of SPOT 2 HRV: a comparison of three methods. Proc. SPIE **1493**, 155–162 (August 1991)
5. Feng, L., Li, J., Gong, W.S., Zhao, X., Chen, X.L., Pang, X.P.: Radiometric cross-calibration of Gaofen-1 WFV cameras using Landsat-8 OLI images: A solution for large view angle associated problems. Remote Sens. Environ. **174**, 56–68 (March 2016)
6. Shin, D., Jin, C., Ahn, H., Choi, C.: Radiometriccross-calibration of KOMPSAT-3 with Landsat-8. Proc. SPIE **9644**, 379–382 (October 2015)
7. CzaplaMyers, J.S., Leisso, N.P., Anderson, N.J., Biggar, S.F.: On-orbit radiometric calibration of Earth-observing sensors using the Radiometric Calibration Test Site (RadCaTS). Proc. SPIE **8390**, 379–382 (May2012)
8. Chander, G., et al.: Applications of spectral band adjustment factors (SBAF) for cross calibration. IEEE Trans. Geosci. Remote Sens. **51**(3), 1267–1281 (March 2013)
9. Bruegge, C.J., et al.: The MISR radiometric calibration process. Remote Sens. Environ. **107**(1–2), 2–11 (March 2007)
10. Gao, C.X., Jiang, X.G., Ma, L.L., Huo, H.Y.: A review of sensor cross radiation calibration. Arid Land Geogr. **36**(1), 139–146 (January 2013)
11. Wulder, M.A., et al.: Current status of Landsat program, science, and applications. Remote Sens. Environ. **225**, 127–147 (May2019)

Validation of ZY1-02D Satellite Multispectral Sensor on-Orbit Radiometric Calibration Accuracy

Hongzhao Tang(✉), Honggeng Zhang, and Chen Hui

Land Satellite Remote Sensing Application Center, Ministry of Natural Resources of P. R. China, Beijing 100048, China
tanghz@lasac.cn

Abstract. ZY-1 02D satellite, carries a 9 bands multispectral camera (VNIC) and a 166 bands hyperspectral imager (AHSI), was successfully launched on 12 September 2019. As a replacement of ZY-1 02C satellite, ZY-1 02D provides more accurate and spectrally dependent radiometric characterization of Earth's surface for Natural Resources Applications. Radiometric characteristics monitoring of ZY-1 02D satellite data is a very essential step that enables further quantitative study and application of the data. This paper describes the method of a reflectance-based vicarious calibration and validation campaign that was conducted in April 2020 at Inner Mongolia Province of China. The results show the accuracy of ZY-1 02D satellite sensors VNIC calibration coefficient is within 7%. ZY-1 02D VNIC has good and stable on-orbit radiometric performance, and ZY-1 02D satellite data is highly reliable and can be used in quantitative natural resources applications.

Keywords: ZY1-02D · VNIC · On-orbit test · Radiometric calibration accuracy · Validation

1 Introduction

On September 12, 2019, China used a Long March 4 carrier rocket to send a 5-m optical operational satellite (ZY-102D) into a predetermined orbit from Taiyuan Satellite Launch Center [1].The 5-m optical service satellite is China's first civilian hyperspectral service satellite, with international leading technical indicators [2].The payload on the satellite focuses on the spectrum subdivision of visible near-infrared and short-wave infrared spectra, and the spectral remote sensing characteristics are prominent, which can realize the fine spectral information investigation of ground objects and meet the needs of natural resources monitoring and investigation in the new era. The ZY-102D satellite is equipped with a 9-band visible near infrared camera, and the absolute radiometric calibration of the payload is an important basic work to ensure the radiation quality of the image and the quantitative application of the data. Validation is absolute radiation calibration precision monitoring satellites orbiting load performance and an important means of radiation characteristics, it can be calibrated during the satellite on-orbit space

L. Wang et al. (Eds.): CHREOC 2022, LNEE 969, pp. 303–312, 2023.
https://doi.org/10.1007/978-981-19-8202-6_27

environment such as the change of the influence caused by the radiation calibration value, improve the accuracy of radiation calibration coefficient, timely improve the quality of remote sensing data products radiation, so as to improve the quality of the application of satellite data, Meet the requirements of quantitative application of remote sensing data [3–5].Based on the satellite-earth synchronous observation experiment carried out in Baotou Radiometric calibration field in Inner Mongolia in April 2020, the absolute radiometric calibration accuracy of visible near-infrared cameras and hyperspectral cameras in the on-orbit site was verified and analyzed using the reflectivity basis method to verify the reliability of radiometric calibration coefficients.

2 Zy1 02d Satellite Parameters

The ZY1-02D satellite adopts a sun-synchronous circular orbit with a designed orbit altitude of 778 km, which can achieve seamless image coverage of the earth within 84 degrees north and south latitude. The satellite's regression cycle is 55 days, and the revisit cycle is 3 days, and it can complete image coverage of China's territory and the whole world once every 55 days. The satellite is equipped with a 9-band visible near infrared camera and a 166-band hyperspectral camera. The visible near infrared camera has a width of 115 km and a spatial resolution of 2.5 m/10 m in panchromatic/multi-spectral bands, enabling simultaneous earth observation in eight multi-spectral bands. The satellite works in a sun-synchronous orbit of 778 km, with a regression period of 55 days and a revisit period of less than three days, which fully guarantees the continuous observation ability of the same target and provides time guarantee for the observation of land reclamation and conversion of farmland to forest.

The main technical specification of ZY-1 02D Satellite are shown in Table 1.

3 Accuracy Verification Method for on-Orbit Radiometric Calibration the Satellite is Introduced

The on-orbit absolute radiometric calibration of satellite payload is the process of establishing the quantitative relationship between the image DN value of sensor response and the radiation energy, that is, obtaining the absolute radiometric calibration coefficient [6]. ZY—102—d satellite payload parameters in front of the satellite launch is for precise measurement in the laboratory, but the satellite launch process and the orbit microgravity condition and state of the change of environmental temperature change can cause the technical parameters of payload, so need to regular the load of the satellite on-orbit absolute radiation calibration accuracy validation [7]. In view of the characteristics of visible near infrared camera and hyperspectral camera of ZY-102D satellite, an on-orbit absolute radiometric calibration accuracy verification method based on baotou radiometric calibration field in Inner Mongolia is proposed in this paper. The natural scene of uniform sandy land in Baotou field is selected as the target object. When ZY-102D satellite passes over Baotou radiometric calibration field in Inner Mongolia, Synchronous measurement using SVC wild terrain spectrometer spectral reflectance data of ground targets by baotou field atmosphere and meteorological data, automatic observing system for the air, ground and atmospheric environmental data, calculate the atmospheric

Table 1. The technical specification of ZY-1 02D Satellite.

Orbital	Orbital type	Synchronous solar regression orbit
	Orbital altitude	778.099 km
	Angle	98.5°
	Descending intersection local time	10:30 AM
	Lateral swing capacity (rolling)	±26°
	Regression cycle	55Days
Satellite weight	Weight	1840 kg
VNIC	Spectral	Panchromatic: 0.45 ~ 0.90 μm Multispectral: 0.45 ~ 0.52 μm 0.52 ~ 0.59 μm 0.63 ~ 0.69 μm 0.77 ~ 0.89 μm 0.40 μm ~ 0.45 μm 0.59 μm ~ 0.625 μm 0.705 μm ~ 0.745 μm 0.860 μm ~ 1.040 μm
	Ground pixel resolution of subsatellite point	Panchromatic 2.5 m, Multispectral 10 m
	Width	115 km
	Quantitative figures	12 Bit
Life		5 Years

extinction coefficient, the calculation of atmospheric water and ozone content, analysis of spectral reflectance data, load spectrum response function and the geometric parameters and the time when the imaging,The obtained and calculated parameters were put into the atmospheric radiation transfer model to calculate the radiation brightness at the entry pupil of the satellite payload, and then compared and analyzed with the radiation brightness observed by the satellite payload to obtain the absolute radiation calibration accuracy of the in-orbit site.The in-orbit radiation accuracy verification and analysis process of ZY1-02D satellite payload is shown in Fig. 1:

3.1 Baotou Radiative Calibration Field

The test area is selected in Baotou National High-resolution Remote Sensing Comprehensive Calibration Field (Baotou Field), which is a comprehensive calibration field led by The Academy of Optics and Electronics of Chinese Academy of Sciences under the continuous support of the key project of "UAV Remote Sensing Comprehensive Verification System" of the Ministry of Science and Technology during the 12th Five-Year Plan

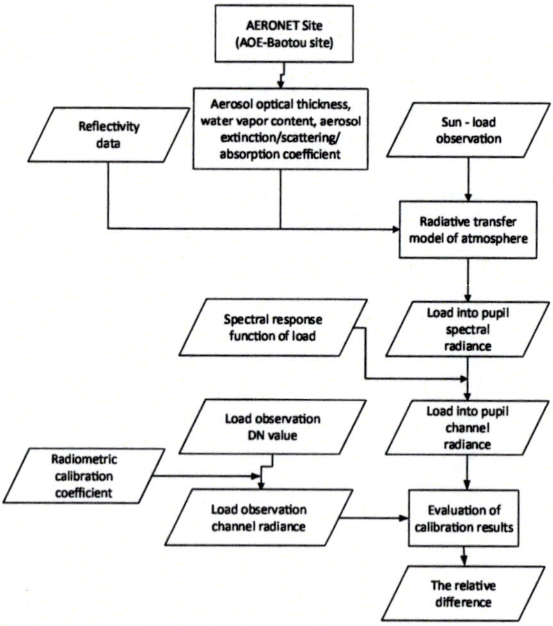

Fig. 1. Flow chart of the reflectance-based calibration and validation approach.

period and subsequent relevant projects. At present, It has been officially listed as "National High-resolution Remote Sensing Comprehensive Calibration Field" by National Remote Sensing Center of Ministry of Science and Technology of China. At the same time, Baotou field was included in the first demonstration sites of THE CEOS Global Autonomous Radiation Calibration Field Network (RadCalNet) and USGS global MTF Evaluation ground standard Target library [8, 9]. Relying on the construction of the test base of North Heavy Industry Group, Baotou Site is located between Daqingshan and Wulashan, with coordinates of 40.85° NORTH latitude and 109.62° east longitude, about 70 km away from baotou city. The terrain is flat, and the altitude of the site is between 1037–1289 m, with an average altitude of 1270 m. The farthest distance of Baotou Field is 32 km from east to west and 16 km from north to south, covering an area of about 300 square kilometers (as shown in Fig. 2).The construction of Baotou field has wide dynamic multi-gray scale fixed targets, fan-shaped targets, strip targets and other artificial targets, as well as large area of uniform sand and other natural scenes. In this radiation calibration accuracy verification test area, the reflectivity of large area uniform sandy land in Baotou field is approximately 20–30%, with an area of 300 m × 300 m. In the natural scene of uniform sandy land, Baotou Field is equipped with CimelCE318 (AOE-Baotou site) of AERONET global Aerosol observation network, automatic weather station, sounding radar, whole sky imager and other equipment, which can obtain atmospheric and meteorological parameters (as shown in Fig. 3).

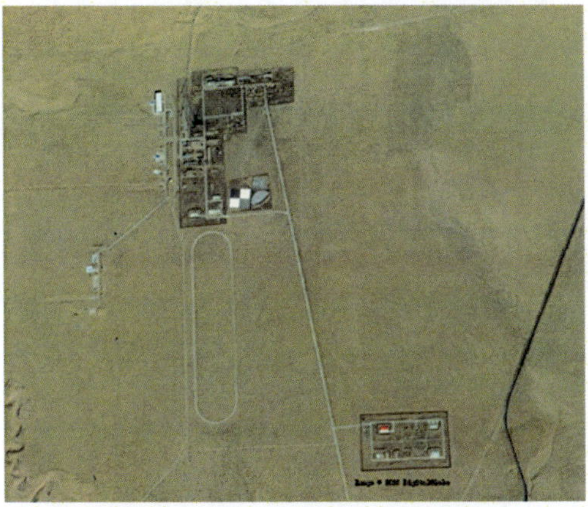

Fig. 2. The RadCalNet Baotou site

Fig. 3. Automatic observation instrument for atmospheric

3.2 Surface Reflectivity Data

In April 2020, zy1-02D satellite transit surface synchronous observation was carried out in Baotou field, and the time was half an hour before and after the satellite transit to ensure the synchronization of the measurement data.At the same time of satellite transit, field feature spectrometer SVC was used to measure the uniform 300 m × 300 m sandy land (Fig. 4), 100 points were uniformly selected for measurement, and encrypted measurements were carried out in some areas in the middle of the sandy land. The average reflectance obtained in the sandy land is shown in Fig. 5:

Fig. 4. The sand of Baotou site

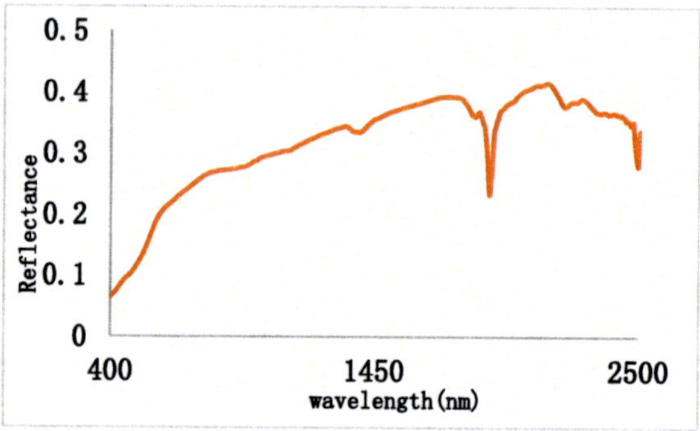

Fig. 5. The reflectance of sand

3.3 Atmospheric Parameter Acquisition

During the satellite overpass, a CE318 sun photometer in Baotou field continuously monitor solar spectral information and earth radiation budget information, and to monitor air visibility, aerosol type, particle size composition and aerosol optical thickness. AOE Baotou Site has joined AERONET global aerosol automatic observation network, collected daily data and sent them to the network for centralized processing, and released the processed atmospheric aerosol optical thickness (level1.5), aerosol physical property parameters (level2.0) and other inversion products. The physical properties of atmospheric aerosol include aerosol particle size, particle shape and particle spectral distribution (including particle scale spectrum and mass spectrum).In general, aerosol models are determined by the physical properties of aerosols. In order to reduce the systematic error caused by aerosol model approximation in the atmospheric radiative transfer model, the physical properties of atmospheric aerosols retrieved by synchronous observation at AERONET AOE-Baotou station in the calibration field were used to determine the self-defined aerosol model. The automatic meteorological observation equipment based on Baotou field continuously monitors surface air temperature, humidity and precipitation, and uses balloon sounding to monitor atmospheric temperature profile, humidity profile and water vapor content at the time of satellite transit.

3.4 Visible NEAR-Infrared Camera VNIC Spectral Response Function

The spectral response function of each band of VNIC is shown in the figure below (Fig. 6):

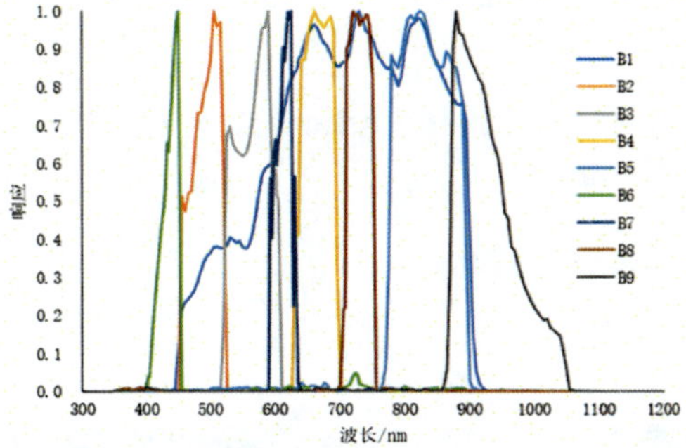

Fig. 6. Normalized spectral response function of the VNIC

3.5 Radiative Transfer Simulation Computations of Incoming Pupil Radiance

The radiation brightness at the entry pupil of zy1-02D satellite load $L_s(\lambda)$ is calculated according to Formula (1):

$$L_s(\lambda) = \frac{1}{\pi} \cdot T_g(\lambda) \cdot \rho(\lambda) \cdot \left[\mu_s E_s(\lambda) e^{\frac{-\tau'}{\mu_s}} + E_d(\lambda) \right] \cdot e^{\frac{-\tau''}{\mu_v}} + L_p(\lambda) \tag{1}$$

where: $L_s(\lambda)$ refers to wavelength λ, which denotes the radiation brightness of satellite load at the entry pupil, in unit of $W \cdot m^{-2} \cdot sr^{-1} \cdot \mu m^{-1}$; $T_g(\lambda)$ refers to the total atmospheric absorption transitance in the upward and downward directions at wavelength λ. $\rho(\lambda)$ is the surface reflectance factor at wavelength λ; $E_s(\lambda)$ is the solar radiation at wavelength λ, unit: $W{\cdot}m^{(-2)}{\cdot}\mu m^{(-1)}$; μs is the cosine of solar zenith Angle θs ($\cos\theta s$); μv is the cosine of the observed zenith Angle θv ($\cos\theta v$); $E_d(\lambda)$ Is the atmospheric diffuse irradiance incident to the surface at wavelength λ (unit: $W \cdot m^{-2} \cdot \mu m^{-1}$); τ' is the vertical atmospheric scattering optical thickness from the sun to the earth; e-$\tau'/\mu s$ is the atmospheric scattering transmittance from the sun to the ground; τ'' is the optical thickness of vertical atmospheric scattering from ground to remote sensor; e-$\tau''/\mu v$ is the atmospheric scattering transmittance from the ground to the remote sensor; $L_p(\lambda)$ refers to the atmospheric path radiation brightness at wavelength λ (unit: $W{\cdot}m^{-2}{\cdot}sr^{-1}{\cdot}\mu m^{-1}$), and the equivalent radiation brightness at satellite payload band is obtained $L_e(\lambda_e)$.

3.6 Image Radiance Inversion

Based on the ZY-102D satellite image and the official radiometric calibration coefficient, the apparent radiance of image inversion is calculated.

$$L_i = Gain_i \times DN_i + Bias_i \tag{2}$$

where L_i is the apparent radiance of satellite payload band (unit: $W \cdot m^{-2} \cdot Sr^{-1} \cdot \mu$), DN is the gray value of satellite image, $Gain_i$ and $Bias_i$ are the gain and bias in radiometric calibration coefficient respectively.

4 The Accuracy of on-Orbit Radiometric Calibration Verifies the Analysis Results

4.1 ZY1-02D Satellite Image Data

See Fig. 7.

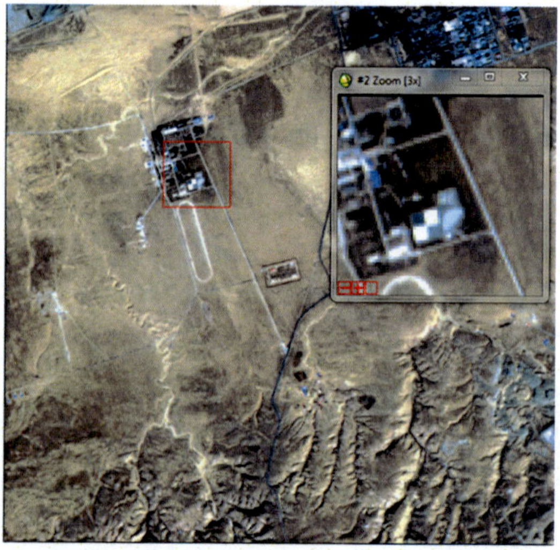

Fig. 7. The ZY-1 02D VNIC data of Baotou site

4.2 Visible Calibration Accuracy Verification Results of Near-Infrared Camera

DN of uniform sand in Baotou site from ZY-102D satellite VNIC image was selected as the value of satellite observation meter, and the radiance of image inversion was obtained by combining with radiometric calibration coefficient.At the same time, surface reflectance, atmospheric parameters and atmospheric profile are input into the radiation

Table 2. The comparison of top-of-atmosphere spectral radiance of ZY-1 02D VNIC

Spectral coverage	DN	Image radiance inversion	Modtran simulates incoming pupil radiance	Relative difference (%)
PAN	1950.66	84.7669	82.614	2.54
1	2128.95	106.3216	104.301	1.90
2	2566.4	109.279	102.995	5.75
3	2388.56	94.4256	89.015	5.73
4	1534.91	65.3677	62.034	5.10
5	1713.39	93.7229	92.064	1.77
6	2692.98	104.1332	98.073	5.82
7	1703.81	80.8719	76.003	6.02
8	1664.61	45.9161	42.454	7.54

transfer model Modtran6 to simulate the radiance of load entering pupil, and the results are shown in the following table (Table 2).

It can be seen from the results in the above table that the apparent radiance obtained from VNIC image and radiometric calibration coefficient is in good consistency with the radiance calculated by simulation based on satellite-earth synchronous observation data. The results show that the difference between them is less than 6%, which shows that the radiometric calibration accuracy of nIR camera meets the requirements of the index. However, the relative difference of band 8 is greater than 7%, which may be due to the large error caused by the band's location in the water vapor absorption channel, and the radiometric calibration coefficient is only obtained through one in-orbit calibration synchronous test. In the future, multiple field tests are needed to fit more accurate calibration coefficients through multiple calibration tests. Due to time and site constraints, this work is only a preliminary verification of the radiation calibration accuracy of multispectral camera during the in-orbit test of ZY1-02D satellite. At present, only a very small number of synchronous observation images are selected for quantitative evaluation, and subsequent tests will continue to verify and monitor the radiation accuracy of the satellite.

5 Conclusion

The ZY1-02D satellite is equipped with a 9-band visible near infrared camera, which is capable of data collection and application with large width, high spatial resolution and high spectral resolution. Radiative calibration precision of the load directly restricts the precision of satellite application, in the subsequent application in work, in order to give full play to the effectiveness of the satellite, need to continue to intensify the radiation calibration and accuracy of monitoring work, for natural resources, such as the typical application scenarios, choose suitable for ZY—102—d satellite multispectral radiation calibration load plan, and to establish and to demonstrate the application of adaptive area, Through long-term research and authenticity verification experiments, the scale capacity of quantitative application is improved.

Acknowledgment. This work was supported by the National Key R&D Program of China, 2018YFB0504800, 2018YFB0504805, 2016YFB0500600, 2016YFB0500602.

References

1. Natural Resources Satellite Remote Sensing Cloud Service Platform. Introduction of ZY-1 02D satellite [EB/OL].[2020–10–09]
2. Hu, Z.: The resolution is better than 5 meters optical remote sensing satellite: revelation "power star" resources ZY-1 02D-xinhuanet [EB/OL]
3. Zhang, Y., et al.: In-flight vicarious radiometric calibration for VIS-NIR channels of FY-1C satellite sensor at Dunhuang site[J]. Acta Meteorol. Sinca **60**(6) (2002)
4. Gao, H.: Study on the calibration and validation of HSI sensor on HJ-1A satellite[D]. In: Institude of Remote Sensing Applications, Chinese Academy of Sciences (2010)
5. Slater, P.N., Biggar, S.F., Holm, R.G., et al.: Reflectance- and radiance-based methods for the in-flight absolute calibration of multispectral sensors[J]. Remote Sens. Environ. **22**(1), 11–37 (1987)
6. Thome, K.J.: Absolute radiometric calibration of Landsat 7 ETM+ using the reflectance-based method[J]. Remote Sens. Environ. (2001)
7. Mccorkel, J., Thome, K., Ong, L.: Vicarious calibration of EO-1 Hyperion[J]. IEEE J. Sel. Top. Appl. Earth Obs. Remote. Sens. **6**(2), 400–407 (2013)
8. Czapla-Myers, J.S., Leisso, N.P., Anderson, N.J., Biggar, S.F.: On-orbit radiometric calibration of Earth-observing sensors using the Radiometric Calibration Test Site (RadCaTS)[C]. Proc SPIE 8390 (2012)
9. Li, C., Ma, L., Gao, C., et al.: Permanent target for optical payload performance and data quality assessment: spectral characterization and a case study for calibration[J]. J. Appl. Remote Sens. **8**(1), 083498
10. Jing, X., Leigh, L., Helder, D., et al.: Lifetime absolute calibration of the EO-1 Hyperion sensor and its validation[J]. IEEE Trans. Geoence Remote Sens. **57**(11), 9466–9475 (2019)

Vibrational Frequency Analysis for the Ziyuan 3-02 Satellite Platform

Mo Fan[1], Xie Junfeng[1], Li Xiang[2](✉), Tang Hongzhao[1], and Zhao Jianyi[3]

[1] Land Satellite Remote Sensing Application Center, Beijing, China
surveymofan@163.com
[2] Wuhan University, Wuhan, China
li.xiang@whu.edu.cn
[3] Liaoning Technical University, Liaoning, China

Abstract. The attitude measurement accuracy of a remote sensing satellite defines the geometric accuracy of satellite mapping. Nevertheless, with the continuous improvement of the ground resolution of high-resolution remote sensing satellites, the accuracy of satellite attitude has become one of the bottlenecks restricting the development of satellite surveying and mapping to expand in scale. Ziyuan 3-02 is the first operational satellite in China's National Medium- to Long-Term Civilian Space Infrastructure Development Plan. Compared to Ziyuan 3-01, Ziyuan 3-02 has hardware optimizations; however, the hardware system for attitude measurement follows the Ziyuan 3-01 configuration with four sets of gyroscopes (each set of three installed in the three axes) and three sets of star sensors. In this regard, this study analyzed the satellite attitude data in the frequency domain and compared the results with those of the Ziyuan 3-01 to better study the attitude jitter of the Ziyuan 3-02 satellite platform. The comparison revealed that Ziyuan 3-02 exhibited the same jitter as the Ziyuan 3-0. In particular, the platform experienced a vibrational frequency of approximately 0.65 Hz.

Keywords: Ziyuan 3-02 · Initial stage in orbit · Attitude data · Frequency domain analysis · Jitter

1 Introduction

Ziyuan 3-02 (ZY3-02) is the second satellite within the Ziyuan-III series satellites. The satellite was launched at Taiyuan Satellite Launch Center on May 30, 2016. In particular, ZY3-02 is equipped with optimized hardware based on Ziyuan 3-01 (ZY3-01), which completed the orbit change during the on-orbit test to form the dual-satellite network. Moreover, ZY3-02 has substantially improved the capabilities of Chinese surveying and mapping applications.

Regarding the satellite attitude measurement hardware, ZY3-02 used the same configuration of ZY3-01, including four sets of gyroscopes (each set of three mounted on the three axes) and three sets of star sensors. The initial satellite attitude filtering determination can be performed using the measurement data from the gyroscopes and the star sensors. Subsequently, the obtained satellite platform attitude can be used for satellite

L. Wang et al. (Eds.): CHREOC 2022, LNEE 969, pp. 313–321, 2023.
https://doi.org/10.1007/978-981-19-8202-6_28

attitude control and preliminary satellite mapping. The geometric positioning accuracy of remote sensing satellites for surveying and mapping is mainly affected by the satellite attitude, orbit, and camera internal parameters [1, 2]. In particular, satellite attitude is a crucial factor (Wang 2001; Wang 2002). Therefore, studies are necessary to analyze the attitude of ZY3-02 during the on-orbit test to explore the change patterns of the satellite platform attitude.

When the satellite travels smoothly on-orbit, it often exhibits similar regular motions [5, 6]. Researchers believe that factors such as the structural composition and temperature of the stars, momentum wheel, and orbit control operations influence the regular motions [7–9]. The short-term attitude jitter of satellite platforms has been detected during the on-orbit operation of various satellites [7, 10], such as a 0.67, 1.6, and 2 Hz jitter of ZY3-01, ASTER, and Landsat-5, respectively. In this study, the jitter of the ZY3-02 platform attitude was analyzed in the frequency domain, and the result was compared with that of ZY3-01. The results revealed that the frequency of the ZY3-02 platform jitter is similar to that of ZY3-01.

2 Hardware Fundamentals

2.1 Star Sensor

Three sets of star sensors were installed on ZY3-02 to improve the measurement of the absolute attitude of the satellite platform. Three star sensors were fixed on the same mounting plate to form the combined star sensor. The schematic setup of the three-star sensor is shown in Fig. 1.

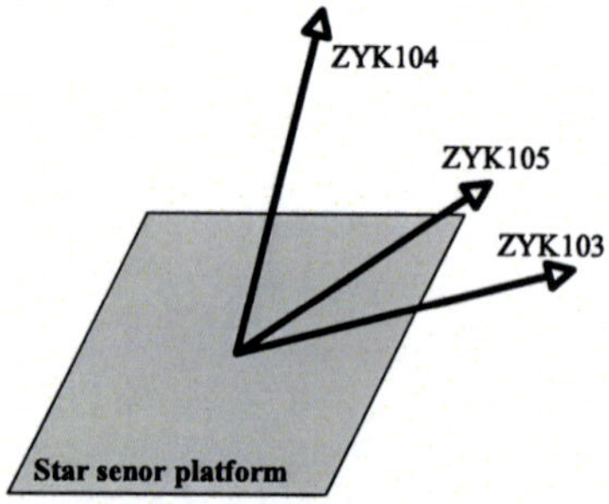

Fig. 1. Schematic setup of the star sensors

The ZYK105 and ZYK104 sensors were equipped with ASTRO10 star sensors. Moreover, the ZYK103 sensor was a Chinese-manufactured APS star sensor. Currently, measurement data of ZYK105 and ZYK104 are mainly used, while ZYK103 is an experimental load that is still in the experimental phase.

2.2 Inertial Gyroscope

The implemented inertial gyroscope had a relatively high-precision measurement capability for the relative attitude and could rapidly record the angular velocity of attitude change. Four sets of gyroscopes were installed on ZY3-02, including two sets of three-floated gyroscopes, one set of dual-floated gyroscopes, and one fiber-optic gyroscope. The complex installation was simplified, as shown in Fig. 2.

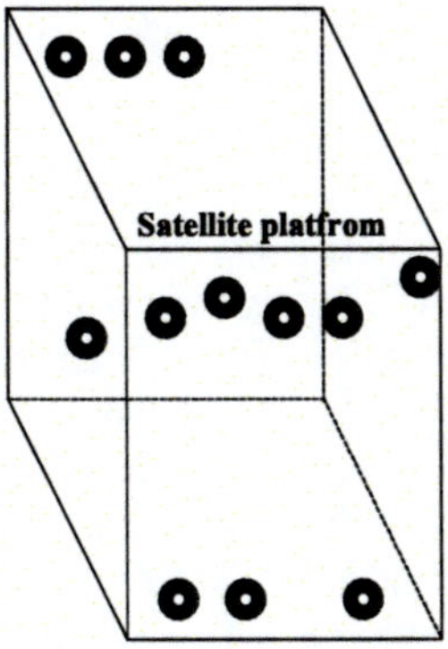

Fig. 2. Schematic setup of the inertial gyroscopes within the entire satellite platform

3 Method

3.1 Extended Kalman Filter

Extended Kalman filter (EKF) is an algorithm that linearizes nonlinear models and performs recursive filtering. This algorithm has been applied extensively to satellite attitude determination. Moreover, EKF take advantage of both star sensor and gyroscope data to achieve optimal accuracy for satellite attitude.

3.2 General Polynomial Fitting

Polynomials can be used to fit data exhibiting a linear response that can be written follows:

$$f(x) = a_0 + a_1 x + \cdots + a_n x^n \tag{1}$$

where $f(x)$ represents the fitting results, $a_0 \sim a_n$ are the coefficients, and n is the polynomial order.

3.3 Fourier Transform

Fourier transform is a typical method implemented for signal frequency-domain analysis. This method properly connects the frequency domain and time domain of signals [11]. The Fourier transform equation is as follows:

$$F(u) = \int_{-\infty}^{\infty} f(t) e^{-j2\pi ut} dt \tag{2}$$

where $F(u)$ represents the frequency-domain values, $f(t)$ represents the time-domain values, t represents time, u represents frequency, and j represents the imaginary part of a complex number.

4 Experimental Results and Analysis

4.1 Experimental Data

In the experiments, the star sensor-gyroscope data was obtained on June 24th, 2016, when the ZY3-02 was on the 382nd orbit during the on-orbit test. The star sensor-gyroscope data were filtered using the EKF to obtain the filtered attitude values, including 946 sets of quaternions. The data exhibit a normal response, showing no errors or anomalies, as shown in Fig. 3. For comparison, the attitude data of the ZY3-01 platform on the 381st orbit during the on-orbit test obtained on February 3rd, 2012, was used. The star sensor-gyroscope data were filtered using the EKF to obtain the filtered attitude values, including 2188 sets of quaternions. The data also exhibit a normal response, showing no errors or anomalies, as shown in Fig. 4.

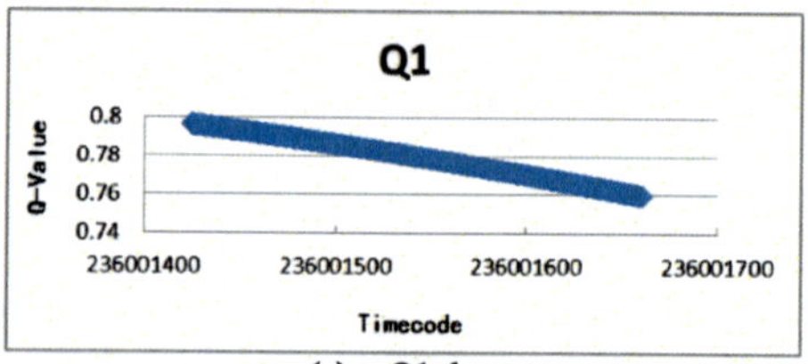

(a) Q1 data

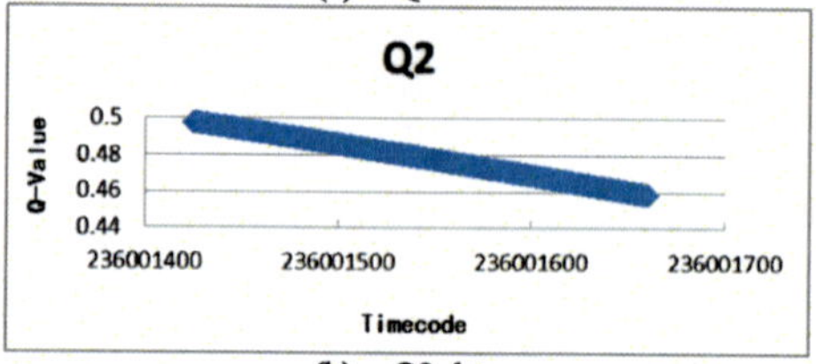

(b) Q2 data

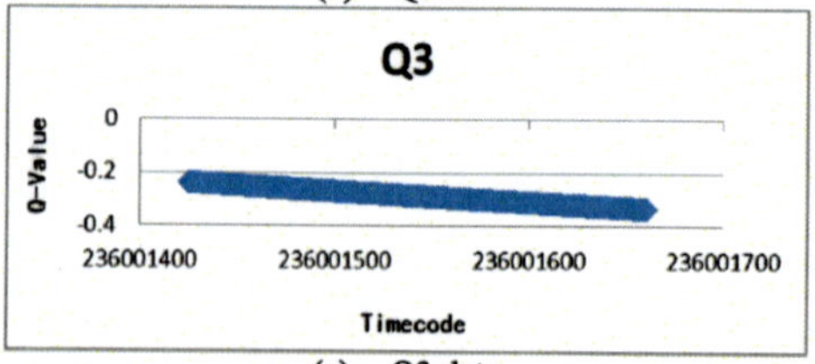

(c) Q3 data

Fig. 3. Q-values of the attitude data of ZY3-02 on 382nd orbit

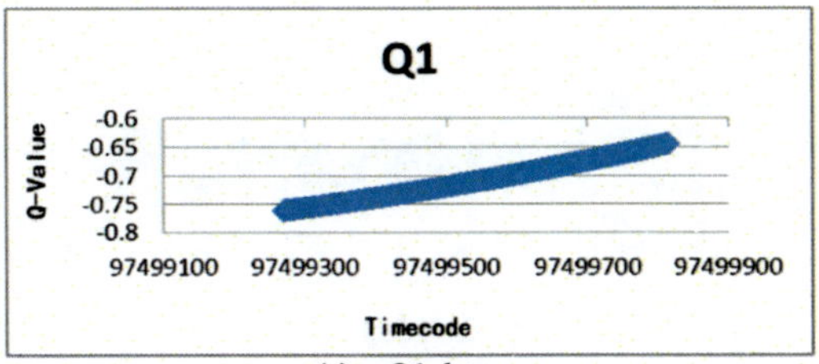

(a) Q1 data

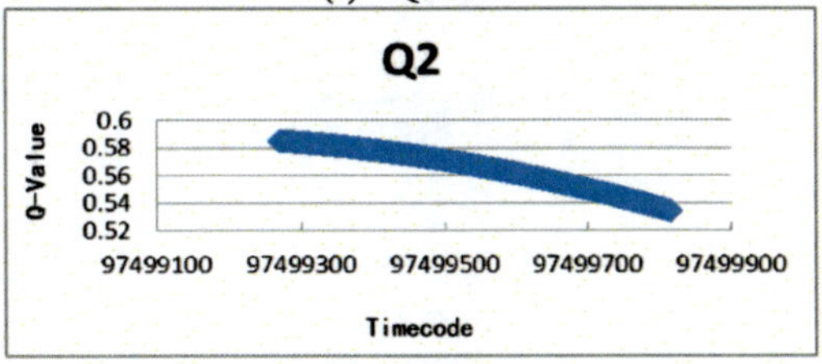

(b) Q2 data

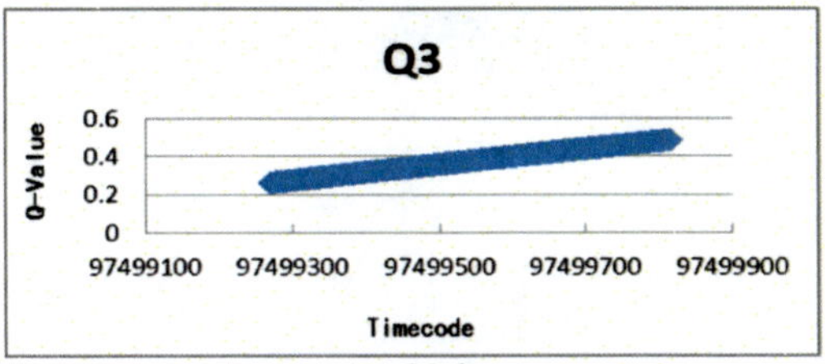

(c) Q3 data

Fig. 4. Q-values of the attitude data of ZY3-01 on 381st orbit

4.2 Experimental Results

The quaternions shown in Figs. 3 and 4 correspond to the attitude values of the satellite body coordinate system relative to the spatial fixed coordinate system. However, due to the earth's rotation, polar motion, and nutation, the values show pseudo-linear changes. Thus, the changes in the satellite platform were analyzed using a general polynomial fitting approach to eliminate the pseudo-linear parts of these quaternions to obtain the attitude data for further analysis. The order of the general polynomial fitting equation was eight. The results obtained using the two sets of data are presented in Figs. 5 and 6.

The Fast Fourier transform was applied to the results of the general polynomial fitting, transforming the time-domain signals to the frequency-domain signals for further analysis. The results are shown in Figs. 7 and 8.

The above frequency-domain distributions of Q-values are quantitatively represented in Table 1.

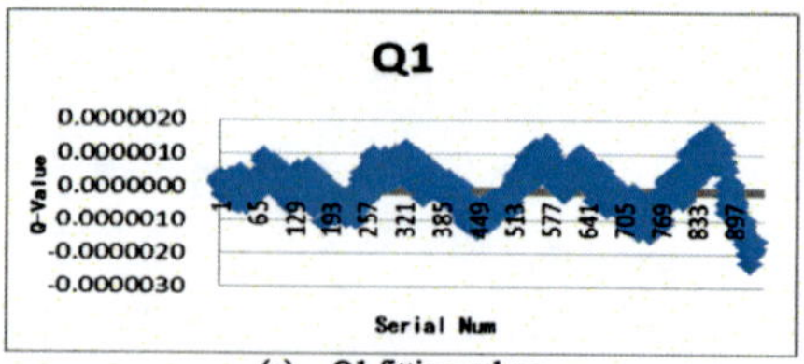

(a) Q1 fitting values

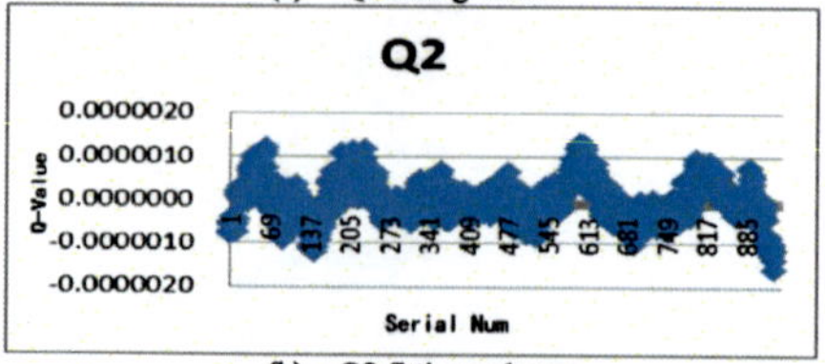

(b) Q2 fitting values

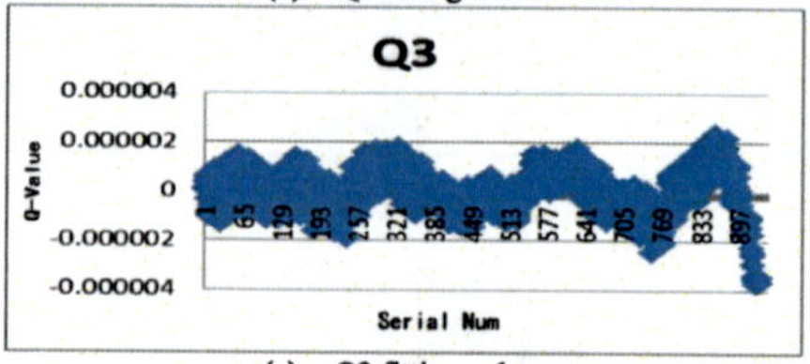

(c) Q3 fitting values

Fig. 5. Fitting values of the attitude data of ZY3-02 on 382nd orbit

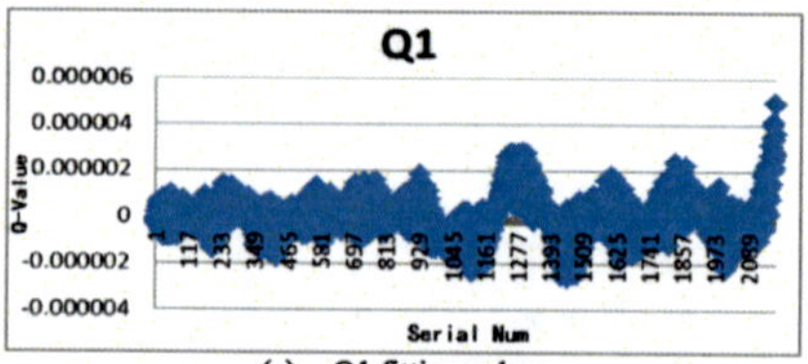

(a) Q1 fitting values

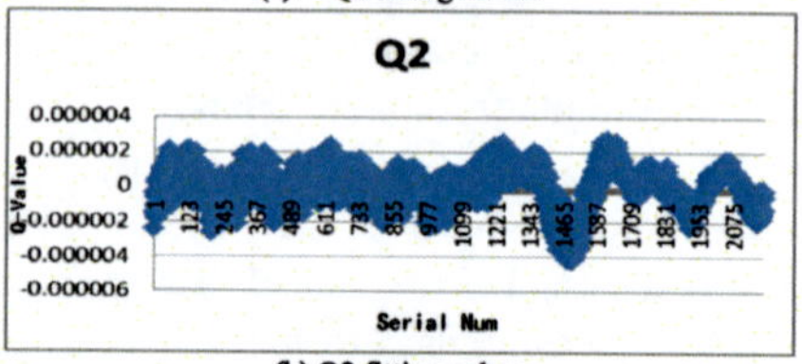

(b) Q2 fitting values

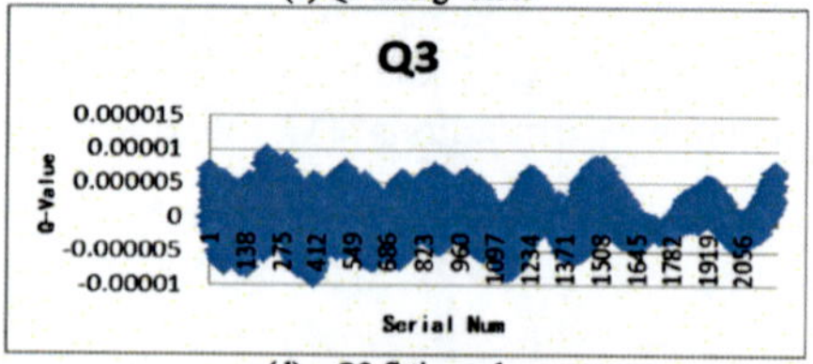

(d) Q3 fitting values

Fig. 6. Fitting values of the attitude data of ZY3-01 on 381st orbit

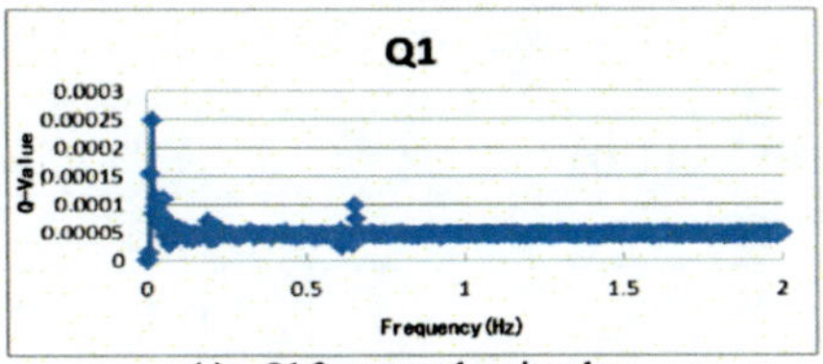

(a) Q1 frequency-domain values

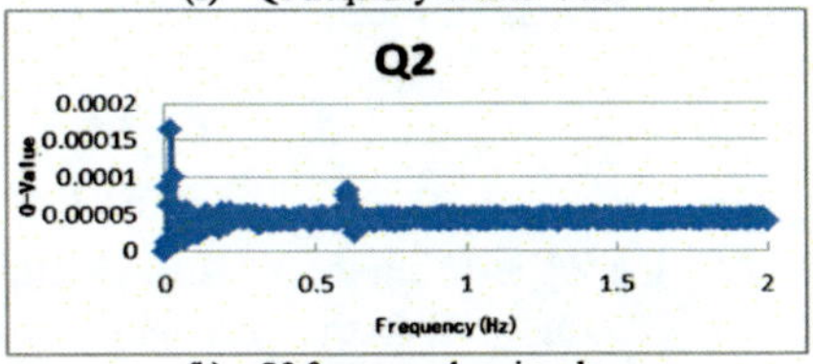

(b) Q2 frequency-domain values

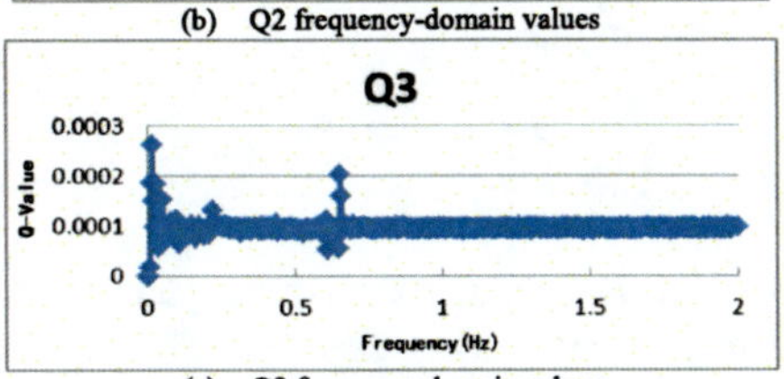

(c) Q3 frequency-domain values

Fig. 7. Frequency-domain distribution of Q-values of ZY3-02 on 382nd orbit

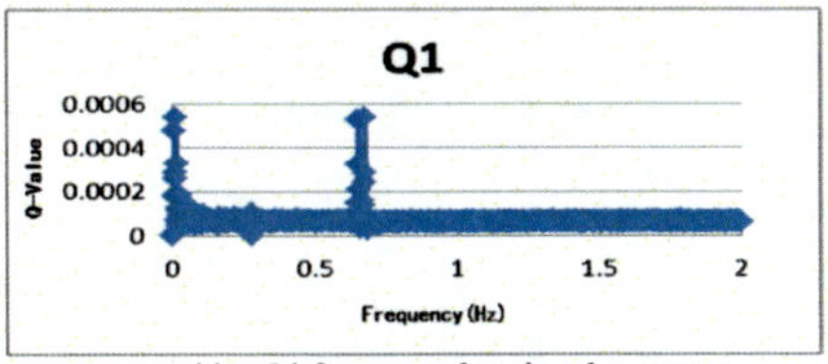

(a) Q1 frequency-domain values

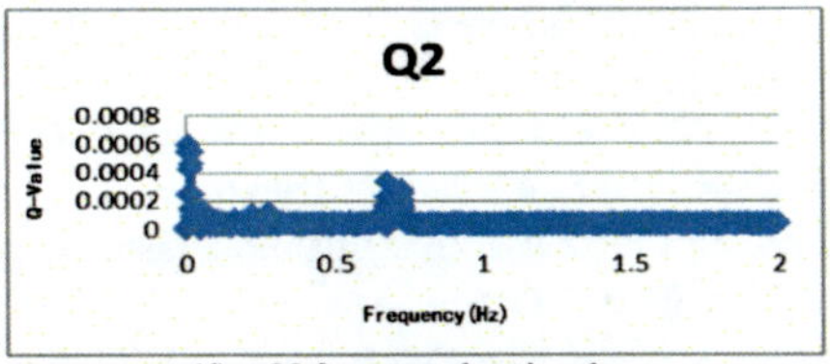

(d) Q2 frequency-domain values

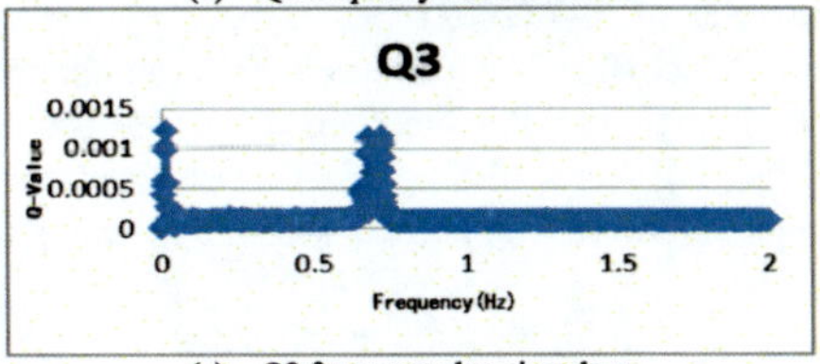

(e) Q3 frequency-domain values

Fig. 8. Frequency-domain distribution of Q-values of ZY3-01 on 381st orbit

Table 1. Jitter frequency of ZY3-01 and ZY3-02

	ZY3–01 (Hz)	ZY3–02 (Hz)
Q1	0.65	0.65
Q2	0.61	0.67
Q3	0.65	0.67

4.3 Result Analysis

Based on the above analysis, the ZY3-02 shows a platform jitter with a frequency of approximately 0.65 Hz during the on-orbit test period, which is similar to the 0.67 Hz of the ZY3–01 satellite platform jitter.

5 Conclusion

This paper analyzed the frequency-domain data of the platform attitude of ZY3-02 and compared the data with the analysis results of ZY3-01. The results showed that the ZY3-02 platform jitter exhibited a frequency of approximately of 0.65 Hz, similar to that of ZY3-01. These results provide a further reference for follow-up analysis.

Acknowledgments. This paper was supported in part by the National Natural Science Foundation of China under Project No. 42001416, No.41971426 and 41771360;in part by the National Key R&D Program of China (2018YFB0504800,2020YFE0200800); Surveying and mapping application demonstration project of GaoFen-2(42-Y30B04–9001-19/21).

References

1. Iwasaki, A., Fujisada, H.: ASTER geometric performance[J]. IEEE Trans. Geosci. Remote Sens. **43**(12), 2700–2706 (2006)
2. Takaku, J., Tadono, T.: PRISM on-orbit geometric calibration and DSM performance[J]. IEEE Trans. Geosci. Remote Sens. **47**(12), 4060–4073 (2009)
3. Wang, R.X.: The method of EFP Aerial triangulation in satellite photography three linear CCD image(1)[J]. Surv. Sci. **04**, 1–5 (2002)
4. Wang, R.X.: The method of EFP Aerial triangulation in satellite photography three linear CCD image(2)[J]. Surv. Sci. **01**, 1–7 (2002)
5. Schwind, P., Müller, R., Palubinskas, G., et al.: An in-depth simulation of EnMAP acquisition geometry[J]. ISPRS J. Photogramm. Remote. Sens. **70**(2), 99–106 (2012)
6. Venkateswaran, N., Goel, P.S., Siva, M.S., et al.: Precision pointing of imaging spacecraft using gyro-based attitude reference with horizon sensor updates[J]. Sadhana **29**(2), 189–203 (2004)
7. Dechoz, C., Lebegue, L.: In-flight attitude perturbances estimation: application to PLEIADES-HR satellites[J]. Proc. SPIE **8866**(2), 206–209 (2013)
8. Tong, X., Ye, Z., Xu, Y., et al.: Framework of jitter detection and compensation for high resolution satellites[J]. Remote Sens. **6**(5), 3944–3964 (2014)

9. Johnston, J.D., Thornton, E.A.: Thermally induced dynamics of satellite solar panels[J]. J. Spacecr. Rocket. **37**(37), 604–613 (2000)
10. Kirk, R.L., HowingtonKraus, E., Redding, B., et al.: High-resolution topomapping of candidate MER landing sites with Mars Orbiter Camera narrow-angle images[J]. J. Geophys. Res. Atmos. **108**(E12), 343–358 (2003)
11. Proakis, J.G., Manolakis, D.G.: The Principle, Algorithm and Application of Digital Signal Processing[M]. Publishing House of Electronics Industry, Beijing (2014)

Land Cover Change Information Extraction of Danjiangkou Wetland Based on GF_1 Data and Softmax Algorithm

Xue Longqin[1,2](✉)

[1] CMA·Henan Agrometeorological Support and Applied Technique Key Laboratory, Zhengzhou, Henan 450003, China
752366302@qq.com

[2] Henan Institute of Meteorological Sciences, Zhengzhou, Henan 450003, China

Abstract. In this paper, based on the GF_1 satellite data of Danjiangkou wetland from 2015 to 2020, the multi-year raster images are classified by softmax classification algorithm, which realized the reuse of wetland classifier based on machine learning method and the dynamic monitoring of land cover change of Danjiangkou wetland. The results showed that the area proportion of Danjiangkou wetland fluctuated and increased from 2015 to 2020, which were 20.44%, 19.10%, 21.50%, 26.23%, 23.58% and 27.35%, respectively, among which the increase was mainly water area. The classification results are verified, and the overall classification accuracy and Kappa coefficient are above 94.0% and 0.898. It is considered that softmax classification algorithm based on GF_1 data is a feasible method to obtain accurate wetland land cover classification information, which can provide research reference for the extraction of wetland land cover information.

Keywords: Danjiangkou wetland · GF_1 · Machine learning · Land cover

Preface

Wetland is a unique ecosystem on the earth, which is closely related to the survival, reproduction and development of human beings. It is the most diverse ecosystem in nature and one of the most important living environments for human beings. It not only provides a variety of resources for human production and life, but also has great environmental functions and benefits. Known as 'kidney of the earth', it is unique to other systems in resisting floods, regulating runoff, storing floods and preventing drought, degrading pollution, regulating climate, controlling soil erosion, promoting siltation and land reclamation, beautifying the environment, etc. [1]. However, in the past 200 years, due to the increasing social and economic activities of human beings and the decreasing wetland area, people are increasingly worried and attached importance to its future development trend [2].

The vast area of wetlands and inconvenient transportation make the field investigation and research take a lot of manpower and material resources, while the continuous development of remote sensing technology provides convenience for wetland monitoring and research. Compared with the traditional investigation methods, using remote

L. Wang et al. (Eds.): CHREOC 2022, LNEE 969, pp. 322–330, 2023.
https://doi.org/10.1007/978-981-19-8202-6_29

sensing technology to investigate wetland resources has the advantages of objectivity, accuracy, economy and high efficiency, and has become the most scientific way to monitor and study wetlands [3, 4]. Gao Jun et al. used Landsat images from 1976 to 2009, and used object-oriented methods on ArcGIS and ENVI platforms to study the annual dynamic changes of wetlands in Tianjin [5]. Zhang Zhijun et al. [6] used TM image data, meteorological observation data and ground survey data, and used wetland type transfer matrix and other methods to analyze the dynamic changes of wetlands in Qinghai Lake Basin in recent 10 years. Zhang et al. [7] used a variety of classification methods to extract coastal wetland vegetation information, and the classification accuracy was much better than the single traditional maximum likelihood method. Lijiang et al. [8] used Landsat TM image data on the Feature Extraction platform of ENVI software to interpret the wetland landscape in the south bank of Hangzhou Bay by object-oriented method, which verified the effectiveness of object-oriented method in extracting wetland information from low and medium resolution remote sensing images.

Machine learning is the core of artificial intelligence, and with the development of artificial intelligence technology, machine learning is widely used in various fields [9–16]. In this study, using remote sensing images from GF_1 satellite data from 2015 to 2020, the softmax classification algorithm was used on the ENVI platform to study the annual dynamic change of Danjiangkou wetland, and the classification results were evaluated and analyzed.

1 Research Area and Data

1.1 Overview of Research Area

Danjiangkou Wetland National Nature Reserve in Henan Province is located in Xichuan County, southwest of Henan Province. Its purpose is to protect the inland wetland ecosystem formed by Danjiang River, Guanhe River and floodplain and its rich biodiversity. Danjiangkou reservoir consists of Hanjiang reservoir area and Danjiang reservoir area in the upper reaches. The reservoir has an average inflow of 39.48 billion cubic meters for many years, and its water source comes from Hanjiang River and its tributary Danjiang River. It is the water source of the Middle Route Project of South-to-North Water Transfer, a national first-class water source protection area and an important wetland protection area in China. In the study area, there are more than a dozen towns in Henan, including Danjiangkou Wetland, Danjiangkou Reservoir and surrounding continuous areas, such as Dashiqiao Township, Taohe Township, Jinhe Town, Xianghua Town and Shangji Town. The study area lies between 111°8′ -111°44′ east longitude and 32°12′-32°35′ north latitude, including water, woodland, construction land, cultivated land and other land types.

1.2 Data and Sources

GF_1 satellite is the first satellite in China's high-resolution Earth observation system, which has the characteristics of high spatial resolution, multispectral and wide coverage. GF_1 satellite is equipped with two 2 m and 8 m high-resolution full-color multispectral cameras, and four 16m medium-resolution wide-width multispectral cameras. Since data distribution in 2013, it has been widely used in various industries in China by virtue of its advantages of large width, high temporal phase and spatial resolution. This study mainly uses GF_1 satellite data to extract the information of land cover change in the study area. The main band characteristics and functions of WFV data from GF_1 satellite are shown in the following table (Table 1).

Table 1. Main features and functions of WFV data band of GF_1 Satellite

Data type	Band number	Band name	Wavelength range (μm)	Key features
Multi_spectral camera	1	Blue band	0.45–0.52	It is penetrating to water bodies. It can distinguish soil, vegetation, forest types, coastline, etc.
	2	Green band	0.52–0.59	Study on plant growth, diseases and insect pests, green reflectance, plant classification and sediment concentration in water
	3	Red band	0.63–0.69	It can distinguish plant types, leaf green absorption rate, suspended sediment, urban outline and water and land boundaries
	4	Near red band	0.77–0.8	It can distinguish water and land boundaries, water systems, roads, residential areas and classify plants

GF_1 satellite data comes from the website of China Resources Satellite Application Center. Six remote sensing images of GF_1 WFV in the study area on October 12th, 2015, September 4th, 2016, September 20th, 2017, September 8th, 2018, September 29th, 2019 and September 7th, 2020 were downloaded. In addition, land use data, DEM data and administrative division data are also used. The administrative division data is used to cut the data and products of the research area, and the land use data and DEM data are used to assist in wetland land use classification. The data comes from the National Satellite Meteorological Center.

1.3 GF_1 Satellite Data Preprocessing

Image preprocessing of GF_1 satellite data in the study area includes geometric correction, radiometric correction, image clipping and other steps to eliminate the interference of deformation, distortion, blur and noise in the original image acquisition process on the extraction process. Geometric correction includes orthorectification and image registration. For the geometric correction of GF_1 satellite image in the study area, orthorectification is carried out by using the orthorectification flow tool provided by ENVI. The geometric correction accuracy error of each image is controlled within one pixel. Radiation correction includes radiation calibration and atmospheric correction. Radiation calibration is to convert the dimensionless DN value recorded by the sensor into the radiation brightness or reflectivity of the atmospheric top layer with practical physical significance, which is realized by the radiation calibration module in ENVI. Its purpose is to correct the error caused by the sensor. The purpose of atmospheric correction is to eliminate the influence of atmospheric and illumination factors on the reflection of ground objects. Atmospheric correction is realized by FLAASH atmospheric correction module in ENVI. After image geometric correction and radiation correction, use ENVI's image clipping function to remove the parts outside the study area.

2 Research Methods

2.1 Establish a Wetland Classification System

According to the general situation and hydrological conditions of the study area and based on the classification system of the Convention on Wetlands, the land classification system of the study area is established. The system is shown in Table 2, in which water bodies and beaches are unified as wetlands, and woodland, grassland, cultivated land, towns and other land are unified as non-wetlands.

2.2 Wetland Extraction Based on Softmax Classification Algorithm

Because the blue band is sensitive to water, the green band is sensitive to vegetation, and the near-red band is sensitive to construction land and bare land, NDVI (Normalized Difference Vegetation Index) is introduced to enhance the vegetation information in the image when GF_1 remote sensing data is used to extract wetlands in the study area. NDWI (Normalized Difference Water Index) is introduced to enhance the information of water in images, and the attribute set of training samples is established by combining these two indices and band information, so as to further enhance the differences among different land cover types.

After the training attribute set is established, according to the classification system, referring to higher-resolution remote sensing images and experiences, training samples with different classifications are selected to obtain the attribute and category values of the

Table 2. Classification of land use in the study area

Code	Category of ground objects	Description
1	Water	Including permanent rivers, seasonal or intermittent rivers and reservoirs
2	Beach	The flat areas on both sides of a flooded river include floodplain, flooded valley and seasonally flooded grassland
3	Woodland	Areas or natural forest areas used for forestry production
4	Grassland	Land dominated by herbaceous and shrub plants
5	plough	Land where crops are planted
6	Urban and other land use	Including towns, industrial and mining areas, places of residence, roads and other construction land and unused land

samples, and softmax classifier is selected to iteratively train the samples. Classifier is an algorithm that iteratively trains the classifier to achieve the minimum error. By iteratively adjusting the internal parameters of the classifier, the error (or loss) converges to the minimum value. Max classification parameters are set as follows: Convergence Criterion is 0.00001, and when the loss difference between this iteration and the previous one is less than 0.00001, the iteration is stopped; The Learning Rate is 100; Maximum Iterations is 100. After training, the classification loss curve is obtained, and the classification error is close to 0.1 as shown in the following figure (Fig. 1).

The trained softmax classifier is used to classify the combined band images from 2015 to 2020, and the classification results of 2019 are shown in Fig. 2.

3 Results and Discussion

3.1 Analysis of Wetland Area Change

The total area of the study area is 1760.2 km^2. According to the classification results, the area percentage of water body, beach, woodland, grassland, town and other land in recent 6 years is shown in the following table. From 2015 to 2020, the proportion of wetland area was 20.44%, 19.10%, 21.50%, 26.23%, 23.58% and 27.35% respectively, showing a fluctuating and rising trend, among which the increased part was mainly water body. In 2018 and 2020, affected by rainfall, the proportion of water body area was high, and other years were relatively stable (Table 3).

3.2 Classification Accuracy Evaluation

In the evaluation of image classification accuracy, the accuracy of classification results is displayed in a confusion matrix, which is used to compare the classification results with the actual measured values. Each column of the confusion matrix represents the

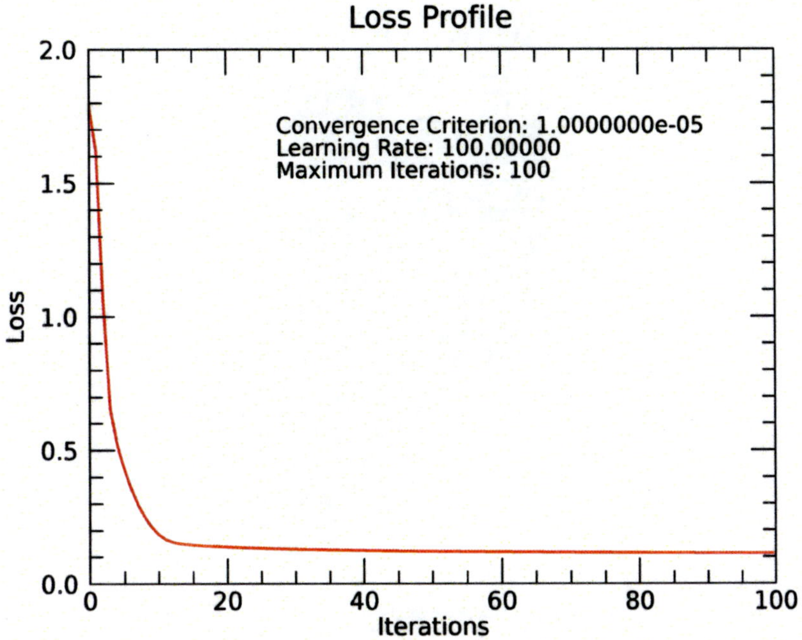

Fig. 1. Loss Profile diagram

actual measured information, and the value in each column is equal to the number of actually measured pixels corresponding to the corresponding category in the classified image; Each row of the confusion matrix represents the classification information of remote sensing data, and the value in each row is equal to the number of remote sensing classified pixels in the corresponding category of measured pixels. The classification results of Danjiangkou wetland land cover from 2015 to 2020 were verified, and the visually discriminated sample patches were used as the reference classification, and the confusion matrix was established to analyze and evaluate the classification accuracy. The reference classification completely corresponds to the statistical types of the classification to be evaluated, including six categories: water body, beach land, woodland, grassland, cultivated land and other land. The verification results in 2019 are shown in Table 4.

From 2015 to 2020, the overall accuracy of classification statistics in the study area reached 95.9% on average, among which the classification accuracy of beach, grassland, town and other land, water, woodland and cultivated land were 83.6%, 69.4%, 83.2%, 99.6%, 96.8% and 69.3% respectively, and the classification accuracy of water was the highest, followed by that of woodland. According to the analysis, the classification accuracy is related to the spectral characteristics of ground objects, and the water features are single and obvious. The spectral differences between woodland, grassland and cultivated land are large and easy to distinguish, while the spectral characteristics of grassland and cultivated land are not obvious and easy to be confused. From the annual point of view, the overall classification accuracy is the highest in 2019, reaching 97.2%, and the lowest

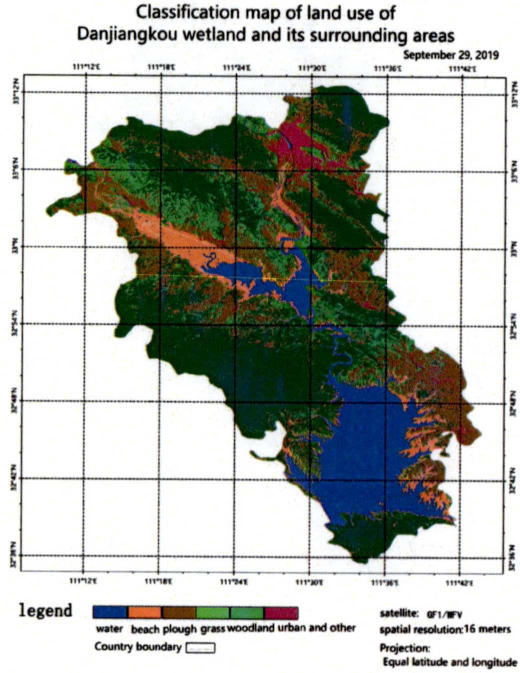

Fig. 2. The classification results of 2019

Table 3. Statistical table of the proportion of land use classification area in the study area from 2015 to 2020

Classify	2015 (%)	2016 (%)	2017 (%)	2018 (%)	2019 (%)	2020 (%)
Water	13.65	15.06	16.01	19.89	15.90	21.22
Beach	6.79	4.04	5.49	6.34	7.68	6.13
Woodland	42.02	44.83	48.06	40.53	40.44	37.96
Grassland	11.46	6.35	7.20	9.71	10.45	9.58
Urban and other land use	1.80	1.94	1.83	1.73	2.32	1.81
Plough	24.28	27.78	21.41	21.80	23.21	23.30

in 2020, reaching 94.0%. According to the analysis, the classification accuracy is related to the quality of satellite data, and the high cloud content of satellite data in 2020 leads to a slightly lower classification accuracy.

4 Conclusion and Prospect

In this paper, using the remote sensing data of Gaofen-1 satellite in Danjiangkou wetland from 2015 to 2020, the reuse of wetland classifier based on softmax machine learning

Table 4. Evaluation table of classification accuracy in 2019

Confusion matrix	Beach	Grassland	Urban and other land use	water	Woodland	Plough	Total
Beach	4956	0	5	91	8	66	5126
Grassland	212	4055	80	73	204	679	5303
Urban and other land use	77	49	5303	5	36	157	5627
Water	258	0	0	110398	0	0	110656
Woodland	85	1933	5	41	97791	510	100365
Plough	51	903	366	0	660	5658	7638
Total	5639	6940	5759	110608	98699	7070	
Overall accuracy = 97.2% and Kappa = 0.953							

method and the dynamic monitoring of land cover change in Danjiangkou wetland are realized. The monitoring results show that the wetland area of Danjiangkou has fluctuated and increased in recent 6 years, and the increased part is mainly water.

The results show that the overall classification accuracy and Kappa coefficient of the study area are above 94.0% and 0.898, and the classification accuracy of beach, grassland, town and other land, water body, woodland and cultivated land are 83.6%, 69.4%, 83.2%, 99.6%, 96.8% and 69.3% respectively, and the classification accuracy of water body. It can be seen from the analysis that the classification accuracy is related to the spectral characteristics of ground objects, and it is also affected by the quality of satellite data.

In the next step, based on the dynamic monitoring of land cover change of Danjiangkou wetland, combined with the changes of vegetation, atmospheric environment and other ecological elements, a comprehensive assessment of ecological quality of Danjiangkou wetland will be carried out, which will provide technical reference for ecological protection of Danjiangkou wetland and "a canal of clear water will be continuously sent to the north" in the middle route of South-to-North Water Transfer Project.

References

1. Zhijun, Z., Qingjun, Y., Wenqi, L., et al.: Remote sensing monitoring of wetland dynamics in Qinghai Lake Basin in recent 10 years [J]. J. Hubei Univ. (Natural Science Edition) **34**(1), 120–124 (2012)
2. Zhenguo, N., Peng, G., Xiao, C., et al.: Preliminary remote sensing mapping of wetlands in China and analysis of related geographical features [J]. Chin. Sci. D: Earth Sci. **39**(2), 188–203 (2009)
3. Bai, Z.: Application of remote sensing technology in Wetland research in China [J]. Remote Sens. Technol. Appl. **1**, 67–71 (1996)
4. Rebelo, L.M., Finlayson, C.M., Nagabhatla, N.: Remote sensing and GIS for wetland inventory, mapping and change analysis[J]. J. Environ. Manag. **90**(7), 2144–2153 (2009)

5. Jun, G., Zhongliang, W.: Analysis of wetland area change characteristics and driving factors in Tianjin from 1976 to 2009 [J]. J. Tianjin Norm. Univ. (Natural Science Edition) **33**(4), 32–38 (2013)
6. Zhijun, Z., Qingjun, Y., Wenqi, L. et al.: Remote sensing monitoring of wetland dynamics in Qinghai Lake Basin in recent 10 years [J]. J. Hubei Univ. (Natural Science Edition) **34**(1), 120–124 (2012)
7. Zhang, Y., Lu, D., Yang, B., et al.: Coastal wetland vegetation classification with a Landsat thematic mapper image[J]. Int. J. Remote Sens. **32**(2), 545–561 (2011)
8. Molijiang, C.Y., Hu, Y., et al.: Object-oriented remote sensing classification of wetland landscape-taking the south bank of Hangzhou Bay as an example [J]. Wetland Sci. **10**(2), 206–213 (2012)
9. Chen, L., Zhao, J., Yi, L., et al.: Research on land use classification based on multi-source remote sensing image fusion based on machine learning [J]. J. Southwest Norm. Univ. (Natural Science Edition) **43**(10), 103–110 (2018)
10. Zhang, S., Hui, L., Xiangren, L.: Wetlands classification method using full convolution neural network and Stacking algorithm [J]. J. Agric. Eng. **36**(24), 257–264 (2020)
11. Liying, F.: Research on Information Extraction of Construction Land from High-Resolution Remote Sensing Images Based on Deep Learning Technology [D]. Zhejiang University, Hangzhou (2017)
12. Gaofeng, Q.: Research on Multi-Spectral Remote Sensing Image Classification and Urban Expansion Based on Machine Learning [D]. Chongqing University, Chongqing (2012)
13. Cao, Z., Lin, N., Xu, W., et al.: Classification of remote sensing images of Dongyu Island based on BP neural network [J]. Mar. Sci. Bull. **35**(5), 587–593 (2016)
14. Zhiqing, Z., Zong, H.: Study on adaptability and quality evaluation of GeoEye-1 multispectral and panchromatic image fusion [J]. J. Southwest Norm. Univ. (Natural Science Edition) **36**(1), 203–208 (2011)
15. Chang, L., Deng, X., Zhou, M., et al.: Convolutional neural network in image understanding [J]. Acta Autom. Sin. **42**(9), 1300–1312 (2016)
16. Wei, W., Xinqi, Z., Zhiyuan, Y., et al.: Research on post-processing method of remote sensing image classification under neighborhood rule [J]. Bull. Surv. Mapp. **S2**, 17–21 (2015)

Deep Learning-Based Matching of Sar Images with Optical Images

Dehua Li[1,2](✉), Canhai Li[1], Hao Han[1,2], and Hui Qu[3]

[1] Land Satellite Remote Sensing Application Center, MNR Beijing, Beijing, China
2321252761@qq.com
[2] Liaoning Technical University, FuXin, China
[3] The Second InStitute of Photogrammetry and Remote Sensing MNR, Harbin, China

Abstract. There are essential differences between optical and SAR remote sensing imaging in terms of mechanism, so the geometric and physical properties of optical and SAR images are extremely different, which makes it more difficult to match images. We designed and improved a deep learning network, firstly, we extracted the high-dimensional features of the image using convolutional neural network, then we used nearest neighbor search for feature matching, in order to reduce the false matching points, we introduced MAGSAC + + to eliminate the outlier points, finally, we compared with SIFT_KNN_BBS and orb_feature_matching matching algorithms. The results show that the matching of heterogenous remote sensing images using deep learning approach has high accuracy and robustness.

Keywords: Multi-sensor · Deep learing · CNN · Image matching

1 Introduction

With the rapid development of science and technology, human beings have gone from face-to-face observation to large-range earth observation with the help of remote sensing satellites, and optical sensors and synthetic aperture radar constitute the two main modes of satellite-based detectors. Optical sensors have the characteristics of intuitive imaging and easy identification, but are greatly influenced by weather and light. Synthetic aperture radar, on the other hand, has the characteristics of penetrating clouds, vegetation and all-weather, but the imaging abstraction is not easy to identify. The advantages and disadvantages of the two complement each other. Therefore SAR image [1] and optical image fusion is of high research value. And the prerequisite for fusion between two heterogenous images is the need for a correct match between the two images. Due to the geometric imaging differences between the two, the matching between them has been a difficult problem in image matching.

Experts at home and abroad have successively proposed different theories to solve this challenge. The most famous one is the SIFT algorithm proposed by Lowe [2], which solves the image rotation, affine transformation, intensity and viewpoint change in matching features to a certain extent. However, due to the key points described based on the

L. Wang et al. (Eds.): CHREOC 2022, LNEE 969, pp. 331–340, 2023.
https://doi.org/10.1007/978-981-19-8202-6_30

gradient distribution of the image local neighborhood, the matching effect is poor for heterogeneous remote sensing images with huge differences in imaging methods. While people are still continuing research along the direction of SIFT, deep learning has gradually become a reality from theory due to the improvement of hardware level, especially convolutional neural network (CNN) [3, 4] has shown great advantages in computer vision such as image recognition, image classification, and image segmentation. Along with the deepening of convolutional layers, the features one can extract are becoming more and more complex and the images that can be processed are becoming more and more complex, which brings the possibility of image matching for heterogenous remote sensing images with relatively large variability, and more and more people have invested in using deep learning to do image matching, such as SuperPoint [5], DELF [6], D2-Net [7], etc., where D2- Net was trained using more than 300,000 pre-matched stereo image pairs and has made significant progress in addressing the impact of matching under scene changes, but these algorithms are applicable to close-up visible images with changing illumination and perspective. In the literature [8], a context-enhanced local descriptor is proposed for the influence matching problem with multimodal heterogeneous sources, and good results are achieved.

After the image features have been extracted, the extracted features need to be matched. The purpose of the matching task is to identify homonyms between two or more images by means of a certain matching algorithm. The matching task is to identify identical points between two or more images by means of a matching algorithm. How quickly and accurately the identical points can be matched is crucial to the whole image matching process, and how to solve the problem of many false matches is also an important part of the image matching process.

Therefore, in this paper, we try to make some small changes on the basis of D2-Net deep learning network to construct a deep learning extraction feature matching method for heterogenous remote sensing images. Triplet Margin With Distance Loss loss function is introduced as a measure of the similarity of two images with good effect. And to solve the problem of many mis-matching points, adaptive Euclidean distance threshold, and MAGSAC++ common constraint purification algorithm are introduced. Through these minor changes, a deep neural network with good robustness is established.

2 Method and Principle

2.1 Feature Extraction and Detection

See Fig. 1.

The feature information that can be extracted in the convolutional neural network is more comprehensive as the convolution deepens and the size of the convolutional kernel changes. Given that VGG-16 has the characteristics of a complex model but fewer parameters, we use VGG-16 as the base model. Although the more abstract the features that can be extracted as the network deepens, the more advantageous it is for matching heterogenous remote sensing images, the deepening of the network is accompanied by the deterioration of the accuracy of feature point locations. The relationship between feature complexity and location accuracy is also very important to be handled.

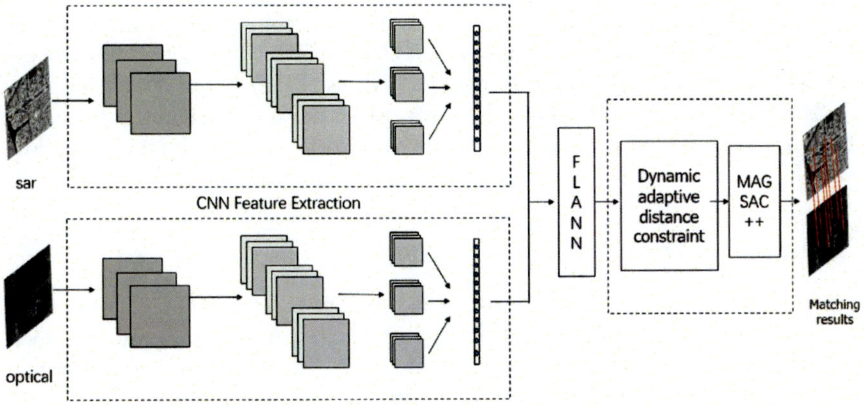

Fig. 1. Algorithm flow of remote sensing image from different sources

VGG-16 has a multi-layer convolutional structure, and we choose the third convolutional layer of the fourth part as the feature mapping. It is able to meet the requirement of localization accuracy while extracting sufficiently complex features. Suppose the size of the input image is h x n and the number of channels is c. After the convolution layer transformation, the size is reduced by 1/6 of the original image and the number of channels is increased to 512. In order to ensure the resolution of the feature map, we modify both the pooling method and the step size, and use hole convolution to expand the perceptual field to improve the generalization ability of feature expression, which is beneficial to the invariant expression of heterogenous image features.

2.2 Optimization of Loss Function

The VGG-16 network model does not have the ability to express the generalization of heterogenous remote sensing images due to the training data used. In order for the CNN network model to learn the expression of common features on heterogenous remote sensing images with large radiometric and geometric differences, new loss functions and relevant datasets need to be introduced to train the model.

We use Triplet Margin With Distance Loss [9] as the loss function because we want to get feature points with some generality when doing feature detection, but at the same time, in the process of feature description, we want the selected feature points to have unique feature vectors that are used to facilitate the search for homonymous image points. Faced with such a problem Triplet Margin With Distance Loss function enhances the uniqueness of the relevant descriptors by penalizing any irrelevant descriptors that lead to false matches, which can be customized by passing in different distance functions, and adding the detection terms to the loss function in order to enhance the repeatability of feature detection, for the batch data containing N samples D (a,p,n). The loss corresponding to the i-th sample is as follows.

$$l_i = \max\{d(a_i, p_i) - d(a_i, n_i) + \boldsymbol{margin}, 0\} \tag{1}$$

where **a**, **p**, **n**, denote anchor points, positive examples (of the same class as anchor points) and negative examples (of a different class from anchor points), respectively. The distance function, **d**, is used to measure the distance between the anchor point and the positive case and the negative case. Margin is considered to be set as a constant. Minimize the loss function so that the distance between the anchor point and the positive example is smaller and the distance to the negative example is larger.

2.3 Reduce False Match Points Use MAGSAC ++

In order to reduce the excessive number of mis-matching points, which affects the alignment effect, we decided to use marginalizing sample consensus ++ (MAGSAC++) to deal with the problem.MAGSAC ++ was proposed by[10] and is based on an improvement of MAGSAC, which introduces a new model quality (scoring) function:

$$Q(\theta, P) = \frac{1}{L(\theta, P)} \tag{2}$$

where

$$L(\theta, P) = \sum_{p \in P} \rho(D(\theta, p))$$

is a loss function of the M-estimator defined by our weight function $\omega(r)$. Function $\rho(r) = \int_0^r x\omega(x)dx$ for $r \in [0, +\infty)$.

Instead of internal outlier decisions, this approach proposes a new marginalization process, the iteratively re-weighted least-squares (IRLS) method. The model parameters in the (i + 1)th step are calculatedas follows:

$$\theta_{i+1} = \arg\min_{\theta} \sum_{p \in P} \omega(D(\theta_i, p)) D^2(\theta, p) \tag{3}$$

where the weight of point **p** is: $\omega(D(\theta_i, p)) = \int \mathrm{P}(p|\theta_i, \sigma) f(\sigma) d\sigma$, and $\theta_0 = \theta$, i.e. the initial model from the minimal sample.

MAGSAC ++ also propose a new sampler, Progressive NAPSAC, for RANSAC-like robust estimators.

2.4 Preventing Over-Fitting

In deep learning, a model is considered to be overfitted if it focuses too much on specific training data and misses the main points. The model provides an answer that is far from the correct answer, i.e., less accurate. This type of model treats noise in irrelevant data as a signal, which negatively affects accuracy. To prevent the overfitting phenomenon, we add a Batch Normalization (BN) layer after the convolutional layer [11], which not only suppresses the overfitting situation to some extent and reduces the necessity of Dropout, but also speeds up the training rate and enables a fast learning rate even for small learning rates.

2.5 Selection of Training Set

For the selection of the dataset, we chose to use the SEN1–2 dataset[12], the first ever dataset that truly contains SAR and optical image pairs and can be called "large scale (over a million)". The dataset contains 282,384 image blocks taken globally and from all four observing seasons, and a particular strength of the dataset is that it can be easily divided into various subsets with clear boundaries (e.g., by scene or season), with the ability to create truly independent training and test sets, so that, depending on the requirements, it can be used as a test set. This allows us to create truly independent training and testing sets, so that training can be targeted to a specific function depending on the requirements. Therefore, this dataset is used for model training. And the data set is divided into two parts according to the ratio of 9:1 for the training set validation set.

3 Results and Discussion

In order to comprehensively evaluate the accuracy and generalization performance of the improved algorithm of heterogenous remote sensing images proposed in this paper, by matching with other three matching algorithms SIFT_KNN_BBS, orb_feature_matching, and FAST_n_BRIEF. Four scenes such as buildings, semi-naked land, and water bodies are processed separately under the same computer configuration. The performance of the different algorithms is analyzed by visualizing the results after processing.

3.1 Test Data

Four types of surface type areas, including Construction, Half naked, Bare land and Water, were selected as test data (Fig. 2).

3.2 Experimental Results and Analysis

Three different algorithms, CNN, SIFT_KNN_BBS, and orb_feature_matching, are used in each of the four scenarios. Among them, SIFT_KNN_BBS and orb_feature_matching are the traditional algorithms. The results are represented in Table1 and Fig. 3.

From Table1, we can see that CNN_Net extracts more feature points, which are evenly distributed in the graph, and has a high correct matching rate and takes less time. Sift_KNN_BBS extracts very few points in the test image, and has a very low correct matching rate and takes longer time. Orb_feature_matching algorithm can be seen in the graph. From Figs. 3, 4, 5 and 6, Although we use the algorithm to eliminate the mismatch points, the result still cannot completely eliminate the mismatch points, so we use the manual method to eliminate the mismatch points and present them. The experiment proves that deep learning has better effect in heterogenous remote sensing image matching.

(a) Construction (b) Half naked

(c) Bare land (d) Water

Fig. 2. Optical images and SAR images in different scenes

Table 1. MATCHING RESULTS

Place	Method	Number of correct matches	Run time[s]
Construction	CNN	113	1.793
	SIFT_KNN_BBS	0	9.617
	orb_feature_matching	86	2.363
	CNN	107	1.375
Half naked	SIFT_KNN_BBS	3	7.296
	orb_feature_matching	64	4.325
	CNN	89	1.446
Bare land	SIFT_KNN_BBS	2	9.538
	orb_feature_matching	51	4.226
	CNN	47	2.251
Water	SIFT_KNN_BBS	2	1.434
	orb_feature_matching	29	7.152

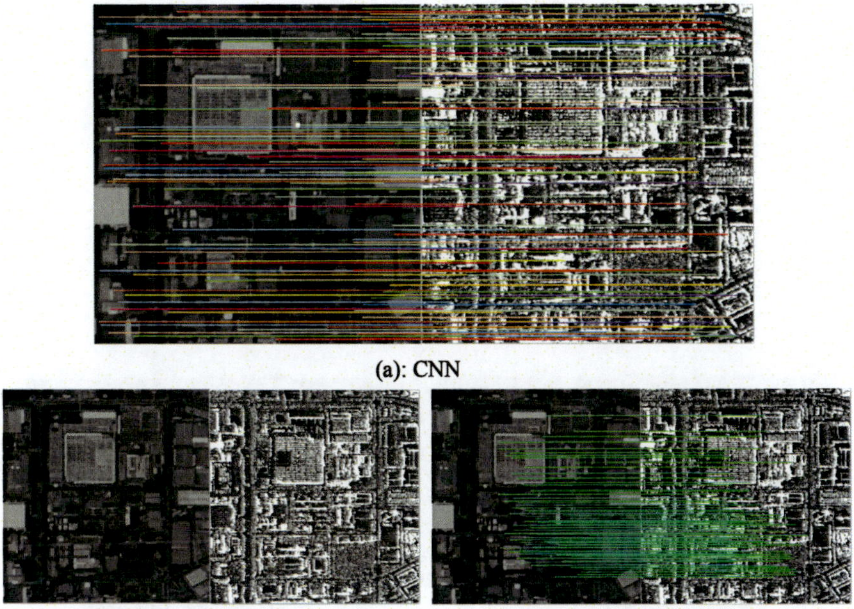

(a): CNN

(b): SIFT_KNN_BBS (c): orb_feature_matching

Fig. 3. Map of correct matching points for three algorithms in building scenes

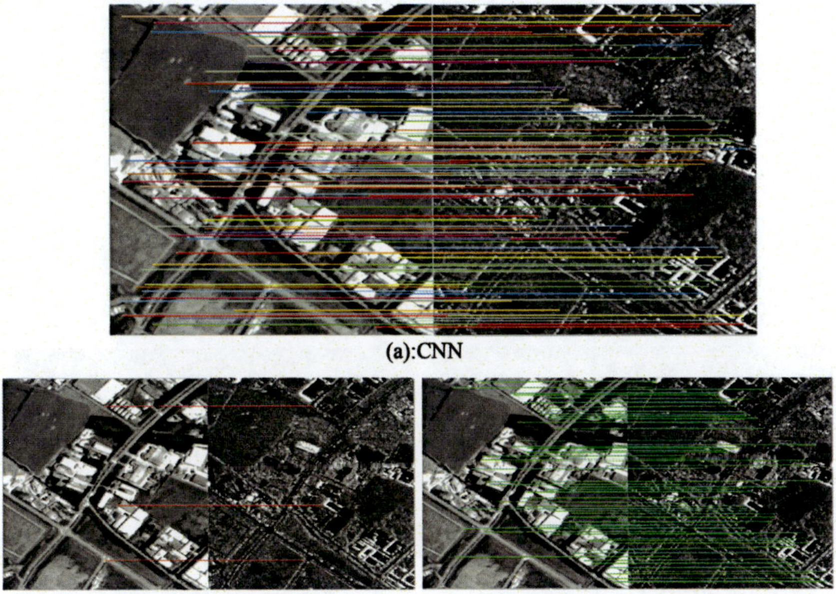

(a):CNN

(b): SIFT_KNN_BBS (c): orb_feature_matching

Fig. 4. Correct matching graph for three algorithms in semi-naked ground scenario

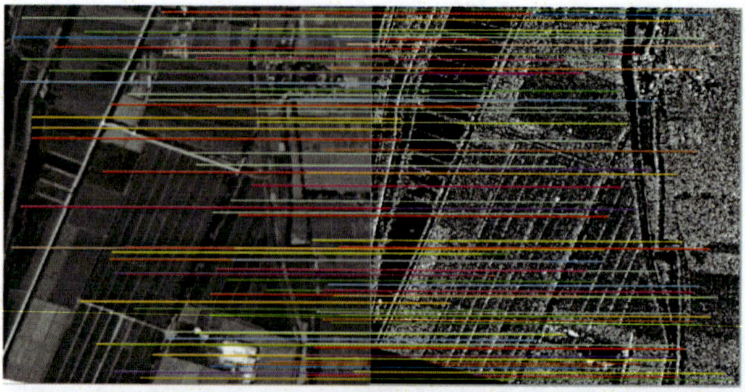

(a): CNN

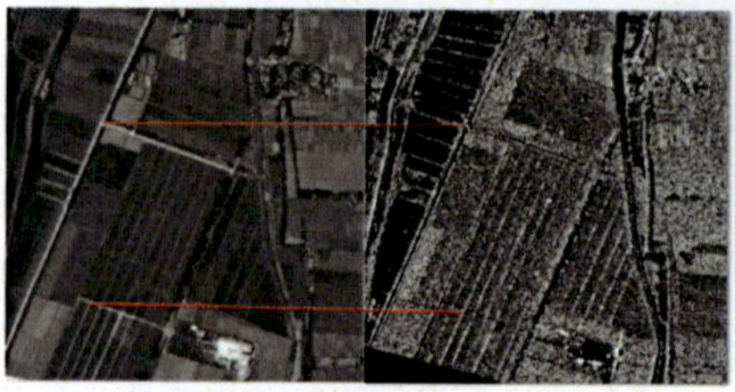

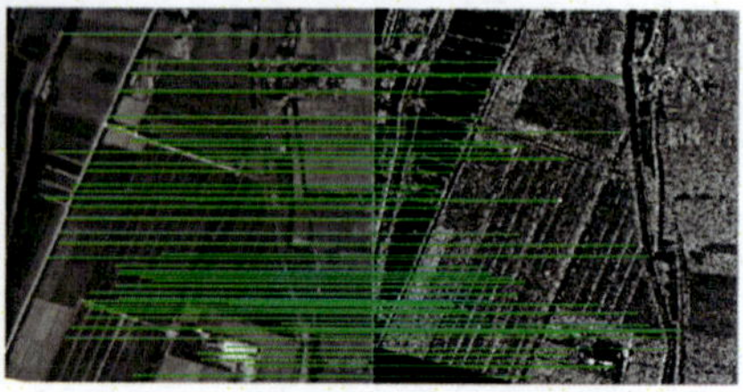

(b): SIFT_KNN_BBS (c): orb_feature_matching

Fig. 5. Correct matching graph for three algorithms in bare ground scenario

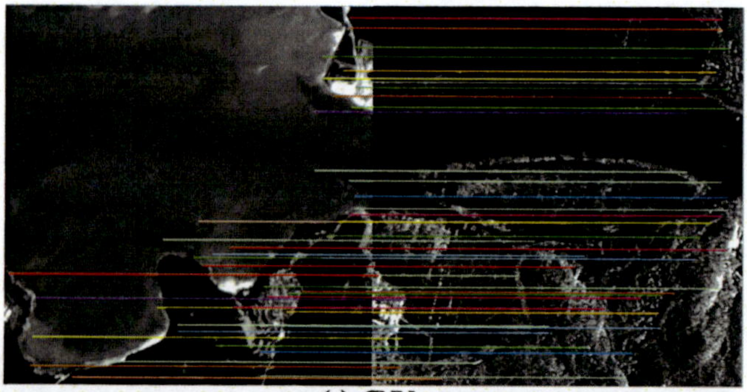

(a): CNN

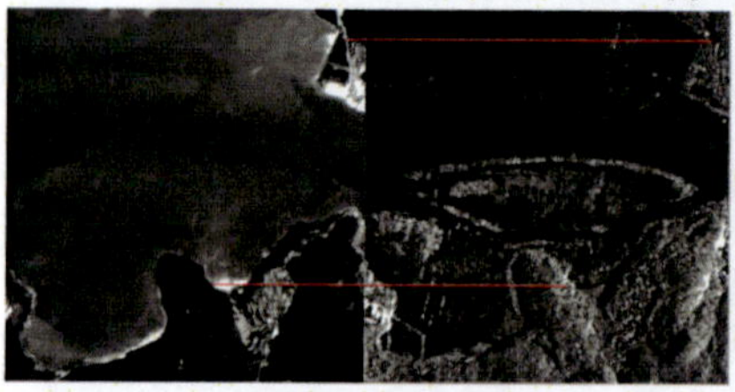

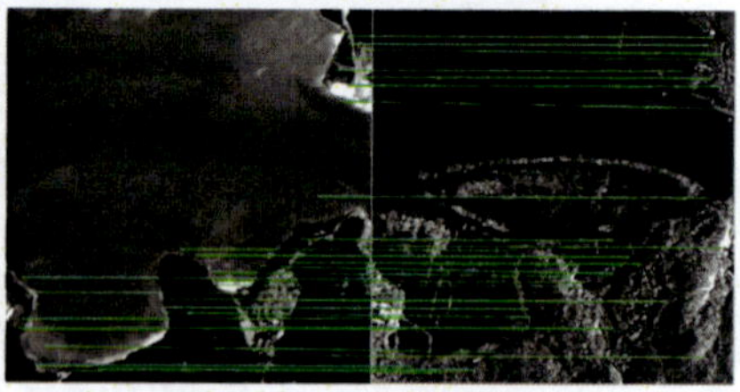

(b): SIFT_KNN_BBS (c): orb_feature_matching

Fig. 6. Correct matching diagram for three algorithms in water scenario

3.3 Number of Correct Matching Points

The correct match point is determined as shown in Eq. (4)

$$Corr(x) : \left\| x_i - x_j \right\| \leq \varepsilon \tag{4}$$

If the pixel distance between a feature point position matched by the algorithm on the reference image and the theoretical point position is less than a given accuracy threshold, it is considered as a correct match.

On the whole image, only the points that satisfy all the above requirements can be defined as correct matching points. Other operations such as image alignment require a certain number of correctly matched points as a prerequisite, and Eq. (4) can reflect the performance of the matching algorithm.

4 Conclusions

With the continuous development of deep learning, we have made considerable achievements in various fields, and it is particularly difficult to match sar images and optical images due to factors such as imaging method and resolution, while deep learning, especially in convolutional networks, has unique advantages in extracting features of images, so we improve the existing classical convolutional networks by making improvements in terms of loss functions and matching constraints. Experiments are conducted in four different natural environment scenes respectively, and the experimental results show that it has adaptability and robustness in different scenes. The experimental results show that it has good matching effect on heterogeneous remote sensing images. With the continuous development of deep learning, there is more room for improvement in feature extraction and feature localization accuracy.

References

1. Geng, J., et al. High-resolution SAR image classification via deep convolutional autoencoders. IEEE Geosci. Remote. Sens. Lett. **12**(11), 2351–2355 (2015)
2. Lowe, D.G.: Distinctive image features from scale-invariant keypoints. Int. J. Comput. Vision **60**(2), 91–110 (2004)
3. Balntas, V., et al.: PN-Net: conjoined triple deep network for learning local image descriptors. arXiv:1601.05030 (2016)
4. Yi, K.M., et al.: Lift: learned invariant feature transform. In: European Conference on Computer Vision. Springer, Cham (2016)
5. DeTone, Da., Malisiewicz, T., Rabinovich, A.: Superpoint: self-supervised interest point detection and description. In: Proceedings of the IEEE Conference on Computer Vision and Pattern Recognition Workshops (2018)
6. Noh, H., et al.: Large-scale image retrieval with attentive deep local features. In: Proceedings of the IEEE International Conference on Computer Vision (2017)
7. Dusmanu, M., et al.: D2-net: A trainable cnn for joint description and detection of local features. In: Proceedings of the IEEE/CVF Conference on Computer Vision and Pattern Recognition (2019)

8. Luo, Z., et al.: Contextdesc: local descriptor augmentation with cross-modality context. In: Proceedings of the IEEE/CVF Conference on Computer Vision and Pattern Recognition (2019)
9. Balntas, V., et al.: Learning local feature descriptors with triplets and shallow convolutional neural networks. In: Bmvc, vol. 1. No. 2 (2016)
10. Barath, D., et al.: MAGSAC++, a fast, reliable and accurate robust estimator. In: Proceedings of the IEEE/CVF Conference on Computer Vision and Pattern Recognition (2020)
11. Ioffe, S., Szegedy, C.: Batch normalization: accelerating deep network training by reducing internal covariate shift. In: International Conference on Machine Learning. PMLR (2015)
12. Schmitt, M., Hughes, L.H., Zhu, X.X.: The SEN1–2 dataset for deep learning in SAR-optical data fusion[J] (2018). arXiv:1807.01569

A Fast Verification Method of Small Building Samples Using Deep Learning

Zhongwu Wang[1], Shucheng You[1], Lei Du[1(✉)], Zhipan Wang[2], Yuhang Gan[1], Zhengyu Luo[1], and Lirong Liu[1]

[1] Land Satellite Remote Sensing Application Center, MNR, Beijing, China
781943277@qq.com
[2] School of Aeronautics and Astronautics, Sun Yat-Sen University, Shenzhen, China

Abstract. Building extraction with deep learning requires a large amount of samples used as train data. Small differences in building features and big differences between building and background features are important for qualified samples. However, in practice, it is difficult to obtain the required massive samples with good quality which restricts the application of building extraction from remote sensing images based on deep learning seriously. To tackle the problem, this paper proposed a fast verification method of small building samples using deep learning extraction with remote sensing images. The main steps include: firstly, combine the morphological building index (MBI) feature with their original bands to be new input images for deep learning network. Secondly, customize network model parameters based on the U-series semantic segmentation framework, use the mini-batch gradient descent method to achieve training parameters and then obtain a suitable deep learning model. Finally, carry out verification of building samples based on model training and accuracy evaluation. Experiments showed that, compared with the traditional U-Net deep learning building sample verification, the proposed method has the advantages of less sample demand, low cost, high accuracy and stable effect.

Keywords: Building sample · Deep learning · Verification · MBI

1 Introduction

Compared with traditional machine learning methods, deep learning technology based on convolutional neural network has improved the building extraction accuracy on remote sensing images significantly [1–4]. However, building extraction with deep learning based on remote sensing images requires a large amount of samples. Building samples should have small differences in building features and big differences between building and background features [5–7]. In practice, it is difficult to obtain required massive samples with good quality which restricts the application of deep learning building extraction from remote sensing images seriously [8–12]. To improve the efficiency and effect of achieving samples, the conventional method is to train the network after a large amount of samples. Firstly, mark a small number of samples. Secondly, train and verify

L. Wang et al. (Eds.): CHREOC 2022, LNEE 969, pp. 341–350, 2023.
https://doi.org/10.1007/978-981-19-8202-6_31

the accuracy quickly and then supplement samples according to verification results, such as supplement positive samples with poor verification and supplement negative samples with false extraction. Thirdly, re-train and re-verify the samples. Finally, repeat the above steps until the sample meets the quantity and quality requirements of the application. Therefore, with a small number of samples, verifying the accuracy for building deep learning extraction is crucial to the efficiency and availability of the whole deep learning samples making.

Deep learning model based on U-series network, such as U-Net, has been a research hotspot of building extraction on remote sensing images because of the advantages of less training samples and faster training efficiency. However, the lack of some types of buildings samples would cause low extraction accuracy, especially on roads, cement pavement and other land types. It weakens the superiority of U-series network greatly which needs relatively few samples. With a small number of samples, this method is difficult to be used for sample data verification and the guidance the sample supplement directly. Therefore, it lacks of suitable methods which can be efficient, high-precision and rapid to the verification of deep learning building samples with remote sensing images when the number of samples is small and building types is not that enough.

2 A Verification Method for Building Samples

2.1 Method

Traditionally, the loop of "sample making-training-verification" technology is used and it is hard to conduct fast verification for a relatively small number of samples with good efficiency and precision. To tackle this problem, this paper proposed a rapid verification method with a small number of samples for building extraction with remote sensing images based on deep learning technology.

The main procedure of this method is shown in Fig. 1, which mainly includes: Firstly, according to image resolution and the average size of the building, the MBI feature is calculated for every image which is used to make the original small building extraction samples. Then, the feature images are added to the corresponding images as separate bands, making new training and testing sample datasets. By adding the training and testing sample datasets of the MBI feature bands, it is beneficial to reduce the internal feature differences in all types of building, enlarge the feature differences between the building and background and help to optimize the parameters of the deep learning semantic segmentation model.

Building samples are obtained from the new synthesized images with a fixed size rectangular window. Calculate the proportion of background pixels in each sample and exclude the sample data as well as their labels with larger proportion than a given threshold. Build the U-series deep learning semantic segmentation model and define a specific network model by setting parameters. The remaining samples are used as the train data of the deep learning network model. The mini-batch gradient descent method is used to train the parameters to obtain the deep learning model of building extraction. Test images are processed to obtain the building extraction results.

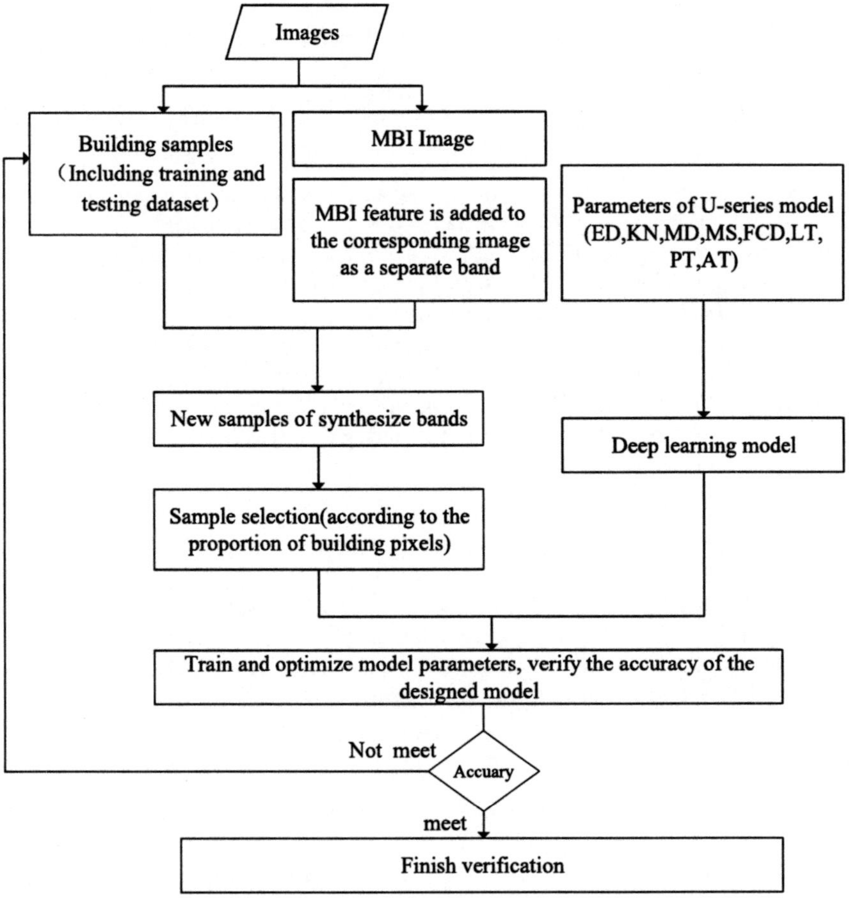

Fig. 1. A fast verification method for building deep learning samples

2.2 Algorithm Procedure

The specific algorithm is as follows:

(1) According to the resolution of images and the average size of buildings, the *MBI* feature TI_{i_MBI} is calculated on each image TI_i (i = 1, 2,…, n, n is the total number of *TI*) in the initial train images (*TI*), and the neighborhood scale is determined according to image resolution. The feature image TI_{i_MBI} is added to the corresponding image TIi as a separate band to synthesize a new train data collection image new $newTI_{i\_} = [TI_{i1}, TI_{i2},\ldots, TI_{ik}, TI_{i\_MBI}]$, *K* is the band number of TI_i, and all new synthesized images $newTI_i$ $(i = 1, 2,\ldots, n)$ are used as a new training sample dataset (*newTI*);

$$\mathrm{TI}_{\mathrm{i\_MBI}} = \frac{\sum_d \sum_s \mathrm{TI\_DMP_i}(d, s)}{D * S} \tag{1}$$

$$\mathrm{TI\_DMP}_{\mathrm{i}}^{(d,s)} = \left|\mathrm{TI}_{\mathrm{i}}^{(d,s)}(b) - \mathrm{TI}_{\mathrm{i}}^{(d,s-\Delta s)}(b)\right| \tag{2}$$

$$\mathrm{S} = \frac{\mathrm{BS}}{\mathrm{PS*2}} \tag{3}$$

(2) For the process image (*PI*), calculate the image *MBI* feature PI_{MBI}. And then the feature image PI_{MBI} is added to the image *PI* as a separate band to synthesize a new process image *newPI* ($newPI = [PI_1, PI_2, \ldots, PI_k, PI_{MBI}]$);
(3) For each train image $newTI_i$ (i = 1, 2,..., n), the corresponding train image TI_Label_i is made by labeling the building samples. Complete contour of building areas needs to be marked, the gray value within the label range is filled with 1 and the gray value outside the label range is filled with 0. All train images TI_Label_i constitute the train image collection dataset *TI_Label*;
(4) Each pair of $newTI_i$ and $\mathrm{TI\_}Label_i$ in *newTI* and *TI_Label* are made by a fixed size *TS* * *TS* rectangular window, and all the cutting results are used as the train dataset *TD* and label dataset LD;

$$\mathrm{TS} = \mathrm{INT}\left(\sqrt{\frac{\mathrm{MGB*1024*1024*1024*8}}{12*(\mathrm{K}+1)*\mathrm{GR}}}/256\right)*256 \tag{4}$$

(5) For each pair of TD_j and LD_j (*j* = *1, 2,..., l, l is the total number of sample dataset*) in train dataset *TD* and label dataset *LD*, the proportion of background pixels in each LD_j sample to the total number of pixels P_j is calculated. Delete the TD_j and LD_j, if $P_j > 0.95$, otherwise keep the sample pair;

$$\mathrm{P_j} = \frac{SP\_Building}{\mathrm{TS*TS}} \tag{5}$$

(6) U-series semantic segmentation deep learning model is constructed, which consists of coding, multi-scale feature construction, skipping, decoding, feature synthesis, classification and other modules. The specific U-series model is defined by eight parameters, including encode depth (*ED*), convolution kernel number (*KN*), multi-Level depth (*MD*), multi-level stride (*MS*), feature concatenation domain (*FCD*), loss type (*LT*), pooling type (*PT*) and activity type (*AT*). All the reserved samples are input into the network model for parameter training to obtain the deep learning model for building extraction, and the new process image *newPI* is processed to obtain the verification results of the building samples.
(7) The deep learning model for building extraction is obtained by all the reserved samples and the mini-batch gradient descent method is used to optimize the parameters. The *newPI* is processed to obtain the verification results of the building samples. The mini-batch gradient descent parameter optimization is to obtain the local optimal value of the parameters by gradient descent with batch small samples.

3 Experiment and Analysis

3.1 Experimental Data

Experiments are carried out on the aerial image datasets, including three-band true color remote sensing images with a spatial resolution of 0.3 meters, 405 km^2 of images, and their ground truth data for testing (https://project.inria.fr/aerialimagelabeling/ [13]). There are five areas such as Austin, Chicago, Vienna, etc., which include densely built cities and sparsely built suburbs, and each area has 36 images with 5,000*5,000 pixels. Among the 180 images, 80 images are used as test images, and their corresponding labels are used as ground truth to test the accuracy of building samples. From the remaining 100 images, 15, 50, 100 images and their corresponding labels are selected separately as the train data, and the remaining samples are taken as the case of few samples by processing samples such as cropping and culling.

The hardware device is a NVIDIA P4000 8GB GPU graphics card (Table 1).

Table 1. Training data and testing data list for samples

NO	Training data		Testing data	
	Quantity/frame	Image dataset name and frame quantity	Quantity/frame	Image dataset name and frame quantity
1	15	Austin1–3 (3 frame) Chicago1–3 (3 frame) Kitsap1–3 (3 frame) Tyrol-w1–3 (3 frame) Vienna1–3 (3 frame)	80	Austin21–36 (16 frame) Chicago21–36 (16 frame) Kitsap21–36 (16 frame) Tyrol-w21–36 (16 frame) Vienna21–36 (16 frame)
2	50	Austin1–10 (10 frame) Chicago1–10 (10 frame) Kitsap1–10 (10 frame) Tyrol-w1–10 (10 frame) Vienna1–10 (10 frame)	80	Austin21–36 (16 frame) Chicago21–36 (16 frame) Kitsap21–36 (16 frame) Tyrol-w21–36 (16 frame) Vienna21–36 (16 frame)
3	100	Austin1–20 (20 frame) Chicago1–20 (20 frame) Kitsap1–20 (20 frame) Tyrol-w1–20 (20 frame) Vienna1–20 (20 frame)	80	Austin21–36 (16 frame) Chicago21–36 (16 frame) Kitsap21–36 (16 frame) Tyrol-w21–36 (16 frame) Vienna21–36 (16 frame)

3.2 Effects and Analysis

Through the technical method described in Chap. 2, the images and their corresponding MBI features of the three sample acquisition data such as Austin1, Chicago1, and Vienna1 are calculated. Taking Austin1 as an example, the origninal image and its corresponding MBI features are shown in Fig. 2.

For the test image, *MBI* characteristic *PIMBI* of the image is calculated, and the characteristic image *PIMBI* is added to the image *PI* as a separate band to synthesize a new test image *newPI*, as shown in Fig. 3.

Label each image in the new training sample collection dataset to generate the corresponding label image. When labeling, mark the complete outline range of the building. The gray value in the label range is filled with 1 while the gray value outside is filled with 0 and all the label images TI_Label_i form the sample label collection data set, as shown in Fig. 4.

Use a rectangular window with a fixed size of 256*256 to cut each pair of sample collection data and label collection data, and use all the cutting results as the training sample dataset and label dataset. Then calculate the proportion P_j of background pixels to all pixels in each sample. If $P_j > 0.95$, delete the sample pair, otherwise keep the sample pair.

Construct a U-series semantic segmentation deep learning model framework composed of encoding, multi-scale feature construction, skip connection, decoding, feature synthesis, classification, and other modules, and use eight parameters such as encoding depth (*ED*), number of convolution kernels (*KN*), multiple-scale depth (*MD*), multi-scale step size (*MS*), feature synthesis domain size (*FCD*), loss type (*LT*), pooling type (*PT*), and activation type (*AT*) to define the specific U-series model. In this example, *ED* is 4, *KN* is 64, *MD* is 1, *MS* is 1, *FCD* is 5, *LT* is *CELoss*, *PT* is *MaxPool*, and *AT* is *Relu*.

The remaining samples are used for parameter training of the input U-series deep learning network model. The maximum number of iterations is 5, the batch sample size is 12, and the learning rate is 0.01. The deep learning model for building extraction is obtained, and the new test image $newPI_i$ is tested. After processing, the verification result of the building extraction sample data of each image is obtained, as shown in Figs. 5, 6, and 7.

Fig. 2. Image (R, G, B) and its corresponding *MBI* feature (after histogram equalization) in Austin1 for traindata collection

Fig. 3. Image (R, G, B) and its new image with added *MBI* feature band (R, G, MBI) in Austin1 for test data collection

Fig. 4. Label image with Austin1 (left) and Austin21 image (right)

In addition, the accuracy of proposed method and traditional U-Net model are compared and the results are shown in Table 2.

4 Conclusion

To effectively represent the differences between buildings and background objects, the addition of the MBI is beneficial to reducing the internal feature differences of buildings, increasing the feature differences between buildings and backgrounds and conducive to the parameter optimization of deep learning semantic segmentation models. Meanwhile, through the design of the semantic segmentation model framework, the specific U-series network can be customized easily as needed, and in the case of a small number of samples, it can also be used to guarantee the extraction accuracy of deep learning so as to realize

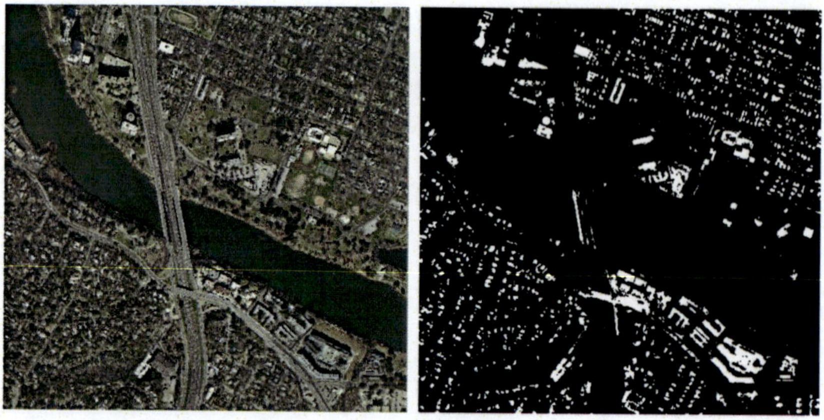

Fig. 5. Verification result of building samples with Austin21 test image

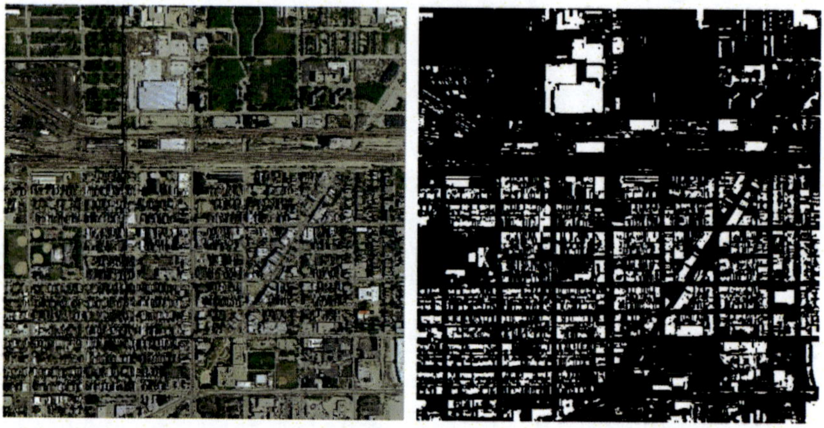

Fig. 6. Verification result of building samples with Chicago21 test image

the rapid verification of high-efficiency and high-precision building sample and provide real and effective reference information for subsequent sample supplements.

Compared with other existing verification methods of deep learning building samples, the method proposed in this paper has the following advantages:

(1) Low cost: U-series neural network deep learning model is used, which adopts a framework design to facilitate the customization of specific networks. It requires few samples, the workload of manually labeling samples is less and the cost is low. It is suitable for rapid accuracy verification of building samples.
(2) High accuracy: MBI feature can narrow the feature differences among different buildings and expand the feature differences between buildings and backgrounds. In the case that the sample is small and cannot cover all building types, the verification

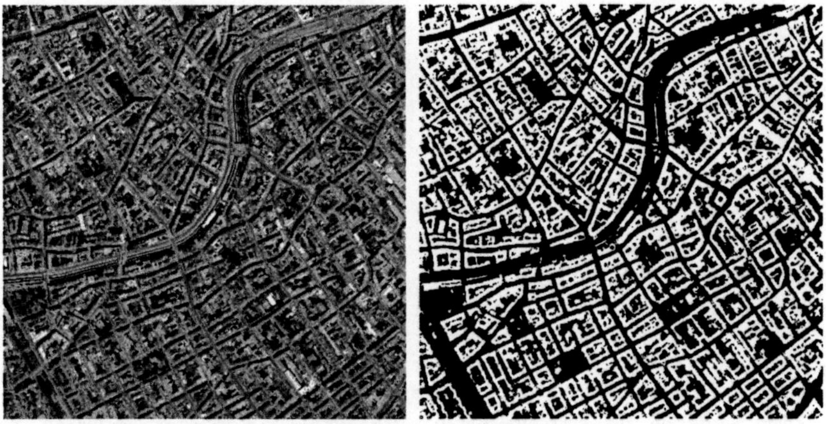

Fig. 7. Verification result of building samples with Vienna21 test image

Table 2. Accuracy comparison between the proposed method and U-Net

Images number used in sample collection	Sample number before selecting	Sample number after selecting	Sample size (pixel)	Method	IOU
15	5,415	3,349	256*256	Proposed method	0.5441
			256*256	U-Net	0.5225
50	18,050	9,803	256*256	Proposed method	0.5814
			256*256	U-Net	0.4395
100	36,100	20,275	256*256	Proposed method	0.6549
			256*256	U-Net	0.6471

accuracy is 0.54–0.65, which is 2–15% higher than the absolute accuracy of the traditional U-Net method.

(3) Stable effect: When 3,349, 9,803 and 20,275 valid samples are used, the accuracy of the proposed method are 0.5441, 0.5814 and 0.6549 respectively. With the increase of the number of samples, the accuracy has been improved steadily which can illustrate the availability of the samples better and truly play the role of sample verification and provide reference for subsequent sample supplements. For the traditional U-Net method, the accuracy is 0.5225, 0.4395 and 0.6471 respectively which are rather unstable (sometimes higher or lower) and cannot meet the need of rapid sample verification.

References

1. Vakalopoulou, M., et al.: Building detection in very high resolution multispectral data with deep learning features. In: 2015 IEEE International Geoscience and Remote Sensing Symposium (IGARSS). IEEE (2015)
2. Yuan, J.: Automatic building extraction in aerial scenes using convolutional networks (2016). arXiv preprint arXiv:1602.06564
3. Guangming, W., Qi, C., et al.: High precision building detection from aerial imagery using a U-Net like convolutional architecture. Acta Geod. Cartogr. Sin. **47**(6), 864–872 (2018)
4. Wang, M., Yuan, S., Pan, J.: Building detection in high resolution satellite urban image using segmentation, corner detection combined with adaptive windowed Hough transform. In: 2013 IEEE International Geoscience and Remote Sensing Symposium, IGARSS. IEEE (2013)
5. Yang, H., et al.: Building extraction in very high resolution imagery by dense-attention networks. Remote Sens. **10**(11), 1768 (2018)
6. Yuan, J.: Learning building extraction in aerial scenes with convolutional networks. IEEE Trans. Pattern Anal. Mach. Intell. **40**(11), 2793–2798 (2017)
7. Ji, S., Wei, S., Meng, L.: A scale robust convolutional neural network for automatic building extraction from aerial and satellite imagery. Int. J. Remote Sens. **40**(9), 3308–3322 (2019)
8. Zhang, L., et al.: An efficient building extraction method from high spatial resolution remote sensing images based on improved mask R-CNN. Sensors **20**(5), 1465 (2020)
9. Lin, X., Zhang, J.: Object-based morphological building index for building extraction from high resolution remote sensing imagery
10. Guo, Z., Shao, X., Xu, Y., et al.: Identification of village building via Google Earth images and supervised machine learning methods. Remote Sens. **8**(4), 271 (2016)
11. Long, J., Shelhamer, E., Darrell, T.: Fully convolutional networks for semantic segmentation. In: Proceedings of the IEEE Conference on Computer Vision and Pattern Recognition, pp. 3431–3440. IEEE, Boston, MA (2015)
12. Maggiori, E., Tarabalka, Y., Charpiat, G., et al.: Convolutional neural networks for large-scale remote-sensing image classification. IEEE Trans. Geosci. Remote Sens. **55**(2), 645–657 (2017)
13. https://project.inria.fr/aerialimagelabeling

Template-Based Universal Adversarial Perturbation for SAR Target Classification

Junfan Zhou, Hao Sun(✉), and Gangyao Kuang

College of Electronic Science, National University of Defense Technology, Changsha 410072, Hunan, China
sunhao@nudt.edu.cn

Abstract. With deep learning models been widely applied to SAR images interpretation, the adversarial robustness problem of models has drawn much attention. Due to the application in military fields, SAR image processing have extremely requirements on security. However, the popular optimized adversarial perturbation algorithms calculated the perturbation for one single image at a time, thus cannot be adapted to real-time adversarial scenarios. Also, the perturbation is designed in full digital space and lacks of sematic information of the targets. Faced with those problems, this paper proposes a SAR universal adversarial perturbation (SAR-UAP) for SAR images adversarial attacks based on template-based lightweight generative model without discriminator. The model has simple structure which consists of six convolutional layers and uses the fixed template as the input of the generator. Template-based UAP ensures that once training is finished, the disturbances remain unchanged in spite of the changing of images chosen to attack. Model performance was tested on the MSTAR dataset for eight classical CNNs. Compared with other GAN-based UAP methods, SAR-UAP gained slightly higher fooling rate which is over 85%, while the illustration showed SAR-UAP are much more target-focused. The transferability of perturbation was also verified.

Keywords: SAR · Adversarial attack · Universal adversarial perturbation

1 Introduction

Deep learning techniques have been developed rapidly in Synthetic Aperture Radar (SAR) image interpretation recently, but it has been pointed out that there exists serious adversarial robustness problem in deep learning models especially for CNN models of classification tasks [1]. The phenomenon of adversarial robustness refers to the fact that constructed but small perturbations added to the benign image can cause the change of result for high-confidence classifier [2]. SAR is widely used in military applications which have extremely requirements for security [3]. Unquestionably, the phenomenon also needs to be taken into consideration for SAR image processing methods.

Subsequently, numerous attack methods were proposed to prove the existence of the phenomenon and have shown high effectiveness. But most of methods hardly can be used in real scenarios because the dependences on input images. Goodfellow et al. [4]

L. Wang et al. (Eds.): CHREOC 2022, LNEE 969, pp. 351–360, 2023.
https://doi.org/10.1007/978-981-19-8202-6_32

first used gradients computed with backpropagation to calculate the noises added to the images separately. Later, PGD [5] was proposed to add a step of initializing the uniform random perturbation, then it runs more iterations which improves the imperceptibility and transferability for perturbed images. Another attack method [6] called C&W perturbs the sample in the direction of the target class by minimizing the distance from the original sample. However, these classical methods mentioned above are all follow the procedure that the perturbations are calculated according to the correlations between the dataset and targeted models. The procedure guarantees the noises is datapoint specific perturbation that are sought to fool a specific image, so it is time costly and not versatile. Due to this characteristic, it becomes impossible for SAR jamming to be achieved in real-world when faced with detectors [7].

Another kind of perturbation which can fool most images from a data distribution can adapt to the real-time constant scene better and it is more suitable for SAR images processing needs. Universal adversarial perturbation (UAP) generates noises only from a sampled vector and doesn't need extra computation. Moosavi-Dezfooli et al. [8] first introduces the concept of UAP into adversarial robustness problem. The method UAP is still an optimized problem which follows the max-norm constraint and achieve at least a fixed amount of fooling rate. Later, Fast Feature Fool [9] was proposed as a novel data-independent approach to generate image-agnostic perturbations for a range of CNNs trained for object recognition. Meanwhile, generative adversarial network is used for learning the distribution of potential disturbances for the classifier [10, 11]. Although UAPs can generate noises for a batch of images at once, for SAR images which can be separated into the target, shadow and speckle noises, the disturbed area of fake images is lack of semantic information and have shown less natural to humans [12]. Structure-preserving attack methods cost more time because the additional modules concerned with semantic information [13]. Additional research is needed to evaluate how to balance the contradiction between structure-consistency and time cost of the perturbations based on UAPs for SAR images.

In this study, we focus on the problem of proposing data-free adversarial attack methods with better for SAR image classifiers. The fooling rate and transferability of our method are verified using MSTAR datasets, which shows great application prospect. Also, the physical interpretability has been improved significantly.

2 Methodology

This paper uses discriminator-free generative adversarial network to learn the distribution of universal noises for the target network. In order to improve perturbation interpretability, the input of the generator is fixed according to the dataset. The overall process of the perturbation generation (see Fig. 1).

2.1 Input Template

Instead of choosing a random normalized vector as the input, input template driven by the dataset was used for injecting prior information of SAR data into the generalization. Simply, the input template was calculated by averaging the sum of all the images in

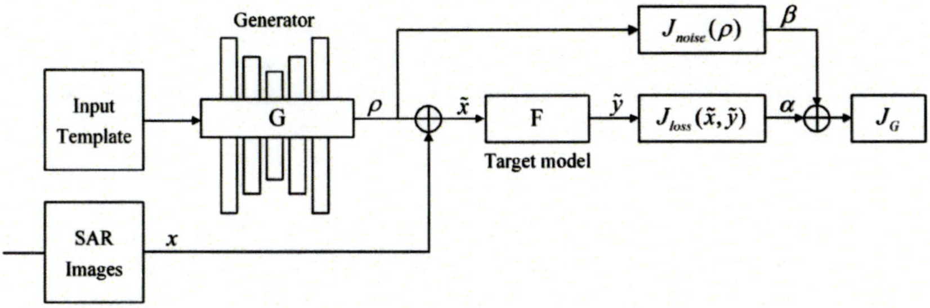

Fig. 1. Overview of the template-based UAP for SAR target classification.

dataset. The input template was gotten as the Fig. 2 shown. The brighter part of the template refers to the area where the target is usually located, while the lower brightness part refers to the area of the sample where the shadow is located. Although the relative positions of shadows and targets change with the azimuth of the SAR target, it can be adjusted in training process.

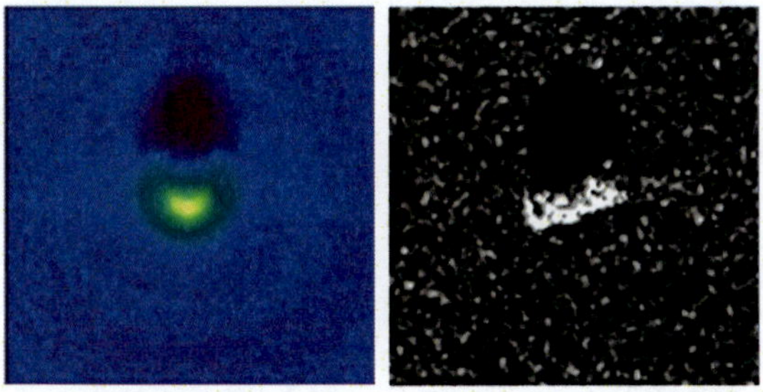

Fig. 2. The input template for MSTAR dataset and a BMP2 sample.

2.2 GAN Architecture

In the implementation of generator, it adopts a combined structure of encoder, decoder and residual module, and the effectiveness of this structure has been verified in image conversion researches. The architecture of the generator is shown as Table 1.

Conv2d represents a 2D convolutional layer, which uses a convolution kernel of size k to perform a convolution operation on the input 3D tensor, and the output channel is a tensor of o. BatchNorm2d represents the batch normalization method. Since the convolution layer has already extracted the features, in order to speed up the training speed and stability, the input tensors are normalized by batches. ReLU is a common activation function that truncates values less than zero. Among them, ReflectionPad2d is called mirror padding. It not only uses zero as the padding value, but copies the values around the width and height of the tensor to pad to the excess width or high school, which can obtain better convolution effect. The main function of the decoding module is to restore the image according to the extracted features. ConvTranspose2d represents the deconvolution operation, and its function is to up-sample the feature matrix. Tanh is an activation function, which is a nonlinear function compared to ReLU, and the output will be smoother. The loss function consists of two parts including C&W loss and perturbation loss.

Table 1. Structure of generator network

Generator		ReflectionPad2d(3,3,3,3)
	Encoder	Conv2d(1,64,7,1), BN, Relu
		Conv2d(64,128,3,2,1), BN, Relu
		Conv2d(128,256,3,2,1), BN, Relu
	Decoder	Convtranspose2d(256,128,3,2,1), BN, Relu
		Convtranspose2d(128,64,3,2,1), BN, Relu
		ReflectionPad2d(3,3,3,3)
		Conv2d(64,1,7,1), Tanh

3 Experiments and Analysis

3.1 Experimental Data and Classifier

In this section, the commonly used public MSTAR dataset is applied to test the proposed method's performance. There are ten types of vehicle targets in the dataset, including 2S1, BMP2, BRDM2, BTR60, BTR70, D7, T62, T72, ZIL131, ZSU234, which are labelled from zero to nine in one-hot vector forms. Also, in order to extensively evaluate the proposed model, eight classical classifiers were chosen. Among them, MobileNet and SqueezeNet represent the lightweight convolutional neural network that may be deployed on airborne edge devices. The train and test accuracy of the eight classifiers are listed in Table 2.

3.2 Fooling Rate and Transferability Test

Table 3 shows the fooling rates of UAP-SAR, NAG and UAN under eight networks. Here, the fooling rate is defined as the proportion of samples with attacking successfully among all the correctly classified sample. UAP-SAR, UAN and NAG are all GAN-based UAP methods. To the contrary of UAP-SAR and UAN, NAG contains a discriminator. The random vector is served as the input of generator in NAG which will be down-sampled to obtain the perturbed images. NAG's function contains two parts, one is the confrontation loss which is calculated using the output of softmax function, another is the diversity loss for shuffling the order in a batch of samples. The UAN feeds random noise into the generator, scales it to the generator and then clips it. The difference between the proposed method and UAN lies in the input vector and the loss function.

Table 2. Training and testing accuracy of SAR target classifiers

Model	Training accuracy (%)	Testing accuracy (%)
AlexNet	100	97.98
ResNet18	100	99.59
ResNet50	100	98.35
ResNet101	99.82	98.89
DenseNet	100	98.60
MobileNet	99.96	99.01
SqueezeNet	100	98.35
VGG16	100	98.68

From Table 3, it can be seen that the proposed method achieves higher fooling rate on most of the models, or performs as the same as the other two comparative methods in only needs 10 epochs training. Consequently, the effectiveness of the methods has been verified.

Several works have indicated that the adversarial examples generated for one model may also serves as the adversarial input for other models which is called transferability. Table 4 demonstrates the transferability across eight models. The first line corresponds to the model used for training generators while the first row corresponds to the model tested on. It shows that half of fooling rates can reach 75% or higher. 93.75% of fooling rates are above 50%. This is expected as the perturbation area is concerned with the input sample semantic features.

3.3 UAPs Illustration

In order to illustrate the characteristics of design perturbation more intuitively, this section will illustrate the effectiveness of the designed method for guiding and enhancing perturbation's semantic information, and also compare the experimental results with the results of methods mentioned in the previous section to further emphasize the advantages of proposed method.

Figure 3 shows the results of the MSTAR data on the test models. As the input template is fixed, for the same training model, the disturbance generated by the generator is constant and does not change with the change of the test data. The first column of Fig. 3 represents the perturbations generated under the corresponding model. It can be seen that the noises are all concentrated in the target center or shadow area of the sample. Taking the adversarial examples of SqueezeNet as an example, the superimposition of strong brightness disturbance on the target may change the attribute scattering center of the sample. Further research can be carried out on how to change the physical properties of the SAR target by adding noise in digital fields.

Compared with SAR-UAP, the noise texture generated by NAG is obvious, which affects the samples increasingly. The noise intensity of NAG is controlled by parameters. There exists irreconcilable contradiction between perturbation invisibility and fooling

Table 3. The fooling rates of three GAN-based UAP on MSTAR

	AlexNet	DenseNet121	MobileNet	ResNet18	ResNet50	ResNet101	SqueezeNet	VGG16
UAP-SAR	89.73	87.18	91.46	88.82	91.34	89.77	85.86	86.31
NAG	85.15	90.02	87.92	88.41	88.25	86.06	87.67	84.41
UAN	85.32	87.05	86.10	84.65	93.65	89.03	90.14	89.28

Table 4. Transferability of non-targeted UAP-SAR on MSTAR

	AlexNet	DenseNet121	MobileNet	ResNet18	ResNet50	ResNet101	SqueezeNet	VGG16
AlexNet	89.73	84.08	91.55	88.78	76.87	90.14	84.74	91.30
DenseNet121	37.94	87.18	35.71	16.58	57.44	61.65	56.08	58.76
MobileNet	60.00	78.80	91.46	44.82	64.16	62.76	62.14	76.08
ResNet18	80.91	86.64	83.75	88.82	84.58	84.78	72.04	74.80
ResNet50	53.69	63.96	48.00	20.62	91.34	53.11	53.90	64.87
ResNet101	69.48	58.60	62.80	70.85	65.73	89.77	68.70	75.67
SqueezeNet	86.72	58.43	58.76	50.64	68.37	57.69	85.86	84.45
VGG16	84.16	83.75	80.37	74.85	78.93	79.92	79.71	86.31

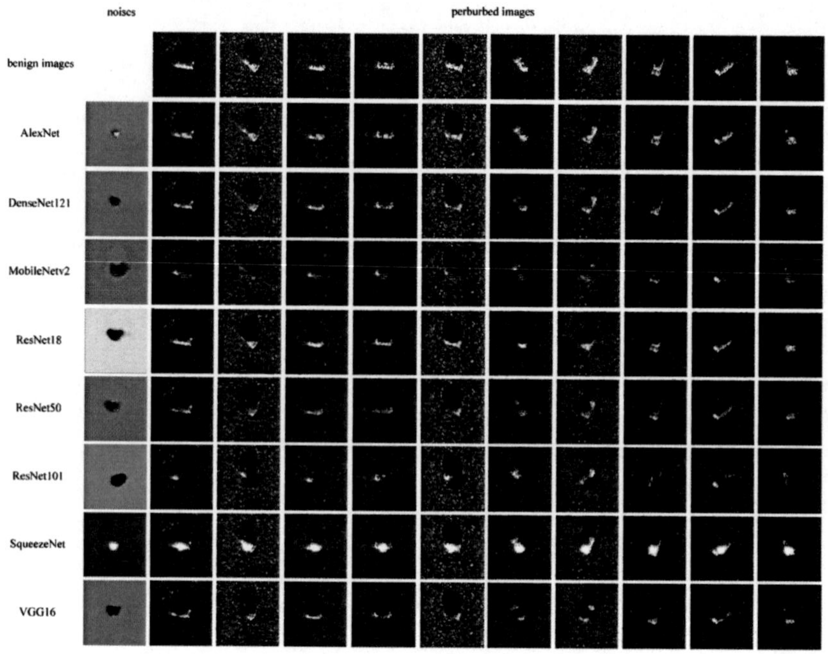

Fig. 3. The results for UAP-SAR on MSTAR samples.

rate. In order to illustrate the prominent feature of the proposed method of giving attention on perturbed regions, the experiment improves the fooling rate of the NAG method by reducing the invisibility of perturbed samples. The results of SAR-UAP-NAG and UAP are illustrated in Fig. 4.

4 Summary

This paper proposes a template-based lightweight generative model without discriminator for SAR image adversarial attacks. The model is simple in structure and only contains 6 layers of convolutional structure. It greatly reduces the amount of calculation and training difficulty, and can also be trained on devices without GPU. The template-based UAP ensures that once the training is completed, the subsequent disturbances can remain unchanged, which is suitable for real-time attack on classification. This design makes generator pay more attention on the target area. With the fooling rate being consistent with other GAN-based UAP methods, the samples generated by the method are more interpretable.

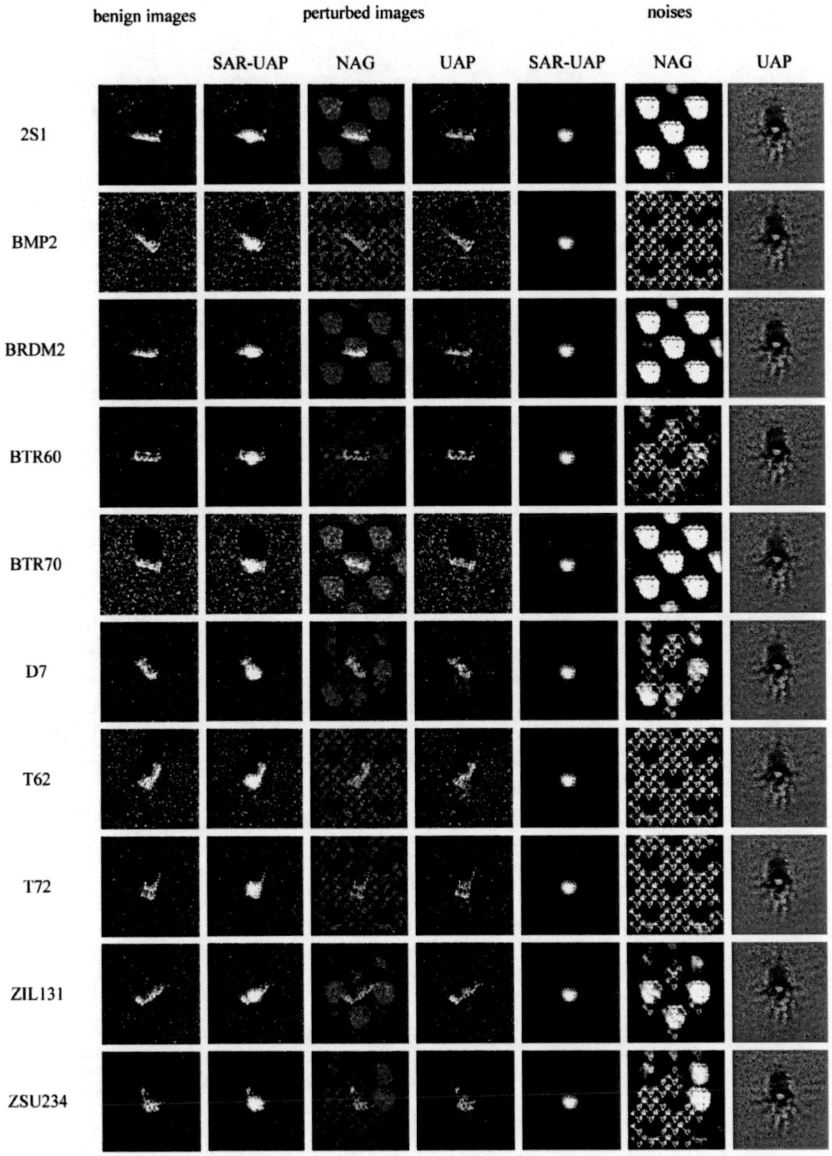

Fig. 4. The results of three GAN-based UAP of MSTAR samples on SqueezeNet.

Acknowledgements. This work was supported by National Natural Science Foundation of China (Project Nos. 61971426).

References

1. Li, H., Huang, H., Chen, L., et al.: Adversarial examples for CNN-based SAR image classification: an experience study. IEEE J. Sel. Top. Appl. Earth Obs. Remote. Sens. **14**, 1333–1347 (2020)
2. Fan, Y., Wu, B., Li, T., et al.: Sparse adversarial attack via perturbation factorization. In: European Conference on Computer Vision, pp. 35–50. Springer, Cham (2020)
3. Xu, Y., Sun, H., Chen, J., et al.: Adversarial self-supervised learning for robust SAR target recognition. Remote Sens. **13**, 4158 (2021)
4. Goodfellow, I., Shlens, J., Szegedy, C., et al.: Explaining and harnessing adversarial examples (2014). arXiv preprint arXiv:1412.6572
5. Madry, A., Makelov, A., Schmidt, L., et al.: Towards deep learning models resistant to adversarial attacks. In: International Conference on Learning Representations (ICLR) (2018)
6. Carlini, N., Wagner, D.: Towards evaluating the robustness of neural networks. In: IEEE Symposium on Security and Privacy, pp. 39–57 (2017)
7. Huang, T., Zhang, Q., Liu, J., et al.: Adversarial attacks on deep-learning-based SAR image target recognition. J. Netw. Comput. Appl. **162**, 102632 (2020)
8. Moosavi-Dezfooli, S.-M., Fawzi, A., Fawzi, O., et al.: Universal adversarial perturbations. In: 2017 IEEE Conference on Computer Vision and Pattern Recognition (CVPR), pp. 86–94 (2016)
9. Mopuri, K.R., Garg, U., Venkatesh Babu, R.: Fast feature fool: a data independent approach to universal adversarial perturbations. CoRR (2017). https://arxiv.org/abs/1707.05572
10. Khrulkov, V., Oseledets, I.: Art of singular vectors and universal adversarial perturbations. In: The IEEE Conference on Computer Vision and Pattern Recognition (CVPR), June 2018
11. Mopuri, K.R., Ojha, U., Garg, U., Venkatesh Babu, R.: Nag: network for adversary generation. In: The IEEE Conference on Computer Vision and Pattern Recognition (CVPR), June 2018
12. Du, C., Zhang, L.: Adversarial attack for SAR target recognition based on UNet-generative adversarial network. Remote Sens. **13**, 4358 (2021)
13. Peng, D., Zheng, Z., Zhang, X.: Structure-preserving transformation: generating diverse and transferable adversarial examples. In: The AAAI Workshop on Artificial Intelligence for Cyber Security, pp. 1–8 (2019)

Semantics-Based Analysis for Multi-source Remote Sensing Image Retrieval

Xiaolu Su[1,3], Lilu Zhu[1,3](✉), Yang Wang[2], and Yanfeng Hu[2,3]

[1] Suzhou Aerospace Information Research Institute, Suzhou, China
zhull@aircas.ac.cn
[2] Aerospace Information Research Institute, Chinese Academy of Sciences, Beijing, China
[3] Suzhou Key Laboratory of Aerospace Big Data Intelligent Application Technology, Suzhou, China

Abstract. With the continuous increasing diversity and volume of the remote sensing data, the urgent need of efficient approaches for multi-source remote sensing data retrieval has attracted a lot of attention in the remote sensing field. However, there are limitations to existing remote sensing image retrieval approaches, such as high use threshold caused by excessive reliance on users' domain knowledge, and imperfect multi-dimensional retrieval ability caused by insufficient data feature utilization. To address these problems, we propose an approach of semantics-based analysis for multi-source remote sensing image retrieval (SBARSIR), to make multi-source remote sensing data more accessible. In more detail, a convolutional neural network (CNN) is exploited for semantic feature extraction, and the global and partial classification results are integrated to improve the accuracy on remote sensing images with differences resolutions. Then, we cross-encode the visual semantic, temporal, and spatial features, and establish the mapping between features and indexing structures to provide retrieval capabilities. Extensive experiments have been carried out on several typical public remote sensing data sets to validate the effectiveness of the SBARSIR proposed. It turns out that the SBARSIR proposed outperforms the main stream approaches under the examined conditions.

Keywords: Multi-source remote sensing image · Remote sensing image retrieval · Visual semantics analysis · Spatio-temporal semantics analysis · Multi-dimensional semantic indexing

1 Introduction

The continuous development of earth observation technology has profound impacts on remote sensing (RS) data, leading to the more abundant RS sources and RS types, the higher spatial and temporal RS data resolutions and increasing RS data volumes [1]. There is no doubt that we have entered an era of remote sensing big data [2], and opportunities and challenges are coming at the same time. The leap of computer technology and communication technology, and the gradual improvement of network infrastructure have greatly broadened the application scope of remote sensing data,

L. Wang et al. (Eds.): CHREOC 2022, LNEE 969, pp. 361–379, 2023.
https://doi.org/10.1007/978-981-19-8202-6_33

including smart city, disaster warning, geological survey, emergency monitoring, etc. As the importance of RS data has become illuminated, it play an important role in the business of the national government and industry sectors [3]. However, these massive and large-scale multi-source remote sensing data pose a substantial challenge to remote sensing data management. Driven by the urgent demand for automatic and effective management of remote sensing data [4], more and more researchers show their interests in remote sensing image retrieval, which aims to find remote sensing images that satisfy the query conditions.

There are generally two different kinds of approaches are always employed to achieve the aforementioned goal metadata-base remote sensing image retrieval (MBRSIR) and content-based remote sensing image retrieval (CBRSIR). In the former one, metadata consist of shooting time, geographic locations, frame of reference, sensor types, waveband information, etc. [2], is always required. The CBRSIR, is born out of image retrieval technology in computer vision field. It aims to analyze and extract the underlying features, and to search through similar remote sensing images in terms of visual content. However, the retrieval capabilities of above approaches have inherent limitations. The metadata-based retrieval implements retrieval capabilities by filtering attribute values such as spatial region, data type, sensor type, etc. Hence, the setting of retrieval conditions is highly dependent on the user's professional knowledge, and can directly affect the performance outcomes. Moreover, due to the strong reliance on artificial generation, it is no doubt that the metadata-base retrieval is time consuming, and will always facing the problem of metadata sparse when the volume of remote sensing images increases. The CBRSIR provides retrieval capabilities based on visual semantic similarity. Yet remote sensing images usually have inseparable temporal and spatial features, which have significant roles in retrieval and analysis of remote sensing image.

In the light of the existing dilemma pointed out above, we put forward a semantics-based analysis approach for multi-source remote sensing image retrieval (SBARSIR). Specifically, the SBARSIR proposed consists of three main parts, visual semantic analysis module, spatio-temporal semantic analysis module and multi-dimensional semantic indexing module. First, visual semantics analysis is conducted on remote sensing images using a convolutional neural network (CNN) based on DenseNet. Considering that the features of the target are usually different in remote sensing images with various resolutions, we integrate the classification results on global image and different image blocks by voting to improve accuracy levels. Then, spatial features and temporal features are extracted from metadata for spatio-temporal semantic analysis, and output the spatio-temporal semantic feature representation of an input remote sensing image. Finally, visual and spatio-temporal semantic feature representation are used to generate indexing structure by means of interleaf coding to provide powerful retrieval capabilities.

The rest of this article is organized as follows. In Sect. 2, we briefly survey related work. In Sect. 3, we describe the SBARSIR proposed in detail. The comprehensive experimental results and analysis are presented in Sect. 4. Finally, we conclude this study and look forward to the follow-up work in Sect. 5.

2 Related Work

This section mainly introduces related work from two aspects, remote sensing image feature extraction, and remote sensing image indexing.

2.1 Remote Sensing Image Feature Extraction

Before the advent of deep learning technology, researchers tended to use artificially designed features to classify remote sensing images. Artificially designed features are usually local or global features designed by professional knowledge. Among them, the low-level features mainly include color histogram [7], color moments [8], GIST [9], etc. These features are simple and easy to implement, but cannot effectively express the high-level semantic information of the image, and the classification accuracy is low. The middle-level features represented by BOVW [10] are obtained by integrating the bottom-level features, which can establish the connection between the low-level visual features and the high-level semantic features, and improve the classification effect of the image.

Since the advent of deep learning technology, the convolutional neural network (CNN) model [11] has gained good results in remote sensing image classification tasks. The CNN model is usually stacked with multiple convolutional layers and pooling layers, and the number of layers of the network is constantly deepening. In the CNN model, the process is regarded as the transformation process from the low-level visual information to the high-level semantic information. Therefore, researchers use the CNN model for high-level feature extraction for image retrieval, avoiding complicated artificial feature design work [11]. Compared with local features, the global features extracted by these CNN model-based methods are more suitable for image retrieval tasks. As a result, feature extraction has gradually transitioned from the way of manually extracting features in the past to using deep neural network technology to automatically learn image features [5]. In [12], the research is conducted on how to extract deep feature representations for high-resolution remote sensing image retrieval (HRRSIR) based on convolutional neural networks; In [13] researchers obtain the deep, compact codes of remote sensing images in several different ways, and introduce PCA to compress the high-dimensional feature learned by traditional DCNNs. In [14], researchers propose a retrieval method based on weighted distance and basic features of CNN, which uses the CNN feature of the image combined with the weight of each image class to calculate the distance between the target and the retrieved image. Nevertheless, these feature extraction methods only consider the extraction of visual features of the data. In addition to semantic features, remote sensing image has corresponding time and space characteristics in the process of shooting and production. It is a description method of remote sensing images from different dimensions. We can measure the correlation by analyzing the internal relations of remote sensing images in different dimensions.

2.2 Remote Sensing Image Indexing

Early researches on various index structures mainly focus on red-black trees, B-trees, hashes, and Bloom filters. Nowadays, indexes of commercial databases occupy 55% of server memory [15, 16]. As the data size increases, the main memory cannot accommodate huge data indexes, and some indexes need to be placed in the external memory, which leads to increases in the number of I/O operations [15]. In addition, the traditional index is designed according to general data structures and does not make use of the data distribution rules in the real world. In the field of image retrieval, the nearest neighbor search technology and the index technology represented by the tree-based structure, which was applied to small and medium-sized databases in the past, have been gradually replaced by the approximate nearest neighbor search method represented by the high-dimensional hash index method. The hashing method is to map the high-dimensional features of the image into the two-dimensional space, generating low-dimensional coding to avoid the "dimension disaster". It uses binary pattern matching or Hamming distance measurement to realize fast search of image coding features and improve retrieval efficiency. With the development of CNN technology, researchers have merged hashing methods with CNN to produce an image retrieval technology based on deep hashing [17, 18]. Learning to hash maps data into the form of binary strings through a machine learning mechanism, to make the hash code maintain the neighbor relationship in the original space as much as possible, reduce data storage and communication overhead, thereby effectively improving the efficiency of data retrieval. Hashing techniques have proven their potential in alleviating the above limitations in the remote sensing field [19]. The hashing-based retrieval method [19] aims to project the original data into a set of compact hash codes to use the fast calculation of Hamming distance to carry out the content-based image retrieval process in a very effective manner. In other words, the input query (along with the image archive) is mapped to a low-dimensional space, and then a bitwise XOR operation can be used to efficiently calculate the ranking process. This method avoids the detailed comparison between the query image and each sample in the database by using approximate nearest neighbor ranking, which can usually achieve good performance in actual remote sensing applications [15]. Besides, it also provides important storage space savings benefiting from the binary compressed representation, showing huge advantages in large-scale RS scenarios. All of the above studies have provided us with new inspirations for designing the indexing scheme.

3 Proposed Approach

As shown in Fig. 1, the overall structure of the SBARSIR consists of three main parts, visual semantic analysis module, spatio-temporal semantic analysis module and multi-dimensional semantic indexing module. To clearly describe the implement of the SBARSIR, we will introduce the approach in detail in the following sections.

3.1 Visual Semantic Analysis Module

In the visual semantic analysis module, we implement remote sensing image classification and semantic feature extraction based on the DenseNet [24] model. Nowadays, commonly used CNN models in the field of remote sensing image classification are Inception [25], AlexNet [26], GoogleNet [27], VGGNet [28] and ResNet [29]. These networks usually have a shallower hierarchical structure and more network parameters.

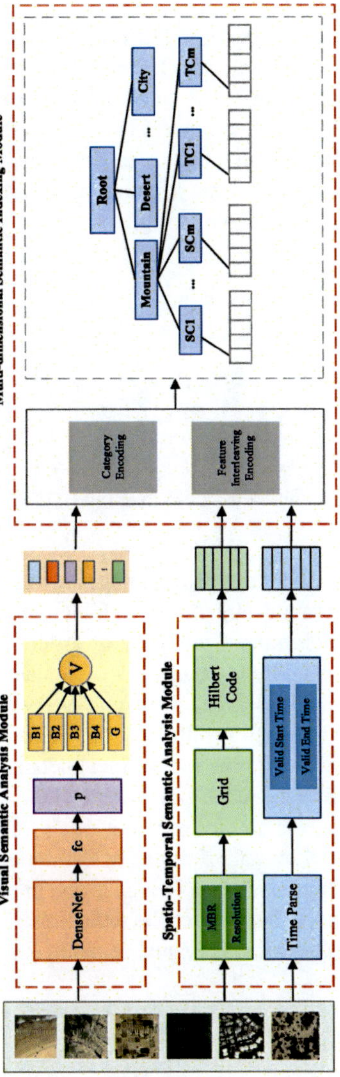

Fig. 1. The architecture of the SBARSIR.

In [28], VGGNet is used to extract features from images, and then the features from different layers are transformed by Discriminant Correlation Analysis (DCA), and the transformed features are fused to construct a new feature, which is classified by SVM and used as Classifiers in other fields [34]. To improve the performance of the network, the usual method is to broaden the network structure and deepen the number of network layers. However, these improvements also brought many problems, such as the disappearance of gradients. DenseNet breaks the rigid thinking of deepening the network layer (ResNet) and broadening the network structure (Inception) to improve the performance of the network [31]. From a feature point of view, through feature multiplexing and bypass settings, the number of network parameters is greatly reduced, and the problem of gradient disappearance is alleviated to a certain extent. Hence, we chose DenseNet network structure to realize the classification function of remote sensing image data. The structure of DenseNet network is shown in Fig. 2.

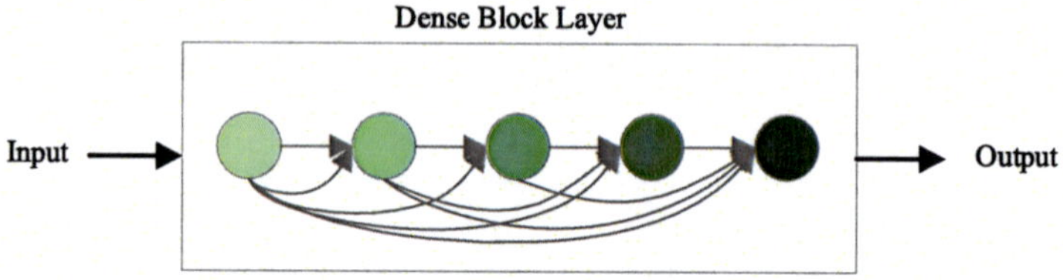

Fig. 2. Structure of the DenseNet model [24].

Considering that multi-source remote sensing images always have different spatial resolutions. That is, the same ground object will exhibit large differences in visual semantics at different spatial resolutions. Continuity of ground objects makes remote sensing image data may contain multiple categories of semantic information, which will cause interference to the final classification results of the remote sensing image. Furthermore, the input data should be uniformly scaled to a smaller size to speed up the training process. This inevitably leads to the loss of local details of the remote sensing image data, which may also lead to a decrease in classification accuracy. Therefore, to optimize the performance of Visual Semantic Analysis Module on remote sensing images of different spatial resolutions and to make full use of the detailed information of the image, we input the remote sensing image data in blocks and integrate the classification results by voting to obtain final classification results with higher accuracies, and output the layer before the classifier as the extracted visual semantic features.

3.2 Spatio-Temporal Semantic Analysis Module

Spatio-temporal semantic analysis is to carry out feature extraction and analysis from multi-source remote sensing images, to provide indexing and retrieval capabilities.

3.2.1 Spatial Semantics Analysis

According to [35, 36], spatial semantics of remote sensing data set is used to describe the spatial range in XY plane. On this basis, the spatial semantic analysis process of remote sensing image data is as follows:

(1) First, extract the minimum bounding rectangle (MBR) of the remote sensing image data.
(2) Then, map the remote sensing image data to a specific pyramid level according to the resolution, and divide it into a set of grids of uniform size.
(3) Next, convert the latitude and longitude information in MBR to corresponding row and column order in spatial pyramid grids.
(4) Finally, fill the grids with the Hilbert space filling curve to realize the orderly organization of the grid in a certain order, and map each grid to a point on the space filling curve to generate the corresponding coding result.

3.2.2 Temporal Semantics Analysis

Temporal semantics of remote sensing data set describes the arrangement order and life cycle of datasets in the time dimension, defines the valid time range of datasets in the form of timestamps, and provides support for sequence analysis and multi-temporal display of datasets [35, 36]. Accordingly, the process of temporal semantic analysis of remote sensing image data is as follows:

(1) First, extract the data production time from the metadata of remote sensing images as the temporal feature of remote sensing image data;
(2) Then, compare the current situation with the RS images in the database;
(3) Next, define the valid time range of the data, where the start time is the production time of the data, and the end time is the production time of the remote sensing image data covering the same time range in the subsequent time.

3.2.3 Spatio-Temporal Clustering

In the previous steps, we have obtained spatial and temporal semantic feature representations of remote sensing images. Now, we input them into cluster methods to divide remote sensing images in time and space dimension respectively. Then we can obtain the results of the division of spatial sub-classes and time sub-classes. Remote sensing data belonging to the same subclass usually has a certain degree of temporal and spatial continuity. Taking the remote sensing image data of the Hawaiian Islands as an example, the results of spatio-temporal sub-classes are shown in the Fig. 3.

3.3 Multi-dimensional Semantic Indexing Module

In the multi-dimensional semantic indexing module, classification results and extracted semantic feature representations of remote sensing data are adopted to build multi-dimensional semantic indexing. The multi-dimensional semantic index has a two-stage structure. In the first stage, we combine the results of visual semantic classification

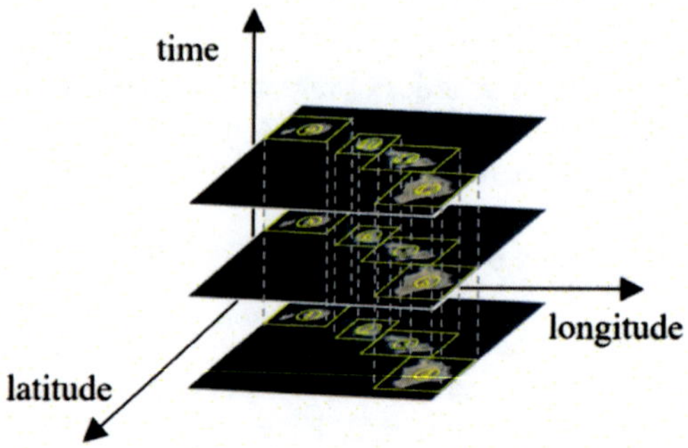

Fig. 3. The results of spatio-temporal subclasses of a remote sensing image.

and spatio-temporal clustering to construct a coarse-grained remote sensing image tree index; in the next stage, we utilize the semantic feature representation of remote sensing data in the visual dimension and spatio-temporal dimension, and implement indexing for each remote sensing data using interleaved encoding.

3.3.1 Remote Sensing Image Tree Index

For remote sensing data v, we quantize and encode it and express it as a three-dimensional vector as

$$\text{code}_v = [MCCode, SCCode, TCCode] \quad (1)$$

where code_v is the quantization code of v, $MCCode$, $SCCode$ and $TCCode$ represent the IDs of the main class, space sub-class and time sub-class to which v belongs, respectively. The name of each class and its corresponding code are shown in Table 1.

Table 1. Examples of classes and corresponding codes

The main class	Code	The subclass	Code
Mountain	0001	Mountain-SC1	0000 0001
		Mountain-SC2	0000 0010
		Mountain-TC1	1000 0001
		Mountain-TC3	1000 0011
Desert	0010	Desert-SC1	0000 0001
Water area	0011	Water area-SC1	0000 0001
Woodland	0100	Woodland-SC1	0000 0001
Grassland	0101	Grassland-SC1	0000 0001
Arable land	0110	Arable land-SC1	0000 0001
City	0111	City-SC1	0000 0001

3.3.2 Multi-dimensional Semantic Interleaving Indexing

IN multi-dimensional semantic interleaving indexing, we use Z-curve to establish a serial relationship between the three dimensions, visual semantics dimension, time dimension and spatial dimension. And we interweave the features of different dimensions in the form of interleaved coding to realize the index construction. The specific implementation is shown in Fig. 4.

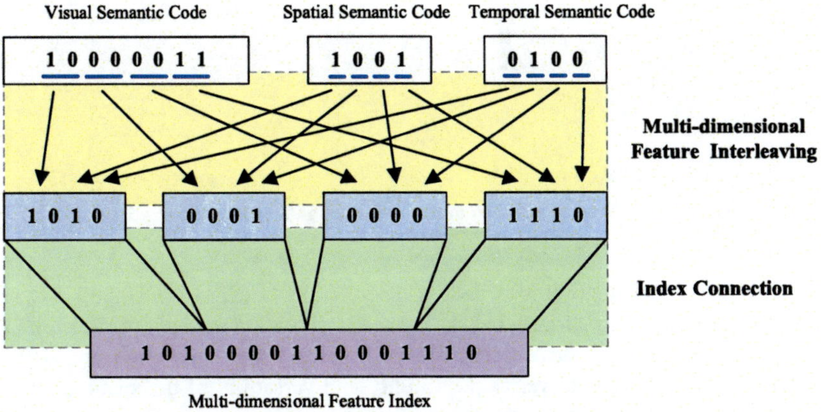

Fig. 4. The specific implementation of multi-dimensional semantic interleaving indexing.

3.3.3 Remote Sensing Image Retrieval

On basis of multi-dimensional spatio-temporal indexing produced, specific process of remote sensing image retrieval is as follows:

(1) Classification prediction. Input query conditions into the index construction process designed in this paper, and get multiple spatio-temporal classifications corresponding to them, including main class, spatial subclass, and time subclass;
(2) Rough distance calculation based on quantitative results. Read the ID and quantization code value of all remote sensing data in the sub-class, and calculate the distance between the query vector and the data;
(3) Sort and selection. Sort the data in the sub-class according to the distance, as shown in Fig. 5, and select the top k data.

4 Experimental Results and Discussion

4.1 Experimental Materials

The main sources of data used in this study include two types, one is public typical remote sensing image training dataset with semantic labels, including UC-Merced (UCM) [37],

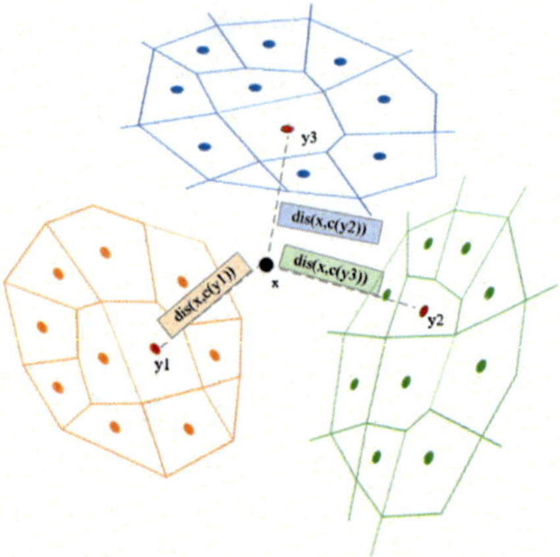

Fig. 5. The distance between query condition and remote sensing data.

RSSCN7 [38], The Google image dataset of SIRI-WHU (SIRI-WHU) [39–41], the Tiangong-2 remote sensing image natural scene classification dataset (TG-2) [42]. The other is a dataset that we made using remote sensing image data downloaded from the Living Atlas platform, to verify the validity and versatility of the spatio-temporal retrieval method in multi-source remote sensing images.

4.1.1 Custom Spatio-Temporal Remote Sensing Image Data Set

We construct a custom spatio-temporal remote sensing image dataset, the Living Atlas. It consisting of images of 12 classes selected from the ArcGIS Earth Platform, including the coastline, desert, forest, farmland, meadow, mountain, parking, resident, rivers, saltern, sea, and snow-mountain. And 300 images measuring 256 × 256 pixels at different times are manually selected for each of all the classes. Each image has three attributes: category, spatial, and time attributes.

The production process of the Living Atlas dataset is as follows:

(1) Locate the positions of objects of different classes on the ArcGIS Earth Platform.
(2) Query all the time versions with local changes on the ArcGIS Earth Platform, and get the corresponding time code.
(3) Convert the location of the object to the level, row, and column order of the tile in the tile pyramid. And combine it with time code to generate the WMTS request.
(4) Send WMTS request to the ArcGIS Earth Platform to obtain the remote sensing tiles in JPG format.
(5) Bind the data with its features such as scene category, spatial attributes, and time attributes. The scene category is manually labeled; the time attribute is the data release time provided by the ArcGIS online earth platform; the spatial attribute is the geographic location of the image, which can be represented by the level, row, and column order of the tile pyramid. The coordinate system of the tile pyramid is the EPSG:3857, and the origin is at the upper left corner (Fig. 6).

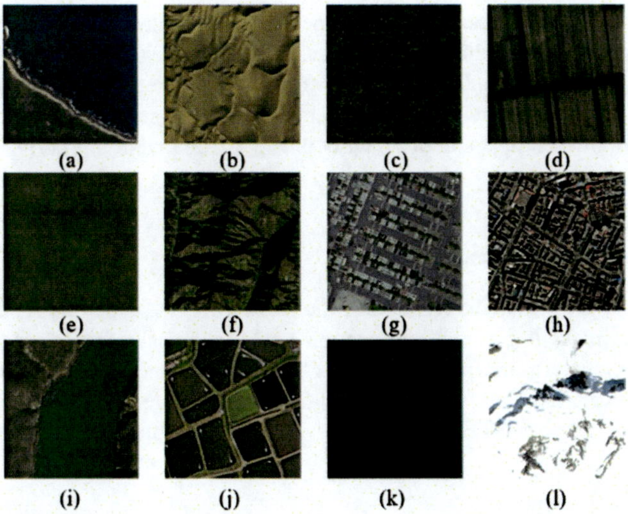

Fig. 6. Example images from the Living Atlas dataset. **a** Coastline; **b** desert; **c** forest; **d** farmland; **e** meadow; **f** mountain; **g** parking; **h** resident; **i** river; **j** saltern; **k** sea; **l** snow-mountain.

4.1.2 Public Remote Sensing Image Data Set

(1) UC-Merced (UCM) [37]: The UC-Merced dataset is a high-resolution remote sensing image proposed in 2010. These images are collected from the National City Map of the United States Exploration Agency and manually extracted remote sensing images of different regions. The classification rate is about 0.3 m. The UC-Merced dataset contains a total of 21 scene categories. In each category, there are 100 images with the size of each image being 256 × 256. Images in different categories are very similar, and in the same category are quite different, which poses a challenge to scene classification. It is also the most frequently used dataset in the field of remote sensing image classification.

(2) The Google image dataset of SIRI-WHU (SIRI-WHU) [39–41]. The SIRI-WHU is a 12-class Google image dataset of SIRI-WHU meant for research purposes. There are 200 images for each of the following classes: agriculture, commercial, harbor, idle land, industrial, meadow, overpass, park, pond, residential, river, water. The pixel size of each image in this dataset is 200*200 pixels, with a 2-m spatial resolution.

(3) The Tiangong-2 remote sensing image natural scene classification dataset (TG-2) [42]. The wideband imager carried by Tiangong-2 is a new generation of wideband, wide-field, and "map-in-one" optical remote sensors, which has acquired many high-quality ground observation images during its orbit. Aiming at the limited number of currently open Earth observation image scene classification datasets, similar scenes, and single data sources, natural scene categories are extracted from Tiangong-2 broadband images, covering image data in different regions and at different times. The extracted data are cropped, selected, and marked to form the

Tiangong-2 natural scene data set. There are RGB three-band true-color image (JPG) and 14-band multispectral images. The scene categories consist of the beach, circular farmland, cloud, desert, woodland, mountain, rectangular farmland, building area, river, and snow mountain. There are 2000 images in each category and the pixel size of each image in this dataset is 128 × 128 pixels (Figs. 7, 8, and 9).

4.2 Experimental Settings

To more fully verify the effectiveness of the SBARSIR, we adapt the model used for comparison to make it more in line with our goal. Specifically, we select three neural network models such as Inception, ResNet, VGG, and combine them with several spatio-temporal indexing methods, such as STHI [43], ST+ [44], and ZMF [45]. Then we get Inception-ZMF, ResNet-STHI, VGG-ST+ to carry out comparative experiments.

4.3 Evaluation Metrics

In our experiments, we adopt Average Precision (AP) as the metric to evaluate the retrieval performance to make a comprehensive evaluation of the SBARSIR we proposed.

As for Average Precision (AP), in the query phase, we generate lots of query conditions with time, space, and semantic categories, and obtain all data that meet the query conditions. Then we can calculate the precision (P) of the query conditions. Finally, the AP can be computed via averaging the P of all query conditions, which is defined as

$$\mathrm{AP} = \frac{1}{|Q|}\sum_{i=1}^{|Q|} P(i) \tag{2}$$

4.4 Experimental Results

4.4.1 Indexing Results of the SBARSIR

We constructed the SBARSIR index model on the remote sensing image dataset. In Fig. 10, we draw the key of the data and the spatial code and time code of the data into a three-dimensional graph and use symbols with different colors and shapes to represent the remote sensing image of different categories. It can be seen from Fig. 10 that the code generated by the multi-dimensional semantic indexing module can guarantee a certain degree of hash ability, and data with similar characteristics in each dimension has the value of a close key.

4.4.2 Performance Evaluation of Remote Sensing Image Retrieval

We generate a set of different time, space and category query conditions, and obtain all data that meet the query conditions. The comparison result of AP is shown in Figs. 11 and 12. Figure 11 shows the comparison results of the retrieval accuracy of each model on different datasets. In Fig. 12, the radar chart is used to show the retrieval accuracy

Fig. 7. Example images from the UCM dataset. **a** Agricultural, **b** airplane, **c** baseball diamond, **d** beach, **e** buildings, **f** chaparral, **g** dense residential, **h** forest, **i** freeway, **j** golf course, **k** harbor, **l** intersection, **m** medium residential, **n** mobile home park, **o** overpass, **p** parking lot, **q** river, **r** runway, **s** sparse residential, **t** storage tanks, **u** tennis court.

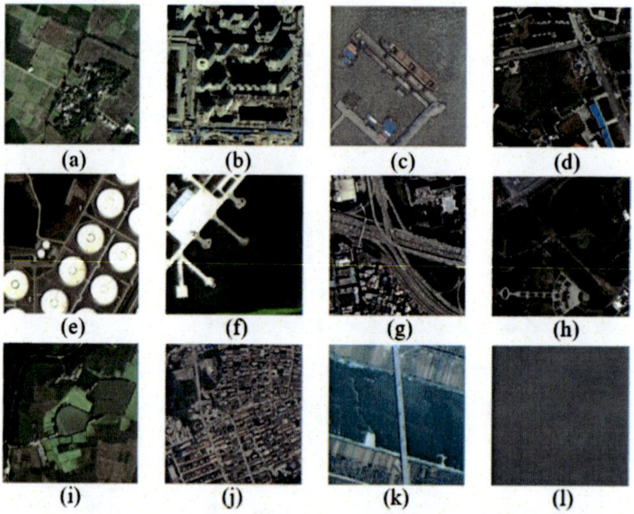

Fig. 8. Class representatives of the SIRI-WHU dataset. **a** Coastline; **b** desert; **c** forest; **d** farmland; **e** meadow; **f** mountain; **g** parking; **h** resident; **i** river; **j** saltern; **k** sea; **l** snow-mountain.

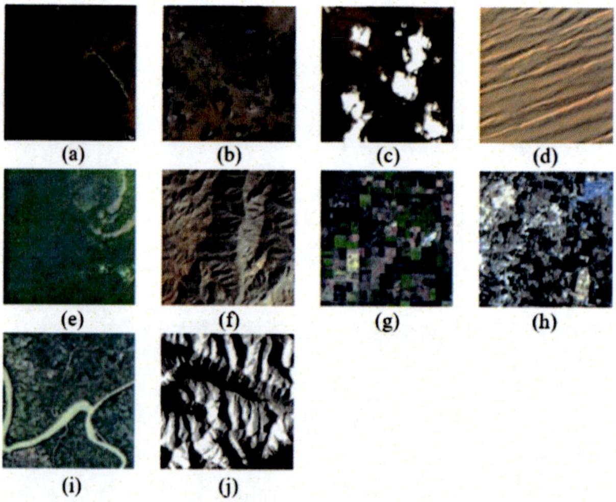

Fig. 9. Example images from the TG-2 dataset. **a** beach, **b** circular farmland, **c** cloud, **d** desert, **e** woodland, **f** mountain, **g** rectangular farmland, **h** building area, **i** river, **j** snow mountain.

results of various data in the Living Atlas dataset. As the figure shows that the SBARSIR can guarantee higher retrieval accuracy for each dataset, especially for data categories with certain retrieval difficulties such as rivers and farms, which can still maintain better results.

In summary, the above results show that the SBARSIR we proposed can accurately extract semantic features from remote sensing image data and fuse semantic features

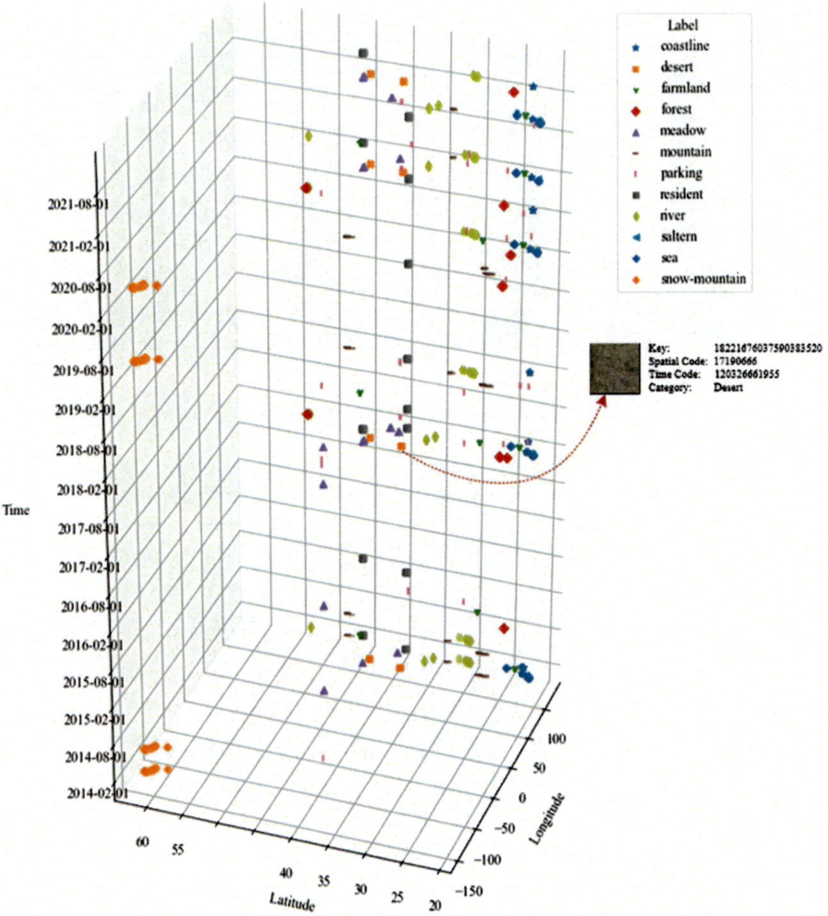

Fig. 10. The multi-dimensional diagram of the SBARSIR indexing results over the Living Atlas dataset.

with spatio-temporal features to build model indexes adaptively. Besides, the SBARSIR also has a good efficiency in retrieval of remote sensing image data in the spatial and temporal dimensions. To be specific, compared with other indexes, the advantages of the SBARSIR are not obvious in the small interval retrieval. With the expansion of the retrieval scope, the retrieval performance can be maintained in a relatively stable state, and the advantages gradually become prominent.

5 Conclusions

Aiming at the problems in remote sensing data applications, this article promotes a semantic-base analysis approach, SBARSIR, for multi-source remote sensing image retrieval. Taking full advantage of CNN techniques, we realize the feature extraction

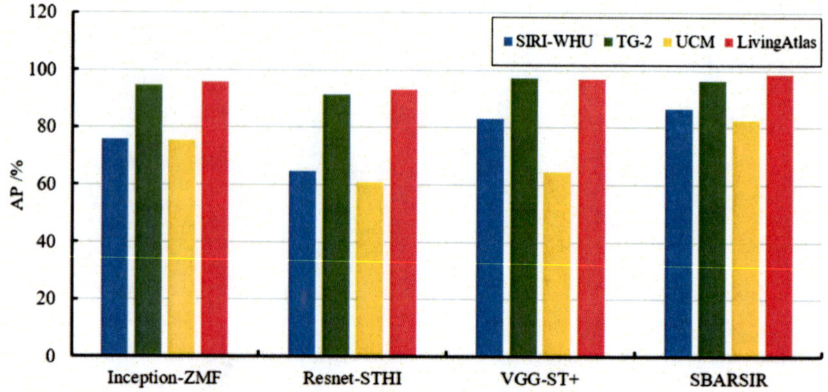

Fig. 11. Comparison of retrieval precision on different datasets.

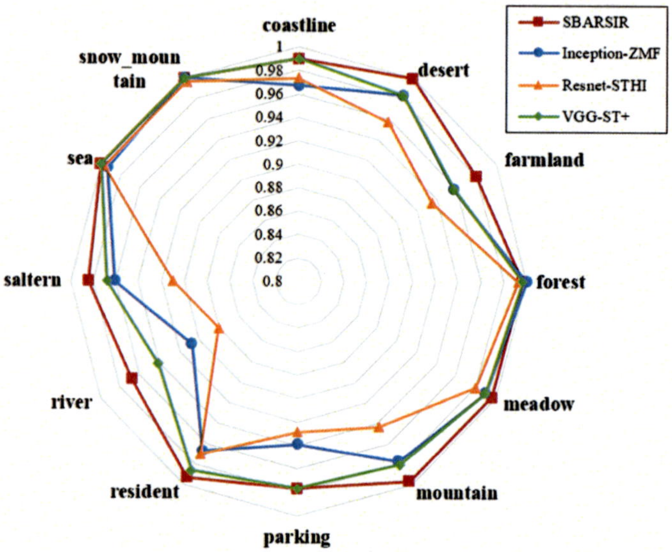

Fig. 12. Comparison of retrieval precision on the Living Atlas dataset.

and the classification of remote sensing data, respectively. Spatial features and temporal features are extracted from metadata for spatio-temporal semantic analysis, and output the spatio-temporal semantic feature representations. Then, visual and spatio-temporal semantic feature representation are used to generate indexing structure by means of interleaf encoding to provide powerful retrieval capabilities. Experiments show that the SBARSIR proposed in this paper has good performance in multi-dimensional retrieval of spatial, time, and data features. It surpasses previous approaches in terms of quality in multi-source remote sensing retrieval. Therefore, the SBARSIR can build access channels for the remote sensing data in various fields.

In future work, on the one hand, we will consider further optimizing the structure design and construction process of the multi-dimensional spatio-temporal index, reduce training costs and improve the stability and portability of the index ulteriorly. On the other hand, we will expand the scope of use of the SBARSIR, seek for its compatibility with remote sensing data analysis and mining, and further develop its practical value.

References

1. Deren, L., Jun, M., Zhenfeng, S.: Spatio-temporal big data and its application. Satell. Appl. **9**, 7–11 (2015)
2. Jiayao, W.: Spatio-temporal big data and its application in smart cities. Satell. Appl. **5**, 10–17 (2017)
3. Deren, L.: The intelligent processing and service of spatio-temporal big data. J. Geo-Inf. Sci. **21**(12), 1825–1831 (2019)
4. Deren, L.: Towards geo-spatial information science in big data era. Acta Geod. Cartogr. Sin. **45**(4), 379–384 (2016)
5. Li, J.Y.: Research on key techniques of fast image search. Shanghai Jiaotong University (2017)
6. Zhang, K.: Research on methods for fast image retrieval. University of Science and Technology of China (2018)
7. Swain, M.J., Ballard, D.H.: Color indexing. Int. J. Comput. Vision **7**(1), 11–32 (1991)
8. Stricker, M.A., Orengo, M.: Similarity of color images. Proc. SPIE - Int. Soc. Opt. Eng. **2420**, 381–392 (1995)
9. Oliva, A., Torralba, A.: Building the gist of a scene: the role of global image features in recognition. Prog. Brain Res.: Vis. Percept. **155**, 23–36 (2006)
10. Sivic, Zisserman: Video Google: a text retrieval approach to object matching in videos. IEEE (2003)
11. Shao, K.: Research on distributed remote sensing image retrieval based on deep learning. Huazhong University of Science and Technology (2019)
12. Zhou, W., Newsam, S., Li, C., et al.: Learning low dimensional convolutional neural networks for high-resolution remote sensing image retrieval. Remote Sens. **9**(5), 489 (2016)
13. Xiao, Z., Long, Y., Li, D., et al.: High-resolution remote sensing image retrieval based on CNNs from a dimensional perspective. Remote Sens. **9**(7) (2017)
14. Ye, F., Xiao, H., Zhao, X., et al.: Remote sensing image retrieval using convolutional neural network features and weighted distance. IEEE Geosci. Remote Sens. Lett. 1–5 (2018)
15. Gao, Y.N., Ye, J.B., Yang, N.Z., Gao, X.F., Chen, G.H.: Middle layer based scalable learned index scheme. Ruan Jian Xue Bao/J. Softw. **31**(3), 620–633 (2020)
16. Zhang, Z., Jin, P.Q., Xie, X.K.: Learned indexes: current situations and research prospects. Ruan Jian Xue Bao/J. Softw. **32**(4), 1129–1150 (2021)
17. Lu, X., Li, S., Rong, X., et al.: Deep hash learning for efficient image retrieval. In: 2017 IEEE International Conference on Multimedia & Expo Workshops (ICMEW). IEEE (2017)
18. Liu, Y., Cheng, M., Wang, F.P., Li, D.X., Liu, W., Fan, J.L.: Deep hashing image retrieval methods. J. Image Graph. **25**(07), 1296–1317 (2020)
19. Li, W.J., Zhou, Z.H.: Learning to hash for big data: current status and future trends. Chin. Sci. Bull. **60**, 485–490 (2015)
20. Kraska, T., Beutel, A., Chi, E.H., et al.: The case for learned index structures. In: Proceedings of the International Conference on Management of Data, pp. 489–504 (2018)
21. Lu, H., Dai, Y., Shen, C., Xu, S.: Indices matter: learning to index for deep image matting. In: Proceedings of the IEEE International Conference on Computer Vision, pp. 3266–3275 (2019)

22. Kraska, T., Alizadeh, M., Beutel, A., Chi, E.H., Ding, J., Kristo, A., et al.: SageDB: a learned database system. In: Proceedings of the Biennial Conference on Innovative Data Systems Research (2019)
23. Kristo, A., Vaidya, K., Çetintemel, U., Misra, S., Kraska, T.: The case for a learned sorting algorithm. In: Proceedings of the International Conference on Management of Data, pp. 1001–1016 (2020)
24. Huang, G., Liu, Z., Laurens, V., et al.: Densely connected convolutional networks. IEEE Comput. Soc. (2016)
25. Szegedy, C., Liu, W., Jia, Y., et al.: Going deeper with convolutions. IEEE Comput. Soc. (2014)
26. Krizhevsky, A., Sutskever, I., Hinton, G.E.: ImageNet classification with deep convolutional neural networks. In: Advances in Neural Information Processing Systems (2012)
27. Xia, G.S., Hu, J., Hu, F., et al.: Formance evaluation of aerial scene classification. IEEE Trans. Geosci. Remote Sens. **55**(7), 3965–3981 (2017)
28. Simonyan, K., Zisserman, A.: Very deep convolutional networks for large-scale image recognition (2014). arXiv preprint arXiv:1409.1556
29. He, K.M., Zhang, X.Y., Ren, S.Q., Sun, J.: Identity mappings in deep residual networks. In: Proceedings of the 14th European Conference on Computer Vision, pp. 630–645. Springer, Amsterdam (2016)
30. Li, M., Zhu, X.Y., Duan, L.: A case-based reasoning approach for task-driven remote sensing image discovery under spatial-temporal constrains. Geomat. Inf. Sci. Wuhan Univ. **42**(6) (2017)
31. Egenhofer, M.J., Al-Taha, K.K.: Reasoning about gradual changes of topological relationships. In: Theories and Methods of Spatio-Temporal Reasoning in Geographic Space, International Conference GIS - From Space to Territory: Theories and Methods of Spatio-Temporal Reasoning, Pisa, Italy, 21–23 Sept 1992. Springer-Verlag (1992)
32. Li, X., Li, J., Wang, X.: ASLM: adaptive single layer model for learned index. In: Database Systems for Advanced Applications (2019)
33. Li, P., Hua, Y., Zuo, P., et al.: A scalable learned index scheme in storage systems (2019)
34. Lin, J.S.: Fuzzy clustering using a compensated fuzzy hopfield network. Neural Process. Lett. **10**, 35–48 (1999)
35. An optimal retrieval method of multi-theme image tiles considering the spatio-temporal semantics
36. Zhu, Q., Li, H., Yang, X.: Hierarchical semantic constraint model for focused remote sensing information services. Geomat. Inf. Sci. Wuhan Univ. **34**(12), 1454–1457 (2009)
37. Piao, S.Z., Yu, J.: Cluster validity indexes for FCM clustering algorithm. Pattern Recognit. Artif. Intell. **28**, 451–461 (2015)
38. Yang, Y., Newsam, S.: Bag-of-visual-words and spatial extensions for land-use classification. In: Proceedings of the 18th SIGSPATIAL International Conference on Advances in Geographic Information Systems, pp. 270–279 (2010)
39. Qin, Z., Ni, L., Tong, Z., et al.: Deep learning based feature selection for remote sensing scene classification. IEEE Geosci. Remote Sens. Lett. **12**(11), 1–5 (2015)
40. Zhao, B., Zhong, Y., Xia, G.-S., Zhang, L.: Dirichlet-derived multiple topic scene classification model fusing heterogeneous features for high spatial resolution remote sensing imagery. IEEE Trans. Geosci. Remote Sens. **54**(4), 2108–2123 (2016)
41. Zhao, B., Zhong, Y., Zhang, L., Huang, B.: The Fisher kernel coding framework for high spatial resolution scene classification. Remote Sens. **8**(2), 157 (2016). https://doi.org/10.3390/rs8020157
42. Zhu, Q., Zhong, Y., Zhao, B., Xia, G.-S., Zhang, L.: Bag-of-visual-words scene classifier with local and global features for high spatial resolution remote sensing imagery. IEEE Geosci. Remote Sens. Lett. (2016). https://doi.org/10.1109/LGRS.2015.2513443

43. Zhou, Z., Li, S., Wu, W., et al.: NaSC-TG2: natural scene classification with Tiangong-2 remotely sensed imagery. IEEE J. Sel. Top. Appl. Earth Obs. Remote. Sens. (2021)
44. Wu, Y.H., Cao, X.F.: Hilbert code index method for spatio-temporal data of virtual battlefield environment. Geomat. Inf. Sci. Wuhan Univ. **45**, 1403–1411 (2020)
45. Zhao, Y.H., Lü, L., Xu, Q.: A multidimensional retrieval strategy for massive spatio-temporal data. Sci. Surv. Mapp. **45**, 203–208 (2020)
46. Xu, A.P., Wang, B., Zhang, X.: Optimization of spatial and temporal big data association query based on HBase. Comput. Appl. Softw. **34**(6) (2017)

A High-Precision Orographic Dataset Suitable for Global Numerical Weather Prediction Spectral Models Was Developed Based on Earth Observation Data

Yingjie Wang[1,2], Jianping Wu[2(✉)], and Dazheng Liu[2]

[1] College of Computer Science and Technology, National University of Defense Technology, Changsha, China
[2] College of Meteorology and Oceanography, National University of Defense Technology, Changsha, China
wjp@nudt.edu.cn

Abstract. The basic orographic dataset in numerical weather prediction (NWP) is critical for accurate weather and climate predictions. However, the original high-resolution orographic elevation data cannot be used directly in NWP models because of the presence of some scale features that are detrimental to NWP. Therefore, high-resolution orographic elevation data requires some pre-processing operations to filter out harmful information before it can be used to NWP. The global NWP spectral models have the advantages of high calculation accuracy and fast convergence speed. However, due to the large amount of computation and long computation time in the NWP spectral model, most of the research and development on orographic datasets for NWP is aimed at grid-point models. Aiming at the autonomously construction, this study uses the earth observation data from two sources to make a basic high-precision orographic dataset suitable for the spectral model. These methods are useful for producing orographic datasets in the global NWP spectral model.

Keywords: Orographic dataset · NWP · Spectral model · Filter

1 Introduction

Orography affects not only local weather but also large-scale circulation through turbulent orography form drag, low-level airflow blocking drag and orographic gravity wave drag. Complex orography features may lead to heavy rainfall [1]. Heavy rainfall may lead to road collapse, landslide, debris flow, urban waterlogging and other geological disasters, causing great damage to the lives of local residents. At present, numerical weather prediction (NWP) technology has been developed rapidly and become more and more mature. In recent years, numerical weather prediction in complex orographic areas has become a hot topic. However, most numerical weather prediction models still face an important and unsolved problem, that is, their low prediction accuracy in complex orographic regions. Weather prediction of complex orographic areas is still a huge

L. Wang et al. (Eds.): CHREOC 2022, LNEE 969, pp. 380–391, 2023.
https://doi.org/10.1007/978-981-19-8202-6_34

challenge. Orography, as one of the key inputs of numerical models, is a vital factor that affects the prediction accuracy of models [2, 3]. Improving the representation of terrain in NWP models is extremely important for improving weather prediction capabilities.

Earth observation satellites play an indispensable role in the national economy, social development and national security. Its application fields include weather forecasting, land census, crop yield estimation, forest survey, geological prospecting, marine forecasting, environmental protection, and disaster monitoring. Its data and information have become the basic and strategic resources [4]. However, at the beginning of the development of numerical weather prediction, there were no orographic data from satellite remote sensing observations, the model's orography of early numerical weather prediction was derived from aerial topographic maps. For example, Berkofsky and Bertoni [5] produced a global orographic data set based on aeronautical charts at a rather unrefined scale. Orographic heights are artificially estimated to produce an average over one-degree squares. What is more noteworthy is that the aeronautical chart was not complete. The regions without values on the aeronautical chart were obtained by interpolation of adjacent values, and then averaged, as shown in Fig. 1. It can be seen that the orographic datasets at that time were very difficult to obtain, and certainly it was not the digitized versions that are usually provided in multiple resolutions today. Even as late as the 1970s, orographic datasets published in tables form were common [6]. During this period, the quality of orographic datasets was uneven, and global coverage was low, especially in areas of South America and Africa. Therefore, it was an urgent need to develop a set of high-resolution and global orographic datasets.

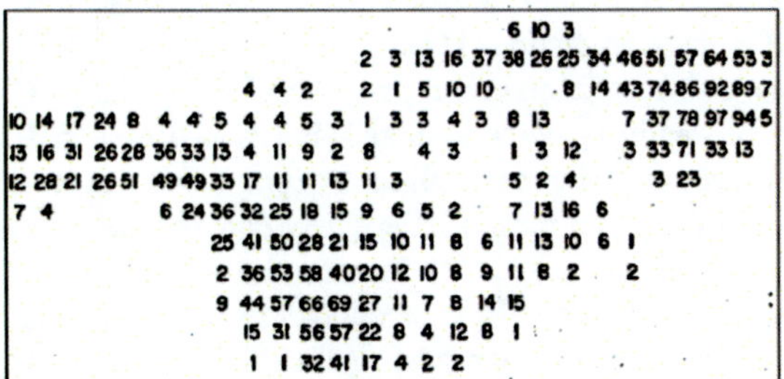

																	6	10	3							
												2	3	13	16	37	38	26	25	34	46	51	57	64	53	3
								4	4	2		2	1	5	10	10			8	14	43	74	86	92	89	7
10	14	17	24	8	4	4	5	4	4	5	3	1	3	3	4	3	8	13			7	37	78	97	94	5
13	16	31	26	28	36	33	13	4	11	9	2	8		4	3		1	3	12		3	33	71	33	13	
12	28	21	26	51	49	49	33	17	11	11	13	11	3				5	2	4			3	23			
7	4				6	24	36	32	25	18	15	9	6	5	2		7	13	16	6						
							25	41	80	28	21	15	10	11	8	6	11	13	10	6	1					
							2	36	53	58	40	20	12	10	8	9	11	8	2		2					
							9	44	57	66	69	27	11	7	8	14	15									
								15	31	56	57	22	8	4	12	8	1									
								1	1	32	41	17	4	2	2											
										4	56	34														
4	12										19	54	3	3			6	10								
													9	19	11					4						
													1		6	8	21	22	10	3						
															1	3	66	13	29	18	7					
														3		33	28	7	9	8	8	1				
																42	14	5	4	1	2	6	7	10	3	
				4												20	63	9	7	9	11	10	15	17	4	
																	51	117	25	15	18	23	25	18		
1		3																78	101	11	14	21	37	11		
		1								1								77	80	5	11	24	3			
			1															89	26	2	8	10				
																3	1	67	17	5	2					
																	8	50	8	3						

Fig. 1. Section of a table showing the hand-digitized orography of North America at a resolution of 5 degrees latitude and longitude from [5]. The data are tabulated in units of 100 ft.

With the hot application of satellite remote sensing, some global orographic datasets are emerging. ETOPO5 is one of the oldest global orographic elevation datasets, first published in 1988 with a resolution of 5 arc min. Some better regional orographic datasets did exist during this period, but they were difficult to obtain due to many limitations. In 1996, the global elevation data ushered in a major leap, that is, the GTOPO30 dataset was publicly released [7]. The GTOPO30 dataset has two significant advantages, one is the global coverage of the dataset, and the other is the resolution of the dataset up to 30 arcseconds. Subsequently, based on the development of satellite and remote sensing technology, the orographic elevation data are constantly updated, and the resolution and accuracy are also constantly improved. The Space Shuttle Radar Topography Mapper (SRTM) mission is a joint effort between the National Aeronautics and Space Administration (NASA), the National Imagery and Mapping Agency (NIMA), and the German and Italian space agencies. The SRTM system was carried on board the space shuttle launched by the United States. After a long time of data processing, the SRTM digital orographic elevation model was made [8]. In 2009, the NASA and the Ministry of Economy, Trade and Industry (METI) of Japan jointly released a more accurate global digital electronic orographic dataset ASTER GDEM [9], which has a resolution of up to 1 arcsecond (about 30 m). So far, an elevation model with uniform resolution and uniform accuracy for most of the earth's surface had been completed. Subsequently, these high-resolution digital elevation datasets, such as SRTM and ASTER GDEM, are used as inputs to various numerical weather prediction models to significantly improve the performance of numerical weather prediction models [10–14]. In addition to global orographic datasets, there are some higher resolution regional orographic datasets, but they are only limited to specific areas and cannot be used for weather predictions in other regions and global climate predictions [15].

However, these high-resolution orographic elevation data cannot be directly introduced into numerical weather prediction models before processing. This is because the numerical model cannot describe the small scale topographic effects of high-resolution topographic elevation data, and they will seriously damage the model performance if they occur in the model dynamic [16, 17]. In general, high-precision orographic elevation data can be obtained, but the NWP model can only describe the influence of the orographic scale suitable for the NWP model resolution on the atmosphere. Therefore, in order to ensure the accuracy and robustness of the numerical weather prediction models, the high-resolution orographic elevation data should be processed by some smoothing and filtering operations before being used as the input of numerical weather prediction models [18].

2 Methods and Datasets

2.1 Background

Orographic processing in NWP models has a long history. Before the original high-resolution orographic elevation data are used as input in the numerical weather prediction models, in order to ensure the accuracy and robustness of the numerical weather prediction models, there were some researches related to handling of orography [19–27]. These studies all show that it is extremely important to strike a balance between orographic authenticity and smoothness in NWP models. However, most of above studies are concerned with NWP grid-point models. The NWP spectral models have plenty of outstanding strengths such as high calculation accuracy, excellent stability, and clean model program [28], which makes it become one of the most widely used models for numerical weather prediction business. The spectral models are primarily used for simulation of the large-scale atmospheric circulation, so the basic model's orography needs to be global in scope. In addition, in the numerical weather prediction spectral model, orography needs to be transformed from grid-point space to spectral space. However, this process will lead to Gibbs phenomena, mainly represented by overshoot and undershoot near steep areas, which will affect the performance of the numerical weather prediction model. Hence, the novel and significant challenge is to make high-precision spectral orography for the numerical weather prediction spectral model. Navarra et al. [29] described a variety of spectral coefficient weighting schemes for constructing spectral model orography and show results using one of these methods that significantly reduce Gibbs oscillations in atmospheric variable fields. Lindberg and Broccoli [30] described a regularization method based on variational constraints to solve the problems of false Gibbs oscillations, false valleys and mountains, and then false small-scale extreme values in the distribution of simulation results when spectral coefficients represent model orography. The results show that this regularization technique can reduce the distortion of features such as mountain heights and continental boundaries, and produce more realistic simulation results.

This study is mainly aimed at the numerical weather prediction spectral model, using the Earth observation datasets to create model's orography applicable to numerical spectral models, and optimize the description of the orographic effect in the numerical weather prediction spectral model.

2.2 Earth Observation Datasets

The latest version of the ASTER GDEM elevation dataset covers the land part between 83° north latitude and 83° south latitude, reaching 99% of the earth's land surface. It is currently the only high-resolution elevation image data covering the global land surface. ASTER GDEM was created by stereo correlation of more than 1.2 million individual ASTER stereo scenes contained in the archive. The GDEM had a vertical accuracy of approximately 10 m and a grid spacing of one arc second latitude and longitude in the horizontal direction (~30 m) [31]. Since the release of the V1 version of ASTER GDEM data on June 29, 2009, it has been widely used in global Earth observation research [32, 33]. In this paper, the latest version of ASTER GDEM, ASTER GDEM V3, is used when constructing the basic orographic dataset of numerical weather prediction global spectral model.

In areas not covered by ASTER (i.e., south of 83S° and north of 83N°), the GTOPO30 dataset is used. GTOPO30 is also a global digital elevation model, and its horizontal grid spacing was 30 arcseconds (~1 km). DEMs are derived from multiple raster and vector sources of terrain information. Completed in late 1996, GTOPO30 was a collaborative effort led by the U.S. Geological Survey's Earth Resources Observation and Science Center (EROS) and took three years to develop.

Firstly, the ASTER data were averaged from 1 to 30 arc-second resolution and merged with the GTOPO30 dataset into a global orographic height dataset. This initial global orographic height dataset grid spacing was 30 arcseconds (~1 km). Regarding the resolution of the final high-precision spectral model's orography, we selected the latest T1279 resolution of NWP spectral model, with a grid spacing of about 16 km at low latitudes. Of course, this method of constructed the spectral model's orography has certain flexibility and robustness, the specific model's resolution can be changed according to actual requirements.

2.3 Methods

When constructing the base spectral model's orography dataset, we use two types of filters, that are a grid-point filter and a spectral one respectively. Besides, we also use the Fourier transforms and Legendre transforms (i.e. spectral transforms) to convert the orography field into spectral coefficients. The grid filter mainly adopts the idea of Gaussian filtering that the weight value of grid points far away from the center is getting smaller and smaller, as shown in (1).

$$\begin{cases} w = \frac{G_1}{P_1} & |R| \le \frac{\Delta}{2} - \delta. \\ w = \frac{G_2}{P_2} & \frac{\Delta}{2} - \delta < |R| \le \frac{\Delta}{2}. \\ w = \frac{G_3}{P_3} & \frac{\Delta}{2} < |R| \le \frac{\Delta}{2} + \delta. \\ w = 0 & |R| \ge \frac{\Delta}{2} + \delta. \end{cases} \tag{1}$$

In Eq. 1, w is filter weight of the grid points. R is range from points to the center, Δ is the filter scale, and δ is the filter edge scale. P1, P2, P3 are number of grid-points that satisfy |R| condition respectively. G1, G2, G3 are the total weight of each part. After the orography field is converted into spectral coefficients, filtering in the spectral space is also required, as shown in (2).

$$f(n) = \frac{1}{1 + 4 \times \left(\frac{n}{mmax} \times \frac{n+1}{mmax}\right)^8} \tag{2}$$

In Eq. 2, the n is total wave number index, *mmax* is maximum truncated wavenumber. Filtering in spectral space is performed by multiplying the spectral coefficient O of the terrain field by Eq. 2. In spectral space, the orographic spectral coefficient O_n^m is multiplied by Eq. 2 for filtering.

3 Orography Processing Steps

3.1 The Pre-filter

The original orographic dataset's resolution was 30 arc seconds (about 1 km at low latitudes). First, the initial orography is processed using filter (1) with a filter scale Δ of 5 km to get rid of small-scale topographic characteristics below 5 km. In NWP, orographic effects below 5 km are often considered to be detrimental to the dynamics of NWP, and these orographic features are generally characterized using physical process parameterization.

The filter scale unit is km, but the dataset resolution unit is radian. Therefore, when filtering, we choose one-dimensional processing. For example, the initial global orographic dataset is processed along the longitude direction first, and then processed along the latitude direction. The filter parameters are different in different longitude directions. Δ is 5 km, G_1 is 0.6, G_2 is 0.34, G_3 is 0.06, δ is 1 km, Eq. 1 will become as follows:

$$\begin{aligned} w_{i,j} &= \frac{0.6}{P_1^{i(or\,j)}} && |R| \leq 1.5. \\ w_{i,j} &= \frac{0.34}{P_2^{i(or\,j)}} && 1.5 < |R| \leq 2.5. \\ w_{i,j} &= \frac{0.06}{P_3^{i(or\,j)}} && 2.5 < |R| \leq 3.5. \\ w_{i,j} &= 0 && |R| > 3.5. \end{aligned} \tag{3}$$

3.2 The Smooth Operation

The dataset obtained in Sect. 3.1 was smoothed. Its current resolution is 30 arc-seconds, and then averaged over a 5 × 5 grid to obtain the orography dataset with a resolution of 2′30″. 2′30″ longitude or latitude is about 5 km near the equator.

3.3 The Second Filter

The second filter was applied to the global orographic dataset with the 2′30″ resolution. The filter scale is 16 km. Since the target resolution of the spectral model was assumed to be 16 km, this filtering is designed to get rid of topographic characteristics between 5 km and the model target resolution. The filtered orographic effects are also represented in the parameterization of physical processes, such as sub-grid scale orographic (SSO) drag parameterization.

The filter scale unit is still km, and the dataset resolution unit is still radian. Therefore, the method in 3.1 should also be used for single-dimension filtering. As when Δ is 16 km, G_1 is 0.6, G_2 is 0.34, G_3 is 0.06, δ is 1 km, there may not grid points satisfy condition $\frac{\Delta}{2} - \delta < |R| \leq \frac{\Delta}{2} + \delta$ near the equator, so that the total weight of filter cannot equal to 1. Therefore, the filter needs to be slightly modified.

When there are grid points satisfy condition $\frac{\Delta}{2} - \delta < |R| \leq \frac{\Delta}{2} + \delta$ (i.e. P_2 is not 0), the filter is as follows:

$$
\begin{aligned}
w_{i,j} &= \frac{0.6}{P_1^{i(or\, j)}} && |R| \leq 7. \\
w_{i,j} &= \frac{0.4}{P_2^{i(or\, j)}} && 7 < |R| \leq 9. \\
w_{i,j} &= 0 && |R| > 9.
\end{aligned} \tag{4}
$$

When no grid points satisfy condition $\frac{\Delta}{2} - \delta < |R| \leq \frac{\Delta}{2} + \delta$ (i.e. P_2 is 0), the filter is as follows:

$$
\begin{aligned}
w_{i,j} &= \frac{1.0}{P_1^{i(or\, j)}} && |R| \leq 9. \\
w_{i,j} &= 0 && |R| > 9.
\end{aligned} \tag{5}
$$

3.4 The Spectral Transform

The resolution of the orographic dataset at this time is 2′30″, and the orographic features below 16 km have been filtered out. Since global spectral models' prediction is carried out in spectral space, all initial field variables need to be converted into spectral coefficients. The grid point orography is also need to be converted into the spectral space of the target resolution. In general, the spherical harmonic is chosen as the basis in NWP spectral model. The spherical harmonic is as follows:

$$Y_n^m(\lambda, \mu) = e^{im\lambda} P_n^m(\mu). \tag{6}$$

where λ is the longitude, $\upsilon = \sin\varphi$, φ is the latitude, $P_n^m(\mu)$ is the associated Legendre polynomial. For the orographic height, the relationship between the grid-point field $O(\lambda, \mu)$ and the spectral coefficient O_n^m is:

$$O(\lambda, \upsilon) = \sum_{m-\infty}^{\infty} \sum_{n=|m|}^{\infty} O_n^m Y_n^m(\lambda, \upsilon) = \sum_{m-\infty}^{\infty} \sum_{n=|m|}^{\infty} O_n^m e^{im\lambda} P_n^m(\mu). \tag{7}$$

The above equation is divided into two steps, that is, the Fourier transform is performed first, and then the Legendre transform is performed, as shown in Eqs. 8 and 9.

$$C^m(\mu) = \frac{1}{2\pi}\int_0^{2\pi} O(\lambda, \mu)e^{-im\lambda}d\lambda. \quad (8)$$

$$O_n^m = \int_{-1}^{1} C^m(\mu)P_n^m(\mu)d\mu. \quad (9)$$

In Eqs. 8 and 9, $C^m(\mu)$ is called Fourier coefficient. There are an infinite number of orographic spectral coefficients included in solution process. The m was called zonal wavenumber and n was called total wavenumber. The values of m and n determine the total number of spectral coefficients. The larger their values are, the more spectral coefficients there are, and the more refined the orographic scale is represented. Whereas, in the actual situation, due to the limitation of computing resources, it is impossible to calculate infinite number of spectral coefficients, so we must choose appropriate values of m and n. The values of m and n also determine the global spectral model's resolution. In this paper, the spectral model's resolution we use is T1279, that is, the maximum of m is 1279 and a triangular truncation form is used for the maximum of n.

3.5 The Spectral Filtering

After converting the orographic field into spectral coefficients, it needs a final filtering. The orographic spectral is gradually tapered by multiplying the orographic spectral coefficient O_n^m by the filter (2) to avoid spectral confusion. The finally calculated orographic spectral can be used as the spectral model's orography data set for numerical weather prediction.

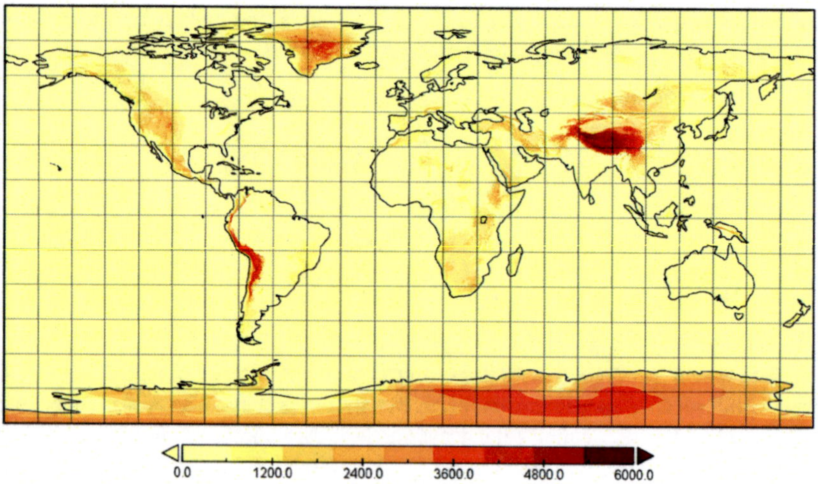

Fig. 2. Orography before processing

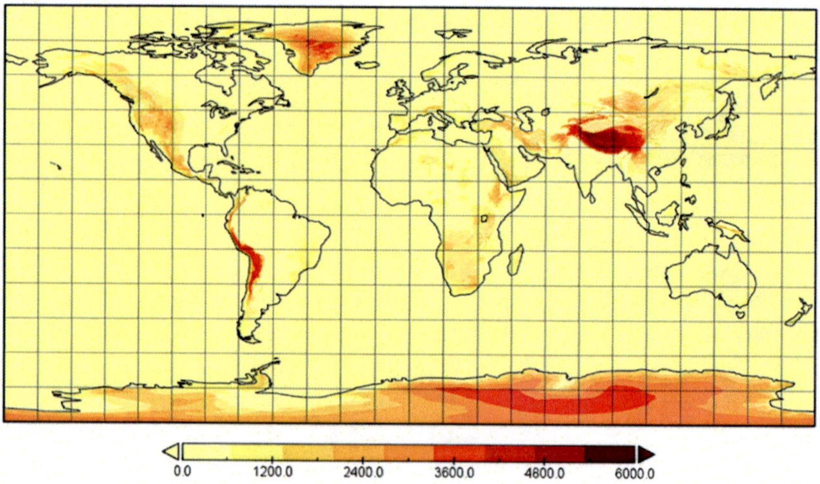

Fig. 3. Orography after all processing operations

4 Summary

This study introduces a production method of the high-precision orography data set applicable to the global NWP spectral model based on Earth observation data. To avoid spectral aliasing in the process from the initial high-resolution elevation data to the global NWP spectral model resolution, and for filtering, a one-dimensional filter is used in grid point space. The filter scale is in kilometer rather than radian, so the filter scale is consistent across the globe. Spectral filters are used in spectral space to make orographic data more suitable for models. The Fig. 2 shows the orography that has just been spliced into global orography but has not been processed, and the Fig. 3 shows the orography after the last step of spectral space filtering. It can be seen from Figs. 2 and 3 that after all orography processing is completed, many small-scale orography is filtered out from the initial global orography, especially in some complex orographic areas, such as the Qinghai-Tibet Plateau in Asia, Greenland in the northern polar region of North America and the Andes Mountains on the western edge of South America. As a result, the newly constructed global orography is relatively smooth. During the operation of the NWP models, the orography with these small-scale orographic features may seriously affect the accuracy and stability of the numerical model, and even cause the model to stop operation. Therefore, these orographic data processing steps are necessary. These methods are useful for producing orographic datasets in the global NWP spectral model. In addition, the calculation method of the orographic dataset is also universal, and the calculation parameters can be changed according to the actual situation.

This paper focuses on the orography construction technology in the spectral model's dynamic for numerical weather prediction. However, the orographic effects are described not only in the model's dynamic, but also in physical process parameterization schemes, such as the turbulent orography form drag parametrization and the sub-grid orography drag parameterization scheme. In future work, we may study orographic parameters in physical process parameterization schemes, such as the filtering orography standard deviation parameter in turbulent orography form drag parametrization scheme and sub-grid orography standard deviation, orographic angle, slope and anisotropic parameters in sub-grid orography drag parametrization scheme. The accurate calculation method of orography parameters coordinated with the model's orography will be studied to construct more accurate orography parameters.

References

1. Stohl, A., Forster, C., Sodemann, H.: Remote sources of water vapor forming precipitation on the Norwegian west coast at 60°N-a tale of hurricanes and an atmospheric river. J. Geophys. Res.: Atmos. **113** (2008)
2. Berckmans, J., Woollings, T., Demory, M.-E., Vidale, P.-L., Roberts, M.: Atmospheric blocking in a high resolution climate model: influences of mean state, orography and eddy forcing. Atmos. Sci. Lett. **14**, 34–40 (2013)
3. Sandu, I., Niekerk, A.V., Shepherd, T.G., Vosper, S.B., Svensson, G.: Impacts of orography on large-scale atmospheric circulation. Clim. Atmos. Sci. **2**, 1–8 (2019)
4. Gao, H., Chen, H., Liu, H., Shi, W.: Development of overseas Earth-observing satellite technology. Spacecr. Eng. **18**, 84–92 (2009)
5. Berkofsky, L., Bertoni, E.A.: Mean topographic charts for the entire Earth. Bull. Am. Meteorol. Soc. **36**, 350–354 (1955)
6. La-Valle, L., Rogo, M., Marini, M.: The topography of the European Mediterranean region for numerical weather prediction. Riv. Meteorol. Aeronaut. **34**, 13–18 (1974)
7. Nikolakopoulos, K.G.: Comparing a DTM created with ASTER data to GTOPO 30 and to one created from 1/50.000 topographic maps. Proc. SPIE **5574**, 43–51 (2004)
8. Farr, T.G., et al.: The shuttle radar topography mission. Rev. Geophys. **45** (2007)
9. Tachikawa, T., et al.: ASTER global digital elevation model version 2 – summary of validation results (2011)

10. De Meij, A., Bossioli, E., Penard, C., Vinuesa, J.F., Price, I.: The effect of SRTM and Corine Land Cover data on calculated gas and PM10 concentrations in WRF-Chem. Atmos. Environ. **101**, 177–193 (2015)
11. De Meij, A., Vinuesa, J.F.: Impact of SRTM and Corine Land Cover data on meteorological parameters using WRF. Atmos. Res. **143**, 351–370 (2014)
12. Kirthiga, S.M., Patel, N.R.: Impact of updating land surface data on micrometeorological weather simulations from the WRF model. Atmosfera **31** (2018)
13. Wen, X., Dong, W., Yuan, W., Zheng, Z.: Establishment and analysis of a high-resolution assimilation dataset of the water-energy cycle in China. Phys. Chem. Earth **87**, 126–141 (2015)
14. Zhang, X., Yin, Y.: Evaluation of the four PBL schemes in WRF model over complex topographic areas. Trans. Atmos. Sci. **36**, 68–76 (2013)
15. Liu, W., Gao, Y., Ran, Y., Cheng, G.: Contrast analyses of simulation results in Heihe basin utilizing the different resolution DEM data. Plateau Meteorol. **26**, 525–531 (2007)
16. Wang, Y., Wu, J.: Overview of the application of orographic data in numerical weather prediction in complex orographic areas. Adv. Meteorol. **2022**, 1279625 (2022)
17. Zhu, W., Chen, Z., Zhang, Y., Yang, J., Zhang, Y.: The impact of high resolution terrain on the prediction of ground elements from grapes model in south china. J. Trop. Meteorol. **35**, 801–811 (2019)
18. He, G.: Review of studies on terrain disposing methods in numerical models and precipitation simulation of orographic effect. Plateau Mt. Meteorol. Res. **41**, 1–8 (2021)
19. Davies, L.A., Brown, A.R.: Assessment of which scales of orography can be credibly resolved in a numerical model. Q. J. R. Meteorol. Soc. **127**, 1225–1237 (2001)
20. Wood, N., Mason, P.: The pressure force induced by neutral, turbulent flow over hills. Q. J. R. Meteorol. Soc. **119**, 1233–1267 (1993)
21. Belcher, S.E., Wood, N.: Form and wave drag due to stably stratified turbulent flow over low ridges. Q. J. R. Meteorol. Soc. **122**, 863–902 (1996)
22. Webster, S., Brown, A.R., Cameron, D.R., Jones, C.P.: Improvements to the representation of orography in the Met Office Unified Model. Q. J. R. Meteorol. Soc. **129**, 1989–2010 (2003)
23. Raymond, W.H.: High-order low-pass implicit tangent filters for use in finite area calculations. Mon. Weather Rev. **116**, 2132–2141 (1988)
24. Rutt, I., Thuburn, J., Staniforth, A.: A variational method for orographic filtering in NWP and climate models. Q. J. R. Meteorol. Soc. **132**, 1795–1813 (2007)
25. Tu, N., Chen, J., He, G.: Research on application of Chebyshev polynomial filtering method in smooth topography of GRAPES mode. Plateau Meteorol. **31**, 47–56 (2012)
26. He, G., Peng, J., Tu, N.: Terrain construction and experiment for numerical model based on high resolution terrain data. Plateau Meteorol. **34**, 910–922 (2015)
27. Chen, L., Xia, Y., Zhuang, X.: Influence of different terrain smoothing schemes in WRF model on precipitation forecast. Meteorol. Sci. Technol. **48**, 664–674 (2020)
28. Chenghai, W., Xiao, L., Yi, Y.: Atmospheric Numerical Model and Simulation, pp. 109–110. China Meteorological Press, Beijing (2011)
29. Navarra, A., Stern, W.F., Miyakoda, K.: Reduction of the Gibbs oscillation in spectral model simulations. J. Clim. **7**, 1169–1183 (1994)
30. Lindberg, C., Broccoli, A.J.: Representation of topography in spectral climate models and its effect on simulated precipitation. J. Clim. **9**, 2641–2659 (1996)
31. Abrams, M., Crippen, R., Fujisada, H.: ASTER Global Digital Elevation Model (GDEM) and ASTER Global Water Body Dataset (ASTWBD). Remote Sens. **12**, 1156 (2020)
32. Chen, Q., et al.: ASTER and GF-5 satellite data for mapping hydrothermal alteration minerals in the Longtoushan Pb-Zn deposit, SW China. Remote Sens. **14**, 1253 (2022)

33. Hao, G., Su, H., Zhang, R., Tian, J., Chen, S.: A two-source normalized soil thermal inertia model for estimating field-scale soil moisture from MODIS and ASTER data. Remote Sens. **14**, 1215 (2022)

PolSAR Image Classification Based on Deep Features

Ping Han(✉), Lunan Men, and Zetao Chen

Tianjin Key Lab for Advanced Signal Processing, Civil Aviation University of China, Tianjin, China
hanpingcauc@163.com

Abstract. The interpretation of polarimetric synthetic aperture radar (PolSAR) images has wide and important applications in target identification, disaster survey, urban planning, road detection and image classification. In PolSAR image classification tasks, some algorithms often obtain different polarimetric scattering features by different target decomposition methods which can generate feature redundant. To solve the problems of complex processing and excessive time consumption caused by multiple dimensional features, a deep feature-based PolSAR image classification algorithm is given in this paper. The algorithm first extracts the original feature set from the PolSAR image which is used to pre-train the Restricted Boltzmann Machine (RBM) model. The redundant features in the original feature set are rejected according to the error which is calculated by the reconstruction error of the input features. The RBM network is initialized by the remaining features and corresponding weights, then train the Deep Boltzmann Machine (DBM). Finally, the softmax classifier is trained with the training samples containing the deep features to achieve image classification. The experimental results show that the proposed algorithm has good feature selection timeliness and the deep features obtained from the algorithm can improve the image classification accuracy.

Keywords: PolSAR image · Deep feature · Image classification · Deep Boltzmann machine

1 Introduction

In recent years, PolSAR image classification technology has been widely used in agricultural production, natural disaster monitoring, geological exploration, and other fields [1]. PolSAR image classification essentially uses the information of polarimetric characteristics, location and spatial structure of each pixel in the image to determine the category to which the pixel belongs [2]. The effective use of the rich polarimetric features in the image is very important for the study of remote sensing feature classification.

L. Wang et al. (Eds.): CHREOC 2022, LNEE 969, pp. 392–403, 2023.
https://doi.org/10.1007/978-981-19-8202-6_35

Deep learning learns abstractly from data by building a neural network model. The literature [3] proposed a backpropagation algorithm, and later built a variety of deep network models based on this algorithm, such as deep Boltzmann machine, convolutional neural networks, and generative adversarial networks [4–6], all of which have been rapidly developed in the field of image processing. In the literature [7], the polarimetric coherence matrix was transformed into 6-dimensional polarimetric features by linear transformation, and then the feature vector was input to the complex convolutional neural network to complete the PolSAR image classification. In the literature [8], since some parameters of the polarimetric scattering matrix obey Wishart distribution, the classification of PolSAR images was realized by combining with DBN model. Later, some scholars used target decomposition to extract multiple polarimetric features to train classifier to achieve PolSAR image classification. The literature [9] used the combination of feature selection and convolutional neural network to achieve PolSAR image classification, but when there are many decision trees in the random forest, the training time and space will be relatively large; The literature [10] accomplished unsupervised PolSAR image classification through using the classical target decomposition method which combined H/α and Wishart classification algorithm, but this algorithm utilized a relatively single target decomposition method with poor feature characterization. However, increasing the dimensionality of polarimetric features generates feature redundancy, which not only increases the extra computational effort, but also causes the classifier to fall into local overfitting, this situation is not conducive to subsequent classification tasks [11]. Therefore, a reasonable combination of different types of polarimetric scattering features is very important to improve the image classification accuracy.

To address the above problems, a PolSAR image classification algorithm based on deep features is given in this paper. The algorithm can effectively deal with the problem of large number of parameters due to feature redundancy. Experimental comparisons are made with the literature [9] and literature [10] using AIRSAR real fully polarimetric data, and the experimental results show that the method not only has good timeliness in processing feature data with large volume and high dimensionality, but also improves the classification accuracy of PolSAR images.

2 Boltzmann Machine

Boltzmann machine is a stochastic neural network. Instead of adjusting the weights based on some deterministic algorithm like other networks, the stochastic network is modified by some probability distribution. The network is trained once and its weights are updated once to adjust and optimize the network parameters.

2.1 Restricted Boltzmann Machine

Restricted Boltzmann Machines (RBM) [12] is a two-layer feedback neural network composed of random neurons via full connectivity, and its layers present symmetric connections. Figure 1 represents the structure of RBM, which can be considered as an undirected graph model consisting of a visible layer and a hidden layer. In the RBM model, when the state of the visible unit is given, the state of each hidden unit is then determined. The activation states of each hidden unit are independent of each other. Based on the above characteristics of RBM, the RBM model can be trained by Gibbs sampling [13] to obtain samples that match the RBM distribution states.

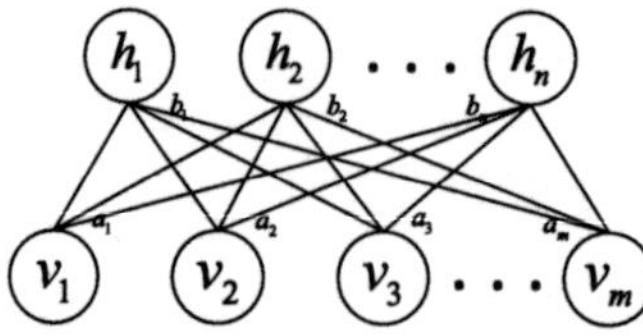

Fig. 1. Restricted Boltzmann machine structure diagram

RBM networks can easily fit input data with known probability distributions. The conclusions of statistical mechanics show that for data with unknown probability distributions, any probability distribution can be transformed into an energy-based model, and the learning process using energy models makes it easy and feasible to learn the distribution of the data. Suppose an RBM has *m* visible units and *n* hidden units, where v_i and h_j denote the states of the first visible unit and the first hidden unit, respectively. For a given set of states (*v*,*h*), the energy possessed by the RBM is defined as

$$E(v, h) = -\sum_{i=1}^{n} a_i v_i - \sum_{j=1}^{m} b_j h_j - \sum_{i=1}^{n}\sum_{j=1}^{m} v_i W_{ij} h_j \tag{1}$$

In the above equation, W_{ij} denotes the connection weight matrix of unit *i* in the visible layer and unit *j* in the hidden layer, a_i denotes the bias of neuron *i* in the visible layer, and b_j denotes the bias of neuron *j* in the hidden layer. From the energy function, the joint probability of the two-layer node (*v*,*h*) can be defined as

$$P(v, h) = \frac{e^{-E(v,h)}}{\sum_{v,h} e^{-E(v,h)}} \tag{2}$$

where the denominator represents the total energy in all possible cases. In a practical task, in order to get the Gibbs distribution [14] represented by the RBM network to fit the distribution of the input samples to the maximum possible extent, the value of the parameter $\theta = \{W_{ij}, a_i, b_j\}$ needs to be found by solving the maximum likelihood function of the input samples. The log-likelihood function for the one-sample RBM model is found as

$$\ln P(v) = \ln\left(\sum_{h} e^{-E(v,h)}\right) - \ln\left(\sum_{v,h} e^{-E(v,h)}\right) \tag{3}$$

and

$$\frac{\partial \ln P(v)}{\partial \theta} = \mathrm{E}_{P(h|v)}\left[-\frac{\partial E(v, h)}{\partial \theta}\right] - \mathrm{E}_{P(h,v)}\left[-\frac{\partial E(v, h)}{\partial \theta}\right] \tag{4}$$

In Eq. (6), $E(\cdot)$ denotes the expectation, the right-hand side of the equation denotes the difference between the expectation of $-\frac{\partial E(v,h)}{\partial \theta}$ with conditional probability $P(v,h)$ and the expectation of $-\frac{\partial E(v,h)}{\partial \theta}$ with joint probability $P(v,h)$, and the left-hand side of the equation denotes the bias (gradient) of the likelihood function with respect to θ. Further biasing the three parameters in θ, the solution of the parameter W_{ij}, a_i, b_j, in the log-likelihood function is obtained by using the Contrastive Divergence (CD) in the iterative process [15].

2.2 Contrastive Divergence

In order to improve the training efficiency of RBM, an approximation algorithm is generally used to estimate the expectation of RBM model. The CD algorithm is an approximation algorithm based on Gibbs sampling, which is applied to the RBM model with the parameter update process of the binary variable input as follows: (1) given the input vector $x^{(0)}$, the activation state $h^{(0)}$ of the hidden layer is calculated by using the connection weight W; (2) compute $x^{(1)}$ and $h^{(1)}$ based on W, and the obtained $(x^{(1)}, h^{(1)})$ is used as the state quantity for the next step of the CD algorithm, which means that the data reconstruction process is completed once. The gradient estimation of the likelihood function can be expressed as

$$\frac{\partial \ln P(v, h|\theta)}{\partial W_{ij}} \approx P\left(h_i = 1|v^{(0)}\right)v^{(0)} - P\left(h_i = 1|v^{(1)}\right)v^{(1)} \tag{5}$$

$$\frac{\partial \ln P(v, h|\theta)}{\partial a_i} \approx v_i^{(0)} - v_i^{(1)} \tag{6}$$

$$\frac{\partial \ln P(v, h|\theta)}{\partial b_i} \approx P\left(h_i = 1|v^{(0)}\right) - P\left(h_i = 1|v^{(1)}\right) \tag{7}$$

$$e_i = \mathrm{E}_{\bmod el}\left[\left(x^{(1)} - x^{(0)}\right)^2\right] \tag{8}$$

The update process of connection weights W, visible layer neuron bias, hidden layer neuron bias and reconstruction error before and after the reconstruction of the RBM model are represented in the above four equations, respectively.

The specific training steps of the CD algorithm are:

Input: training set $\{X_1, X_{2,..}X_n\}$, learning rate: ε , sigmoid activation function: σ , training period: T.

Output: a,b,W.

Training process:

Initialization: a,b,W is set to a random smaller value.

For t=1,2…T

For for each training sample $X_i(1 \le i \le n)$, the initial activation state of the visible layer neurons: $v_1 = X_i$;

State analysis of hidden layer neurons: $P(h_{1j} | v_1) = \sigma(b_j + \sum_i v_{1i} W_{ij})$;

Extracting the activation state of hidden layer neurons: $h_{1j} \in \{0,1\}$;

Reconstruction of the activation state of neurons in the visible layer:

$$P(h_{2j} | v_2) = \sigma(b_j + \sum_i v_{2i} W_{ij}) \tag{9}$$

The parameters are updated as:

$$W \leftarrow W + \varepsilon(P(h_1 = 1 | v_1)v_1^{\mathrm{T}} - P(h_2 = 1 | v_2)v_2^{\mathrm{T}}) \tag{10}$$

$$a \leftarrow a + \varepsilon(v_1 - v_2) \tag{11}$$

$$b \leftarrow b + \varepsilon(P(h_1 = 1 | v_1)v_1^{\mathrm{T}} - P(h_2 = 1 | v_2)v_2^{\mathrm{T}}) \tag{12}$$

End For

End For

2.3 Deep Boltzmann Machine

Deep Boltzmann Machine (DBM) is an undirected graph model that adds multiple hidden layers to the structure of RBM model [16]. The structure of DBM with three hidden layers is shown in Fig. 2.

To simplify the expression, the energy function omitting the bias parameter DBM is expressed as follows

$$E\left(v, h^{(1)}, h^{(2)}, h^{(3)} | \Omega\right) = -vW_1h^{(1)}h^{(1)^{\mathrm{T}}}W_2h^{(2)} - h^{(2)^{\mathrm{T}}}W_3h^{(3)} \tag{13}$$

In the above equation, $(\cdot)^{\mathrm{T}}$ denotes the transpose, $\Omega = \{W_1, W_2, W_3, B, A_1, A_2, A_3\}$ is the model parameter, W_1, W_2, W_3 denotes the connection weights of the display and hidden layers and the hidden and hidden layers, respectively. B is the bias of the display layer, and $A_i(1 \le i \le 3)$ is the bias of the i hidden layer. Similarly, the DBM with three hidden layers is considered as a RBM with two layers. The data learning process of the first RBM is described in Sect. 2.1. For the second RBM, the result of the first hidden layer can be regarded as a visible layer, and then the parameters of the two RBM are fine-tuned, and then the weights are updated by the CD algorithm. The number of neurons in different hidden layers are set enables the DBM to extract more distinguished abstracted features to express deeper structural information of the data [17].

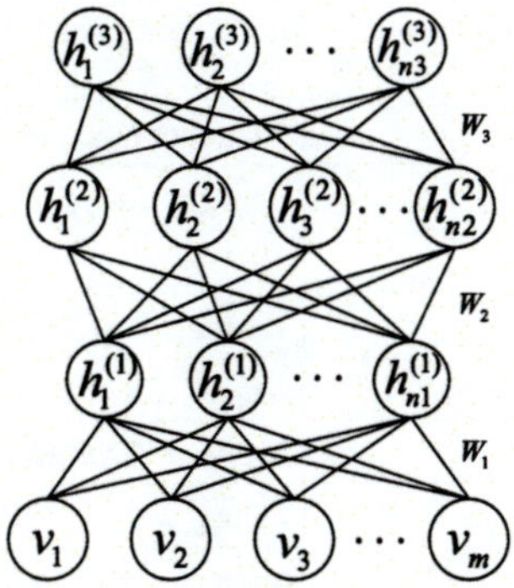

Fig. 2. Deep Boltzmann machine structure diagram

3 Algorithm Processing

The overall flow chart of the algorithm in this paper is shown in Fig. 3, and the algorithm consists of two main parts: training and testing.

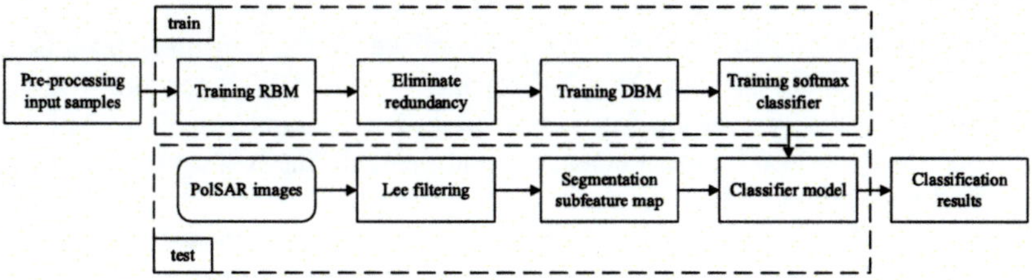

Fig. 3. Flow chart of algorithm

As shown in Fig. 3, the datas are preprocessed. Lee filtering is performed on datas to reduce the influence of coherent speckle noise on the experimental results. Multi-dimensional polarimetric feature parameters are extracted by using different polarimetric target decomposition methods from the datas to form training samples. The polarimetric parameters are shown in Table 1.

Training:

(1) The RBM model is pre-trained with above training samples. The output of each input feature in the RBM hidden layer neuron is calculated, and the display layer feature parameters are reconstructed by using the CD algorithm according to Eq. (8).
(2) De-redundancy. Equation (8) e_i represents the variance of $x^{(1)}$ and $x^{(0)}$, which represents the deviation of features before and after reconstruction. The specific steps of redundancy removal are: (a) The original feature set is used to pre-train the RBM model and calculate the reconstruction error of the input features e_i. (b) A certain number of features from the features of the m visible layer is selected

Table 1. Polarimetric feature parameter set

Feature type	Characteristic parameters
Polarimetric parameters	C_{11}, C_{22}, C_{33}, T_{11}, T_{22}, T_{33}
Cloude decomposition	*H, A, a*
Y-decomposition	*Yamaguchi (Dbl, Hlx, Odd, Vol)*
K-decomposition	*Ks, Kd, Kh*
F-decomposition	*Freeman (Dbl, Odd, Vol)*
Coherence matrix eigenvalues	$\lambda_1, \lambda_2, \lambda_3$
Scattering power	*Span = Freeman (Db + Odd + Vol)*

to be eliminated, and calculate the reconstruction error e'_k for the $k(1 \leq k \leq m)$th feature to be eliminated. (c) If $e'_k < e_k$, the kth feature in step (a) is considered as a redundant feature, and the feature is removed from the input feature set, otherwise the feature is retained. (d) When the maximum number of iterations is reached, a subset of redundant features is obtained and the parameters such as visible layer neuron bias, hidden layer neuron bias and connection weights corresponding to the remaining features are updated.

(3) Train the DBM. The RBM network is initialized by the remaining features and the corresponding weights. Deep Boltzmann Machine (DBM) network which can extract deep features of the data is trained, and the parameters of the DBM network are optimized by the gradient descent method, finally a set of deep feature parameters are outputted.
(4) The output feature parameters from step (3) are used to train the softmax classifier, and the trained image classification model is obtained after parameter tuning and optimization.

Testing:

(1) Lee filtering is performed on PolSAR image and then the deep features are calculated on the image.
(2) The process of splitting the sub-feature map is to divide the feature image obtained in step (1) into sub-image blocks of size 8×8 to form a test set, and finally input the sub-image blocks into the classification model to obtain the classification results.

4 Experimental Results and Analysis

In this paper, the fully PolSAR data acquired by the US airborne AIRSAR system is used for classification experiments, and the classification results are analyzed and compared with those of the literature [10] (T + CNN) and literature [9] (RF + CNN) to verify the effectiveness of the algorithm in this paper. In the training stage, the number of neurons in the visible and hidden layers of the RBM is set to 26, the learning rate is 0.1, and the training times are 10; the number of neurons in the visible and three hidden layers of the DBM is set to 7, the learning rate is 0.13, and the training times are 50; the CD training algorithm are both adopted. The learning rate of the SoftMax layer is set to 0.0002, and the number of iterations is 200. In order to compare the accuracy of classification results, this paper expresses the proportion of correctly classified pixel points to all pixel points, that is the overall classification accuracy OA parameter.

Experiment 1: The image data of L-band San Francisco area acquired by AIRSAR system with image size 900 × 1024. The experimental sample set is 25,000 data of size 8 × 8, 80% of data is used for training and the rest is used for testing. A set of 7-dimensional features after eliminating redundant features by RBM is: T_{22}, C_{11}, λ_3, A, H, *Span*, K_d. The 7-dimensional feature parameters and the corresponding connection weights are used to initialize to the RBM model parameters and the DBM is trained to output a set of deep features after reaching the maximum number of iterations (Fig. 4 and Table 2).

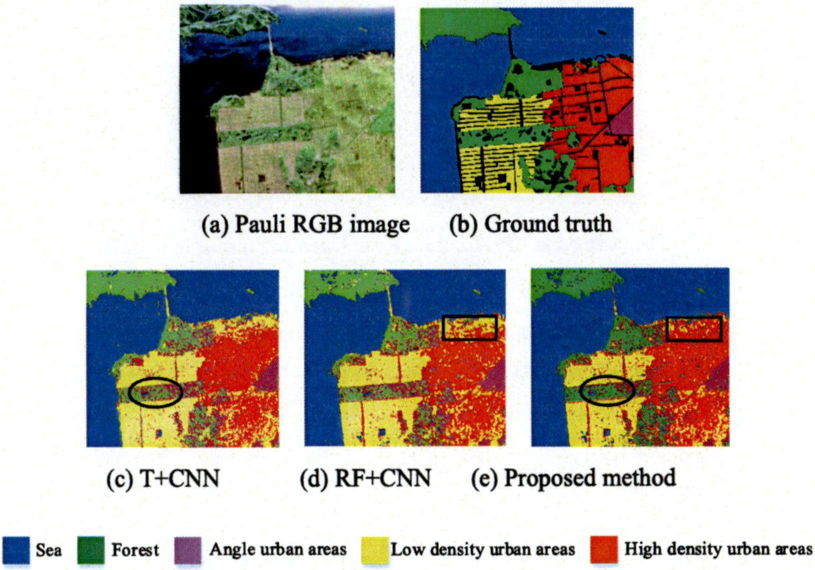

Fig. 4. The image data of L-band San Francisco area and results of experiment.

Table 2. Classification results of the San Francisco area PolSAR image (%)

Categories	Sea	Forest	Angle urban	High urban	Low urban	OA
T + CNN	95.65	77.89	97.28	76.82	94.36	88.40
RF + CNN	97.04	81.58	99.71	77.66	96.12	89.82
Proposed method	99.37	92.07	96.28	86.42	91.79	93.19

Experiment 2: PolSAR image of Flevoland area whose image pixel size is 1024 × 750, the experimental dataset is 20,000 data of size 8 × 8, 80% is used for training and the rest is used for testing. The original feature set extracted for this image is retained after redundancy is removed as: a, λ_1, T_{22}, H, λ_2, A, Y_Dbl (Fig. 5).

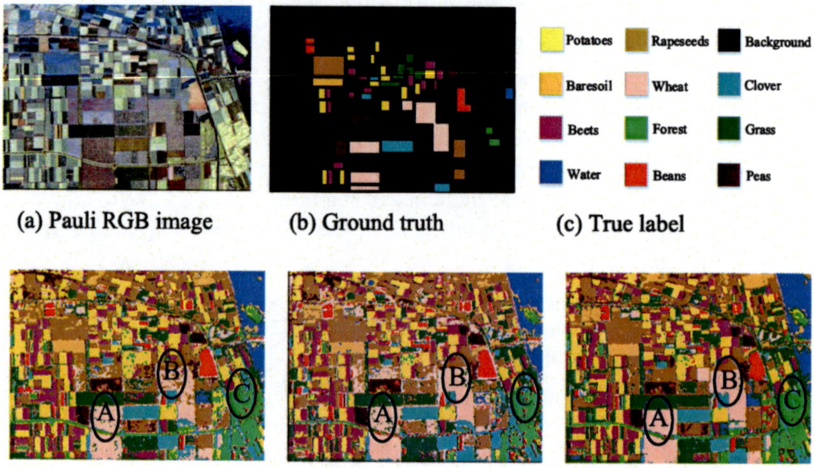

(a) Pauli RGB image (b) Ground truth (c) True label

(d) T+CNN (e) RF+CNN (f) Proposed method

Fig. 5. The image data of L-band Flevoland area and results of experiment.

Table 3. Classification results of the Flevoland area PolSAR image (%)

Categories	Baresoil	Beets	Forest	Grass	Lucerne	Peas
T + CNN	94.30	88.89	97.52	69.94	82.00	90.01
RF + CNN	96.82	94.17	91.30	72.66	93.84	88.70
Proposed method	99.35	91.27	95.69	81.05	95.47	93.75

Potatoes	Beans	Water	Wheat	Rapeseed	OA
87.68	98.24	98.27	68.35	92.31	87.95
92.45	95.85	100.0	76.23	88.91	90.08
95.96	95.07	100.0	82.59	90.80	92.82

From the above two experimental results, it can be seen that the method in this paper has less misclassification compared with the other two traditional feature extraction-based methods. As shown in Experiment 1, the correct classification results of this paper at complex terrain are better than the other two methods, as marked by ellipse and rectangular regions. As shown by the results of Experiment 2, the classification purity of this paper's method at A, B and C is higher than that of the other two methods, and there exist less misclassified regions. From the quantitative results in Tables 3 and 4, it is clear that the method in this paper achieves higher overall accuracy compared with the other two methods. Among them, in Experiment 1, the classification accuracy of this method in forest and high-density urban areas is improved by nearly 10% accuracy than the other two methods. In Experiment 2, the accuracy of this paper's method in baresoil, grass, potatoes, peas and wheat is improved significantly and achieves the maximum overall accuracy of 92.82%.

In order to compare the classification effects of PolSAR images before and after feature selection, this paper illustrates the experiments with image data from the San Francisco area. The experimental conditions such as the number of sample and network parameters were kept basically the same, the polarimetric parameters without feature selection are shown in Table 1, and the comparison results are represented in Fig. 6 and Table 4.

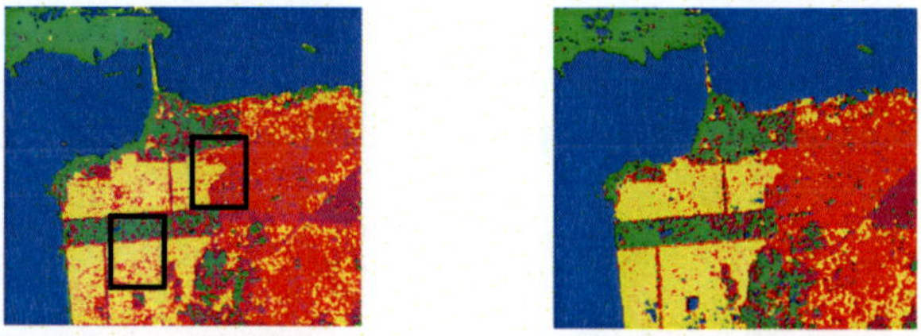

(a) Before feature selection (b) After feature selection

Fig. 6. Classification results before and after feature selection

Table 4. Accuracy of classification results before and after feature selection (%)

Categories	Sea	Forest	Angle urban	High urban	Low urban	OA
Before feature selection	93.12	77.89	86.14	69.36	82.11	80.05
After feature selection	99.37	92.07	96.28	86.42	91.79	93.19

The classification results from Fig. 6 show that the classification effect before feature selection has a significant misclassification situation compared to the effect after feature selection. In Fig. 6(a), both high-density and low-density urban areas have more areas misclassified as angle urban areas, as shown at the rectangular in the figure. The accuracy comparison from Table 4 shows that the classification accuracy of both high-density urban area and low-density urban area before feature selection is lower than that of the classification result of after feature selection, and the total classification accuracy of result after feature selection is higher than that of the classification result without feature selection.

To verify the timeliness of feature selection in this paper, four different feature selection methods were used to conduct time statistics for different size data sets with sample size of 6,000 and 20,000, and the statistical results are shown in Table 4. The experimental platform is Matlab, the environment is Windows 10-64bit, memory 16G, CPU is Intel Core i7-10700, the statistical results are shown in Table 5.

Table 5. Time used by the four algorithms (s)

Algorithms	Time			
	6,000 samples		20,000 samples	
	26 features	48 features	26 features	48 features
FCBC [18]	21.37	39.02	54.71	126.08
RF [9]	32.2	61.7	90.65	158.31
Literature [19]	93.28	227	264.27	587.9
Proposed method	8.11	14.47	18.64	39.83

As can be seen from Table 4, the deep feature selection method has certain advantages over traditional feature selection algorithms (FCBF, RF, optimal feature selection method) in terms of feature selection efficiency and shorter time, which also illustrates that the deep learning-based feature selection method is suitable for handling larger datasets.

5 Conclusion

Aiming at the problems of large computational complexity caused by high-dimensional features extracted by target polarimetric decomposition, repeated computation caused by feature redundancy and slow processing of large number of parameters, this paper gives a deep feature-based PolSAR image classification method by using the features of deep learning that can mine deep information of data and can handle large-scale data. Through experimental verification, the method not only can quickly filter effective features, but also improves the classification accuracy of PolSAR images. However, in the process of redundancy removal using RBM, the number of features to be eliminated needs to be set in advance, and setting different numbers of features to be eliminated has partial influence on the number of output features. In the subsequent study, the adaptive idea needs to be used to determine the optimal number of features to be eliminated in order to improve the adaptiveness of the algorithm.

References

1. Huadong, G.: Radar Theory and Application to Earth Observation. Science Press, Beijing (2000)
2. van Zyl, J.J.: Unsupervised classification of scattering behavior using radar polarimetry data. IEEE Trans. Geosci. Remote Sens. **27**(1), 36–45 (1989)
3. Plaut, D.C., Nowlan, S.J., Hinton, G.E.: Experiments on learning by back propagation. Tech. Rep. Carnegie Mellon Univ. 54 (1986)
4. Salakhutdinov, R., Mnih, A., Hinton, G.: Restricted Boltzmann machines for collaborative filtering. In: Proceedings of the 24th International Conference on Machine Learning, pp. 791–798 (2007)

5. Yang, Z.Z., Kuang, N., Fan, L., et al.: A review of image classification algorithms based on convolutional neural networks. Signal Process. **34**(12), 1474–1489 (2018)
6. Goodfellow, I., Pouget-Abadie, J., Mirza, M., et al.: Generative adversarial nets. In: Advances in Neural Information Processing Systems, pp. 2672–2680 (2014)
7. Zhou, Y., Wang, H., Xu, F., et al.: Polarimetric SAR image classification using deep convolutional neural networks. IEEE Geosci. Remote Sens. Lett. **13**(12), 1935–1939 (2016)
8. Fang, L., Jiao, L., Hou, B., et al.: POL-SAR image classification based on Wishart DBN and local spatial information. IEEE Trans. Geosci. Remote Sens. **54**(6), 1–17 (2016)
9. Han, P., Sun, D.D.: Feature selection combined with deep learning for polarized SAR image classification. Signal Process. **35**(06), 972–978 (2019)
10. Qiyu, H.: Research on contrast enhancement and classification method of polarized SAR images based on target decomposition. University of Electronic Science and Technology (2011)
11. Wang, M.J.: Polarized SAR image feature classification based on sparse learning and metric learning classifier. Xi'an University of Electronic Science and Technology, Shaanxi (2017)
12. Sutskever, I., Hinton, G.E., Taylor, G.W.: The recurrent temporal restricted Boltzmann machine. pp. 1601–1608 (2008)
13. Hinton, G.E.: A practical guide to training restricted Boltzmann machines. In: Neural Networks: Tricks of the Trade, 2nd edn., pp. 599–619 (2012)
14. Cho, K.H., Raiko, T., Ilin, A.: Gaussian-Bernoulli deep Boltzmann machine. In: International Joint Conference on Neural Networks. IEEE (2013)
15. Zhang, J., Jiang, Y., Xuewei, H.U., et al.: Convolutional restricted Boltzmann machine based on fast persistent contrastive divergence. Comput. Eng. (2016)
16. Qian, X., Liu, G.: Nearest-neighbor restricted Boltzmann machine for collaborative filtering algorithm. In: International Conference on Advanced Hybrid Information Processing. Springer, Cham (2017)
17. Shalev-Shwartz, S., Singer, Y., Srebro, N., et al.: Pegasos: primal estimated sub-gradient solver for SVM. Math. Program. **127**(1), 3–30 (2011)
18. Nemirovski, A.S., Juditsky, A., Lan, G., Shapiro, A.: Robust stochastic approximation approach to stochastic programming. SIAM J. Optim. **19**(4), 1574–1609 (2009)
19. Han, P., Chen, Z., Wan, Y., et al.: PolSAR image classification based on optimal feature and convolution neural network. In: IGARSS 2020 - 2020 IEEE International Geoscience and Remote Sensing Symposium. IEEE (2020)

Research on Lake Changes in China Over the Past 50 Years Based on Satellite Remote Sensing Technology

Lei Du, Xu Han(✉), Shucheng You, Ke Liu, Zhongwu Wang, Yun He, and Xinglin Mu

Land Satellite Remote Sensing Application Center, MNR, Beijing, China
wgkwg@163.com

Abstract. Affected by multiple factors such as global warming, melting glaciers and increasing rainfall, there are signs indicating that a trend of warm and humid climate is forming in Northwest China. To monitor and analysis this trend, four typical provinces such as Qinghai, Xinjiang, Ningxia and Gansu are selected as research areas. Based on Chinese ZY and GF serial satellite images and the Landsat images, a 50 years long-time monitoring and analysis is carried out on lakes which are more than 1 km^2 in the above areas. Considering the lake extraction accuracy as well as the workload of data processing, this paper mainly uses a combined method which includes the OTSU threshold method and multiple optimized seed region-growing algorithm to quickly achieve a total of 11 period lake information of the four provinces in China from 1970 to 2021. Results show that there is a significant rising trend in the number and surface area of lakes in the four northwestern provinces in China from 1970 to 2021. The results will contribute to the related research on water resources and climate change in northwest China.

Keywords: Lake change · Long-time monitoring · Global warming · Northwest China

1 Introduction

Affected by multiple factors such as global warming, northwest China is experiencing a new trend of warm and humid climate. As an important part of water resources, lakes are one of the sensitive indicators of climate change. The four provinces of Qinghai, Xinjiang, Ningxia and Gansu are located in northwestern China and covering a vast territory of China. Traditionally this region has a dry climate with little and unevenly distributed rainfall. Though basically the region lacks of lake water resources, it is extremely rich in glacier resources which makes it an ideal area to study long-time monitoring of lake changes in Chinese under the background of global warming [1–5].

A lot of researches have been done on lake automatic extraction. Experts and scholars have conducted continuous research on how to accurately and quickly extract the surface water with different conditions, such as, different types, regions, scales, and scenarios and have successively proposed a variety of surface water extraction algorithms, including: threshold method (including single band, multi-band and water index, etc.), classifier

L. Wang et al. (Eds.): CHREOC 2022, LNEE 969, pp. 404–410, 2023.
https://doi.org/10.1007/978-981-19-8202-6_37

method (including the support vector machine, decision tree and object-oriented, etc.), "global-local" automatic extraction and deep learning algorithms [6–12].

With the rapid development of the satellite remote sensing industry, it is now applicable to carry out large-scale and long-time monitoring of lakes with satellite remote sensing technology. Considering that the spatial resolution of Landsat satellite images is generally only 30 m, only lakes with an area of 1 km^2 or above are monitored in this paper. Unlike other long-time lake monitoring research, this paper firstly establishes a lake monitoring list of the research area based on three important thematic databases, that is, the National Land Survey, the National Geographical Condition Monitoring and China Census for Water. And then carries out 11 periods of lake monitoring with satellite remote sensing technology. Moreover, in terms of the temporal phase of the data, this paper selects as much data as possible from high water season (mainly the second half of a year) from 1970 to 2021 (a total of 11 periods). This has laid a solid foundation for lake long-time monitoring as well as data analysis and guarantees the credibility of the data results.

2 Research Area

The four provinces of Qinghai, Xinjiang, Ningxia and Gansu are located in northwestern China, covering a total land area of more than 2.8 million km^2. Traditionally this region has a dry climate with little and unevenly distributed rainfall. However, the region is extremely rich in glacier resources. According to the latest monitoring data in 2020, the total area of glaciers in Xinjiang, Qinghai, and Gansu provinces is about 27,000 km^2, accounting for about 57% of China's glacier resources. Affected by factors such as global warming, the number and area of lakes in this area are also changing dynamically, so this area is an ideal area to study the changes of lakes in China in the past 50 years.

3 Data and Method

3.1 Data

Based on Chinese ZY and GF serial satellite images of 2 m resolution from 2019 to 2021 (3 periods, all high water season) and the Landsat images of 30 m resolution from 1970 to 2015 (8 periods, mainly high water season), satellite image dataset of the four northwestern provinces from 1970 to 2021 is made, consisting of 11 periods. The dataset is as shown in Table 1.

3.2 Method

Lake extraction method is shown in Fig‘. 1. The main steps include:

(a) Based on Chinese ZY and GF serial satellite images of 2-m resolution and the Landsat images of 30-m resolution, screen out and build 11 periods image dataset of the four northwestern provinces from 1970 to 2021.

Table 1. Satellite image dataset of the study area (1970–2021)

No	Date	Data source	Resolution (m)	Time phase
1	2021	China GF, ZY satellites	2	High water season
2	2020	China GF, ZY satellites	2	High water season
3	2019	China GF, ZY satellites	2	High water season
4	2015	Landsat	30	High water season (Mainly)
5	2010	Landsat	30	High water season (Mainly)
6	2005	Landsat	30	High water season (Mainly)
7	2000	Landsat	30	High water season (Mainly)
8	1995	Landsat	30	High water season (Mainly)
9	1990	Landsat	30	High water season (Mainly)
10	1986	Landsat	30	High water season (Mainly)
11	1970	Landsat	30	High water season (Mainly)

(b) Establishes the lake monitoring list and its buffer areas of the research area based on three important thematic databases, that is, the National Land Survey, the National Geographical Condition Monitoring and China Census for Water.
(c) Carry out 11 periods of automatic extraction of lakes in the study area from 1970 to 2021 through OTSU threshold method for water bodies, combined with multiple optimized seed region-growing algorithms.
(d) Perform binarizing, small and pseudo target automatic removing, smoothing, thinning and merging etc., to get rid of holes, broken-pieces, jagged edges, etc.
(e) Carry out the quality inspection, accuracy assessment and statistical analysis of lakes.

4 Analysis

4.1 The Overall Change of Lakes Over the Past 50 Years

The changes of lakes with the area of 1 km^2 and above in the four northwestern provinces from 1970 to 2021 are monitored and analyzed using satellite image data during the high water season. The monitoring results show that from 1970 to 2021, there is an obvious upward trend in the number and surface area of lakes in the four northwestern provinces (Fig. 2, Table 2). In 1970, the number of lakes in the four northwestern provinces is 374, which increases to 399 in 2021, posting an increase of 25 lakes compared with 1970. Since 1995, the surface area of lakes in the four northwestern provinces has shown an increasing trend, reaching to 21,740.70 km^2 in 2019, the largest value in the past 50 years.

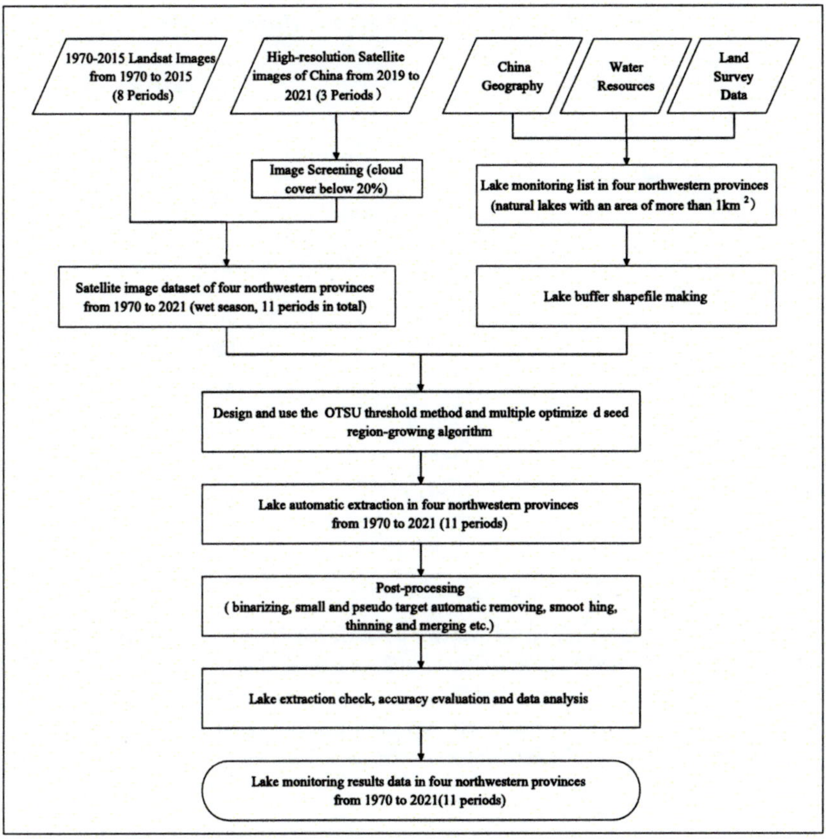

Fig. 1. Method of lake longtime monitoring

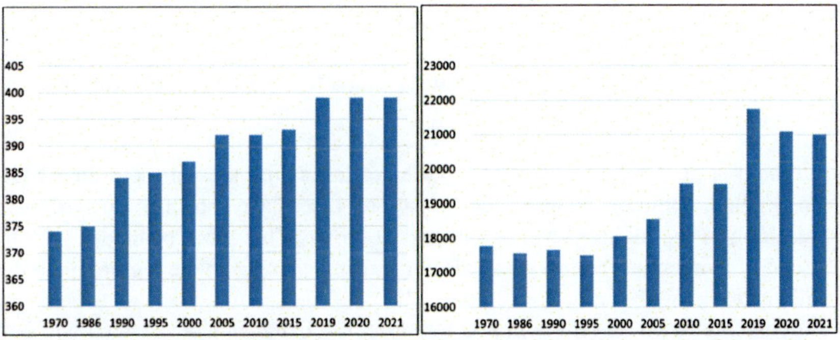

Fig. 2. Histogram of lake number (left) and total lake area (right, km^2) of four northwestern provinces of China from1970 to 2021

Table 2. Lake number and total lake area change analysis

Year	Lake number	Lake area/km^2	Area change rate/%
1970	374	17769.48	
1986	375	17562.73	−1.16
1990	384	17664.42	−0.59
1995	385	17508.63	−1.47
2000	387	18050.18	1.58
2005	392	18546.35	4.37
2010	392	19570.34	10.13
2015	393	19556.59	10.06
2019	399	21740.70	22.35
2020	399	21088.73	18.68
2021	399	21007.91	18.22

4.2 Change of Typical Lakes (Qinghai Lake)

Qinghai Lake is the largest inland saltwater lake in China. It is located in the northeastern part of the Qinghai-Tibet Plateau and to the northwest of Xining City. From 1970 to 2021, the water surface area of Qinghai Lake shows a fluctuating upward trend (Figs. 3 and 4, Table 3). The lake surface area increases continuously from 2005, and reaches to 4,501.27 km^2 in 2021, the largest value in the past 50 years, posting an increase of 255.42 km^2 compared to 2005.

5 Conclusion

Based on Chinese ZY and GF serial satellite images and the Landsat images, this paper mainly uses a combined method which includes the OTSU threshold method and multiple optimized seed region-growing algorithm to quickly achieve a total of 11 period lake information of four northwestern provinces in China from 1970 to 2021. Results show that:

(1) In the past 50 years, the number and area of lakes with area of more than 1 km^2 in Qinghai, Xinjiang, Ningxia and Gansu provinces have shown an upward trend, reaching to the highest level of the past 50 years. This trend coincides with the trend of temperature, rainfall and glacier melting changes in the region in the past 50 years.

(2) The phenomenon of get warm and humid in the Qinghai-Tibet Plateau will have a certain impact on the regional ecological environment. It is necessary to continuously monitor the water resources and ecological changes in the region, analyze the change trend and impact mechanism, to promote sustainable utilization and management of water resources.

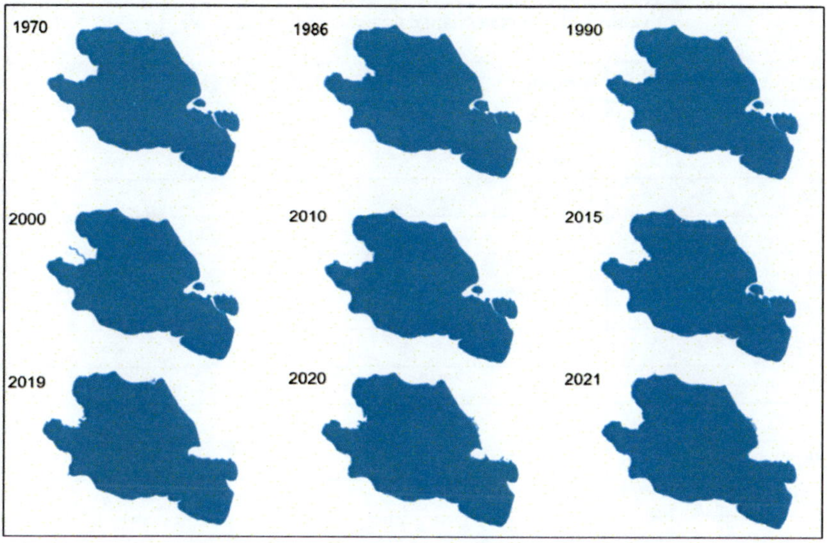

Fig. 3. Changes of Qinghai Lake from 1970 to 2021

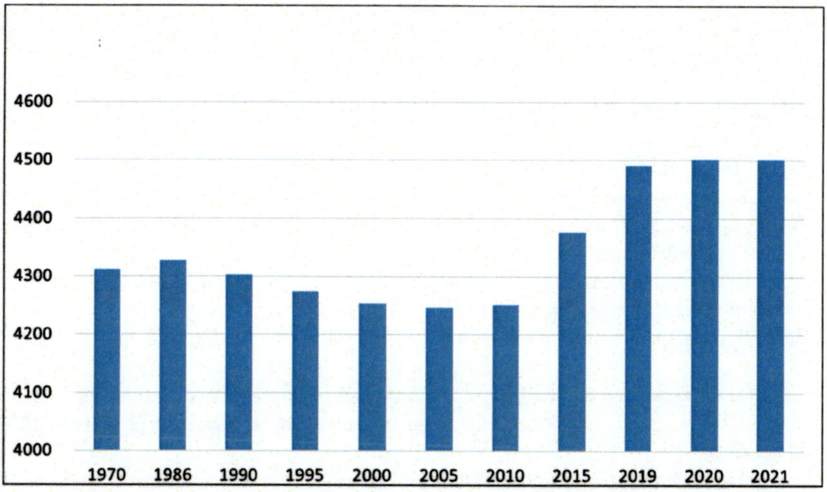

Fig. 4. Histogram of the Change of Qinghai Lake Area from 1970 to 2021 (km^2)

Table 3. Lake area change analysis of Qinghai Lake

Year	Area/km^2	Area change/%
1970	4312.15	–
1986	4326.95	0.34
1990	4303.22	−0.21
1995	4274.09	−0.88
2000	4253.22	−1.37
2005	4245.85	−1.54
2010	4251.06	−1.42
2015	4375.94	1.48
2019	4490.84	4.14
2020	4501.11	4.38
2021	4501.27	4.39

References

1. Zhang, G., Yao, T., Chen, W., et al.: Regional differences of lake evolution across China during 1960s–2015 and its natural and anthropogenic causes. Remote Sens. Environ. **221**, 386–404 (2019)
2. Adrian, R., O'Reilly, C.M., Zagarese, H., et al.: Lakes as sentinels of climate change. Limnol. Oceanogr. **54**(6), 2283–2297 (2009)
3. Feyisa, G. L., et al.: Automated water extraction index: a new technique for surface water mapping using Landsat imagery. Remote Sens. Environ. **140**, 23–35 (2014)
4. Zhang, J., Zhou, J., Zhang, G., et al.: Climate- and human-driven variations in lake area and number in North Xinjiang, China. Int. J. Remote Sens. **42**(2), 469–485
5. Pekel, J.F., Cottam, A., Gorelick, N., et al.: High-resolution mapping of global surface water and its long-term changes. Nature **540**, 418–422 (2016)
6. Hang, W., Fen, Q.: A review of remote sensing image water extraction. Sci. Surv. Map. **43**(5), 23–32 (2018)
7. Liegang, X., Zhanfeng, S., Junli, L., et al.: Automatic interpretation of diverse water bodies from remotely sensed images in complex background. China J. Image Graph. **18**(11), 1513–1519 (2013)
8. Li Dan, W., Baosheng, C.B., et al.: Review of water body information extraction based on satellite remote sensing. J. Tsinghua Univ. **60**(2), 147–161 (2020)
9. Giardino, C., Bresciani, M., Villa, P., Martinelli, A.: Application of remote sensing in water resource management: the case study of Lake Trasimeno, Italy. Water Resour. Manag. **24**(14), 3885–3899 (2010)
10. Huang, X., Xie, C., Fang, X., Zhang, L.: Combining pixel- and object-based machine learning for identification of water-body types from urban high-resolution remote-sensing imagery. IEEE J. Sel. Top. Appl. Earth Obs. Remote Sens. **8**(5), 2097–2110 (2015)
11. Xu, H.: Modification of normalised difference water index (NDWI) to enhance open water features in remotely sensed imagery. Int. J. Remote Sens. **27**(14), 3025–3033 (2006)
12. Wang, G., Wu, M., Wei, X., Song, H.: Water identification from high-resolution remote sensing images based on multidimensional densely connected convolutional neural networks. Remote Sens. **12**(5) (2020)

Author Index

B
Baogang, Geng, 59
Bing, Zhou, 161

C
Cai, Qifa, 79
Chen, Guangming, 79
Chen, Zetao, 392

D
Dong-sheng, Yang, 72
Du, Lei, 341, 404
Du, Lu, 59

F
Fan, Caizhi, 135
Fan, Mo, 313
Fang, Dong, 3
Fang, Panfei, 102

G
Gan, Yuhang, 341
Gao, Feng, 216
Gou, Peng, 124
Guanghui, Wang, 274
Guanglong, Wang, 161
Guo, Liang, 170
Guo, Rui, 170

H
Ha, Changliang, 178
Hai, Zhang, 3
Han, Hao, 331
Han, Ling, 242
Han, Ping, 392
Han, Xu, 404
Hao, Wang, 15
He, Yun, 404
Hongzhao, Tang, 313
Hu, Yanfeng, 361
Hu, Ye, 59
Huang, Long, 229
Hui, Chen, 293, 303

J
Jiaju, Ying, 161
Jiale, Zhao, 161
Jiancheng, Dang, 3
Jianwei, Qi, 274
Jianyi, Zhao, 313
Jie, Shang, 37
Jun, Zhou, 3
Junfeng, Xie, 313
Junyao, Li, 37

K
Kuang, Gangyao, 351

L
Lei, Deng, 161
Lei, Shu-Qi, 116
Li, Canhai, 331
Li, Dehua, 331
Li, Weiqiang, 150
Liu, Aifang, 229
Liu, Dazheng, 380
Liu, Dongsheng, 242
Liu, Jun, 46

L. Wang et al. (Eds.): CHREOC 2022, LNEE 969, pp. 411–412, 2023.
https://doi.org/10.1007/978-981-19-8202-6

Liu, Ke, 404
Liu, Lirong, 341
Liu, Peiwei, 102
Liu, Yang, 124
Liu, Yu, 284
Longqin, Xue, 322
Luo, Zhengyu, 341

M
Ma, Chengzhi, 150
Mei, Zheng, 264
Meibing, J. I. N., 201
Men, Lunan, 392
Mu, Xinglin, 404

N
Nie, Tiening, 46
Nie, Wei, 124

P
Peijie, Liu, 37
Pengrui, C. E. N., 264

Q
Qi, Jianwei, 284
Qian, Y. U., 72
Qianghui, Wang, 161
Qing, Zeng, 15
Qu, Hui, 331

S
Shouye, L. V., 264
Song, Chao, 135
Song, Qi, 150
Su, Xiaolu, 361
Sun, Hao, 351
Sun, Yue, 95

T
Tang, Hongzhao, 293, 303
Tao, Zhang, 274
Tengfei, Zhao, 37
Tianye, Wang, 15
Tongzhi, Yang, 3

W
Wang, Fan, 229
Wang, Feng, 95, 116, 193
Wang, Guanghui, 284
Wang, Guojun, 102
Wang, Hengyang, 293
Wang, Hui, 170
Wang, Jie, 284
Wang, Leiguang, 102
Wang, Mengmeng, 135
Wang, Shuqi, 178
Wang, Yang, 361
Wang, Yegui, 79
Wang, Yingjie, 380
Wang, Yunsen, 150
Wang, Zhipan, 341
Wang, Zhongwu, 341, 404
Wei, Zhang, 274
Wei-guo, Zang, 72
Wu, Jianping, 380
Wu, Panfeng, 150

X
Xiang, Li, 313
Xiao-xue, Yuan, 72
Xu, Hui, 229
Xu, Nuo, 124
Xu, Weiheng, 102
Xu, Yifan, 229
Xuan, Yao, 59

Y
Yang, Hong, 95
Yang, J. I. N., 201
Yang, Zi-An, 95, 193
Yaohai, Dong, 15
Ying, Wang, 201
You, Shucheng, 341, 404
Yu, Guangrui, 178
Yu, Hui, 46
Yu, Lili, 178
Yue, Dong-Xiao, 116
Yunfei, Li, 59

Z
Zhan, Bingguang, 46
Zhang, Honggeng, 293, 303
Zhang, Menglin, 216
Zhang, Tao, 284
Zhang, Xiaofang, 79
Zhang, Yonghe, 135
Zhao, Jun, 46
Zhao, Shen, 37
Zhao, Xiaopeng, 170
Zheng, Nai-Rong, 95, 193
Zheng, Pengfei, 102
Zheng, Yalan, 124
Zhiqiang, Bian, 15
Zhou, Junfan, 351
Zhou, Tianyu, 124
Zhu, Lilu, 361